Mathematische Leitfäden

J. J. Benedetto
Real Variable and Integration

Mathematische Leitfäden

Herausgegeben von
em. o. Prof. Dr. phil. Dr. h. c. G. Köthe, Universität Frankfurt/M., und
o. Prof. Dr. rer. nat. G. Trautmann, Universität Kaiserslautern

Real Variable and Integration

with Historical Notes

by John J. Benedetto
Professor at the University of Maryland

1976

 B. G. Teubner Stuttgart

Prof. John J. Benedetto

Born 1939 in Boston. Received B.A. from Boston College in 1960, M.A. from Harvard University in 1962, and Ph.D. from University of Toronto in 1964. Assistant professor at New York University from 1964 to 1965; and research associate at University of Liège and the Institute for Fluid Dynamics and Applied Mathematics from 1965 to 1966; employed by RCA and IBM from 1960 to 1965. At University of Maryland, assistant professor from 1966 to 1967, associate professor from 1967 to 1973, and professor beginning in 1973. Visiting positions include the Scuola Normale Superiore at Pisa from 1970 to 1971 and 1974 (spring) and MIT in 1973 (fall); also Senior Fulbright-Hays Scholar from 1973 to 1974.

CIP-Kurztitelaufnahme der Deutschen Bibliothek

Benedetto, John J.
Real variable and integration : with histor.
notes. — 1. Aufl. — Stuttgart : Teubner, 1976.
  (Mathematische Leitfäden)
  ISBN 3-519-02209-5

© B. G. Teubner, Stuttgart 1976
Printed in Germany
Setting: William Clowes & Sons Ltd., London
Printer: Johannes Illig, Göppingen
Binder: F. Wochner KG, Horb/Neckar
Cover design: W. Koch, Sindelfingen

Dedicated to the memory of Sam

Introduction

The subject matter in this book is a fundamental part of the basic graduate real variable course as I now teach it. Since there are several excellent texts that generally cover the material here, I'm obliged to render an "apologia" for the present text concerning its content, presentation and existence.

The theme of this book is the notion of absolute continuity and its role as the unifying concept for the major results of the theory, viz., the Lebesgue dominated convergence theorem (LDC) and the Radon–Nikodym theorem (R–N). The main mathematical reason that I've written this book is that none of the other texts in the area stresses this issue to the extent that I think it should be stressed.

Let me be more specific.

The problem of taking limits under the integral sign, that is, "switching limits", is in a very real sense the fundamental problem in analysis. Lebesgue's axiomatization which formulates and proves LDC in an optimal way yields the most important general technique for examining such problems. This material is developed in Chapter 3. Shortly after Lebesgue's initial work Vitali gave necessary and sufficient conditions to switch limits in terms of uniform absolute continuity. Vitali's result led to research which has culminated in Grothendieck's study of weak convergence of measures. This latter material is usually not included in most texts; in particular, its relationship to LDC is not emphasized. This is the reason that I've included Chapter 6.

Knowledge of the structure of measures provides an important tool in modern potential theory, harmonic analysis, and probability theory; its scope of application ranges from establishing a mathematical model for the continuous spectrum of white light to formulating the action of the stock market as Brownian motion in terms of the Wiener measure. The key theorem in this milieu is R–N and the major results involve decompositions of a given measure into various parts with specific properties. R–N is the usual fundamental theorem of calculus (FTC) for the case of functions defined on the real line; and the basic formula by which integration and differentiation are considered as inverse operations is characterized in terms of absolute continuity. We have dwelled on these issues in Chapter 4 and Chapter 5. We give the classical point function results, study the abstract setting, examine their relation, and spend a good deal of time with examples; generally, our treatment is more extensive than usual.

Besides the theme of absolute continuity there are two features of a more secular nature. First, I've included some extensive historical and motivational passages; integration theory did not develop in a vacuum and I've presented information on the development of Fourier series because of its close relation with many of the notions from real analysis. Second, there are large problem sets including certain types of problems which abound in the folklore (e.g. the *Amer. Math. Monthly*) but which are generally omitted from a full year real-variable text whose purpose is to present systematically *n* topics. Some of the problems are quite difficult and will probably challenge even the most mathochistic student. On the other hand I've listed a batch of more or less routine exercises in each chapter. I hope that the historical remarks and problems (especially the harder ones) are read since I believe they provide relevant perspective.

The catalyst for writing this book occurred during the academic year 1970–71 when I was a guest at the Scuola Normale Superiore in Pisa. At that time I discovered Vitali's work at a different level than I had previously known it. He is responsible for the first non-measurable set, the first statement and proof of FTC, the first necessary and sufficient conditions for LDC, the first statement and proof of Lusin's theorem, and more (as you'll see in the text). Essentially, I've decided to engage in some advertising for this most important figure in integration theory.

Now some comments on the format are necessary. The basic text, Chapters 1 through 6, begins with the fundamental classical problems which led to Lebesgue's definition of integral, and develops the theory of integration and the structure of measures in a measure theoretical format. We have viewed this body of information in a unified way; and have relegated to Appendix I, II, and III equally important subject matter which doesn't fit in perfectly with our singular theme of absolute continuity. In Chapter 1 I've inserted details for some of the elementary material, but there is a bloc of more advanced matter, with details omitted, which should be read before moving on to Chapter 2; of course, the instructor may choose to develop this material more fully at any time. Chapters 2 through 5 form the core of the course and are presented in full detail. In Chapter 6 I again discuss things rather briskly; and, although this material could be omitted in a systematic treatment, I think it is important that the student is at least aware of the subject matter.

At the present time I lecture on Chapters 1–4 and most of Appendix I in the first semester. After finishing Appendix I I do the remainder of the book in the second semester. Usually some classroom time remains after I've finished the text. In the past I've lectured on distribution theory or harmonic analysis at this point. Given the student's background, excellent treatments of these topics are found in [57, Chapter 1] (where details must be added) and [16; 35; 45], respectively.

The appendices are meant to serve as outlines for the subject matter treated there. We of course need the functional analysis concepts from Appendix I in Chapters 5 and 6. Fubini's theorem (Appendix II) is certainly one of the most important theorems in integration theory, but doesn't blend in exactly with our absolutely continuous story; although, even in this situation, a connection can be made with R–N from the point of view of conditional probabilities (and, of course, LDC is used to prove Fubini's theorem). In Appendix III our base space is not only a measure space but a topological space, and for this reason we have put the Riesz representation theorem (RRT) outside of the main text. These distinctions are, of course, artificial from most standpoints and I view them as an editorial constraint of no importance. Once again, the instructor may choose to develop the material in the appendices. In fact, I tend to emphasize Appendix III as an introduction to the study of Radon measures and to what is the most suitable approach to use measure theory in harmonic analysis, partial differential equations, and distribution theory.

One last point. There are discussions and name dropping (of concepts) without accompanying definitions and results. Of course, the core material is presented logically; but I think that perspective can be gained by occasional digressions where it would have a negative effect to define every single word.

Acknowledgement

Several friends, both colleagues and students, have generously helped me: A. Beardon critically read parts of the manuscript; S. Espelie and R. Johnson used my notes in their own real variable classes; E. Korn assisted me with some difficult exercises; D. Madan advised me on recent developments in logic; and D. Schmidt provided some editorial assistance.

University of Maryland, Spring 1976   J. J. Benedetto

Table of contents

Notation

Sets and spaces

$A(\mathbf{T})$ A3.1.5b $A'(E)$ A3.1.5b $A'_s(E)$ A3.1.5b
$A'(\mathbf{T})$ A3.1.5b

$\mathscr{B}$ 2.1, 2.3 $\mathscr{B}(X)$ 2.1 $BV[a, b]$ 4.1
BV_{loc} 4.1 $BV(\mathbf{T})$ 4.1 $B(X)$ I.2

C 1.2 C_μ 5.2 $\mathbf{C}$ 1.1
$C(f)$ 1.3.1 $C_0(X)$ III.2 C^∞ P1.16
$C(X)$ I.2 $C_b(X)$ I.2 $C[0, 1]$ 4.1
$C[a, b]$ III.1 $C_c(\mathbf{R})$ I.6 $C_c(X)$ III.3
$C_c^+(X)$ III.4

$D(f)$ 1.3.1 $D^0(X)$ III.3

$F(X)$ 5.5 $\mathscr{F}_\sigma$ 1.3.1 $\mathscr{F}_{\sigma\delta}$ 1.3.1

$\mathscr{G}_\delta$ 1.3.1

$J^+(X)$ III.4

$L_\mu^1(X)$ 3.2 $L_m^1(\mathbf{T})$ 3.3 $L_m^\infty(\mathbf{T})$ 3.3
$\mathscr{L}_m^\infty(X)$ I.2 $L_m^2(\mathbf{T})$ 3.3 $L_\mu^\infty(X)$ I.2
$L(f)$ 4.4 $(L_\mu^p(X))'$ 5.5 $\mathscr{L}_\mu^p(X)$ I.2
$L_\mu^p(X)$ I.2

$\mathscr{M}(\mathbf{R})$ 2.1, 2.3 $\mathscr{M}^n$ 2.2 $M(X)$ 5.1, III.2
$M_c(X)$ 5.2 $M_d(X)$ 5.2 $M_{ac,\nu}(X)$ 5.3

$\emptyset$ 1.1

$\mathscr{P}(X)$ P1.3

$\mathbf{Q}$ 1.1
$\mathbf{R}$ 1.1 $\mathbf{R}^+$ 1.1 $\mathbf{R}^*$ 2.3
$\mathbf{R}^n$ 2.2

$SM(X)$ 5.2 $(S)M(X)$ 5.2 $(S)M_c(X)$ 5.2

T   1.1

Z   1.1   **Z**$^+$ 1.1

E_ξ P1.12 $(X, \mathscr{A})$ 2.3 $(X, \mathscr{A}, \mu)$ 2.3
$(\mathbf{T}, \mathscr{M}, m)$ 3.3 X' I.8, I.9 X'' I.9
$\sigma(X, X'), \sigma(X', X),$ $L(X, Y)$ I.10
    $\sigma(X', X'')$ I.9

Other symbols

$\aleph_0$ 1.1 AC 2.2 $a.e.$ 2.3

card   1.1 c 2.3

D^+, D_+, D^-, D_- 4.3

$ecnd$ 1.3.2   ess sup   2.4

F-T   II.1   FTC   Introduction, 4.3, $B(y)$ 4.2
                              4.4, 4.5
C_E 4.2 χ_A 1.3.1 $N(x)$ 4.1
$P(x)$ 4.1 $Vf(x)$ 4.1 f^+, f^- 3.2
$\hat{f}$ 3.3 $\hat{\mu}$ P5.16 $\Gamma(x)$ 2.2
$W(x)$ 1.3.2 r_y function   1.3.1

int   1.2

LM 2.2   LDC   Introduction, 3.3 $\overline{\lim}, \underline{\lim}$ 1.2
lsc III.4

m 1.2, 2.1, 2.2, 2.3 m^* 2.1 m^n 2.2
μ_E 4.2 δ_x 2.3, 5.2 $\mu\text{-}a.e.$ 2.3
$\mu \times \nu$ II.1 $\mu \perp \nu$ 5.2 $\mu \ll \nu$ 5.2
$|\mu|$ 5.1 μ^+, μ^- 5.1

$\| \ \|_1$ 3.3, I.2 $\| \ \|_\infty$ 2.4 $\| \ \|_p$ I.2

$O(G(x)), o(G(x))$ P3.10
R-N   Introduction, 5.1,   RRT   Introduction, 4.1,
    5.3   III.2, **Theorem III.2**

1 Classical real variable

1.1 Set theory—a framework

"There are mathematicians who claim that there is no difference between mathematics and set theory, but I believe this claim can be dismissed. No mathematician of my acquaintance would abandon his field if an apparently insurmountable contradiction were discovered in the general concept of subset" [44, *Preface*].

The following two articles give readable expositions, including a survey of recent results, in the area of "foundations":

Monk, J. D.: On the foundations of set theory. Amer. Math. Monthly **77** (1970) 703–710;

Henkin, L.: Mathematical foundations for mathematics. Amer. Math. Monthly **77** (1970) 463–486.

Our approach to set theory will be uncritical and incomplete; we refer to the above references for a more thorough treatment, and we assume an acquaintance with naïve set theory and the axioms for the real numbers (as treated in an advanced calculus course).

C, **R**, and **Q** will be the fields of complex, real, and rational numbers, respectively. **Z** is the ring of integers, and the quotient group $\mathbf{T} = \mathbf{R}/2\pi\mathbf{Z}$ is identified with $[0, 2\pi) \subseteq \mathbf{R}$. $\mathbf{R}^+$ and $\mathbf{Z}^+$ are the non-negative elements of **R** and **Z**, respectively. $\emptyset$ denotes the empty set, $A^\sim$ is the complement of the set A, and $A \backslash B = \{x \in A\} \cap \{x \notin B\}$. The cardinality of A is denoted by card A and we write card $\mathbf{Z} = \aleph_0$. "$S_1 \Rightarrow S_2$" is read as "S_1 implies S_2", and "$S_1 \Leftrightarrow S_2$" is read as "S_1 if and only if S_2".

1.2 The topology of R

$x \in \mathbf{R}$ is the limit of the sequence $\{x_n; n = 1, \ldots\} \subseteq \mathbf{R}$ if

$$\forall \varepsilon > 0, \exists N \text{ such that } \forall n \geq N, \quad |x_n - x| < \varepsilon.$$

$x \in \mathbf{R}$ (resp., "$x = \infty$") is a cluster point of $\{x_n : n = 1, \ldots\} \subseteq \mathbf{R}$ if

$$\forall \varepsilon > 0, \forall N, \exists n \geq N \text{ such that } |x_n - x| < \varepsilon \text{ (resp., } x_n \geq \varepsilon).$$

Given $\{x_n : n = 1, \ldots\} \subseteq \mathbf{R}$ we define

$$\overline{\lim}\, x_n = \inf_n \sup_{k \geq n} x_k, \tag{1.1}$$

$$\underline{\lim}\, x_n = \sup_n \inf_{k \geq n} x_k. \tag{1.2}$$

Thus $x \in \mathbf{R}$ (resp., "$x = \infty$") is $\overline{\lim} \, x_n$ if and only if

$$\forall \varepsilon > 0, \exists N \text{ such that } \forall n \geqslant N, \quad x_n < x + \varepsilon$$

and $$\forall \varepsilon > 0, \forall N, \exists n \geqslant N \text{ such that } x_n > x - \varepsilon$$

$$(\text{resp. } \forall \varepsilon > 0, \forall N, \exists n \geqslant N \text{ such that } x_n > N).$$

Naturally the above notions are also defined for "$x = -\infty$".

Intuitively, $\overline{\lim} \, x_n$ is the largest cluster point of $\{x_n : n = 1, \ldots\}$ and $\underline{\lim} \, x_n$ is the smallest cluster point of $\{x_n : n = 1, \ldots\}$. Note, for example, that

$$\overline{} \, x_n = -\infty \quad \text{if and only if} \quad \lim x_n = -\infty.$$

The following facts are proved by routine calculations:

$$\underline{\lim} \, x_n + \underline{\lim} \, y_n \leqslant \underline{\lim} \, (x_n + y_n) \leqslant \overline{\lim} \, x_n + \underline{\lim} \, y_n$$
$$\leqslant \overline{\lim} \, (x_n + y_n) \leqslant \overline{\lim} \, x_n + \overline{\lim} \, y_n, \tag{1.3}$$

provided "$\infty - \infty$" doesn't occur;

$$\overline{\lim} \, (x_n + y_n) = \overline{\lim} \, x_n + \lim y_n \tag{1.4}$$

provided $\lim y_n$ exists.

We shall use the following result in the construction of Cantor sets.

Proposition 1.1 *Let $p > 1$ be an integer.*
a) *If $x \in (0, 1)$ there is $\{a_n : n = 1, \ldots$ and $0 \leqslant a_n < p\} \subseteq \mathbf{Z}$ such that*

$$x = \sum_1^\infty a_n / p^n, \tag{1.5}$$

and the representation (1.5) is unique except when $x = q/p^n$, in which case there are exactly two such representations.
b) *If $\{a_n : n = 1, \ldots$ and $0 \leqslant a_n < p\} \subseteq \mathbf{Z}$ then $\sum a_n / p^n$ converges to some $x \in [0, 1]$.*

"Proof". Just to see what's happening let $p = 3$ and let us write $\frac{1}{2}$ as in (1.5).

$$\frac{1}{2} = \frac{1}{3} + r_1 \quad \text{and} \quad \frac{1}{3^2} < r_1 = \frac{1}{6} = \frac{1}{2} - \frac{1}{3} < \frac{1}{3}$$

$$\frac{1}{2} = \frac{1}{3} + \frac{1}{3^2} + r_2 \quad \text{and} \quad \frac{1}{3^3} < r_2 = \frac{1}{2} - \frac{1}{3} - \frac{1}{3^2} < \frac{1}{3^2}.$$

Therefore,

$$\frac{1}{2} = \sum_1^n \frac{1}{3^j} + r_n, \qquad \frac{1}{3^{n+1}} < r_n < \frac{1}{3^n},$$

and so

$$\frac{1}{2} = \sum_1^\infty \frac{1}{3^j}.$$

Consequently we write

$$\tfrac{1}{2} = .111\ldots \quad (3).$$

(The symbol "(3)" designates that we have expanded the number ("$\tfrac{1}{2}$" in this case) in the base 3.)

In the same way we have

$$\frac{1}{3} = \begin{cases} .100\ldots & (3) \\ .022\ldots & (3). \end{cases}$$

                                                                              "q.e.d."

A set $U \subseteq \mathbf{R}$ is open if for each $x \in U$ there is an open interval I such that $x \in I \subseteq U$. The interior of $X \subseteq \mathbf{R}$, denoted by int X, is the union of all open intervals contained in X. It is easy to prove that

the finite intersection of open sets is open and that the arbitrary union of open sets is open.

Theorem 1.1 *If $U \subseteq \mathbf{R}$ is open then $U = \overset{\infty}{\underset{1}{\bigcup}} I_j$, where I_j is an open interval and $I_j \cap I_k = \emptyset$ for $j \neq k$.*

Proof. U open implies that for each $x \in U$ there is $y > x$ such that $(x, y) \subseteq U$. Set

$$a = \inf\{z \colon (z, x) \subseteq U\}, \quad b = \sup\{y \colon (x, y) \subseteq U\}.$$

Then $x \in (a, b)$, and $I_x = (a, b)$ is an open interval. We shall prove:

a) $U = \bigcup\{I_x \colon x \in U\}$;
b) $\{I_x \colon x \in U\}$ is disjoint (unless $I_x = I_y$);
c) card $\{I_x \colon x \in U\} \leqslant \aleph_0$ (again excluding the case $I_x = I_y$).

(With regard to b) we say that a family of sets is disjoint if the elements of the family are pairwise disjoint.)

It is necessary to make the following preliminary observation: if $x \in U$ and $I_x = (a, b)$ then $I_x \subseteq U$ and $a, b \notin U$.

To see this let $w \in (x, b)$ so that by the definition of b there is $y > w$ for which $(x, y) \subseteq U$ and hence $w \in U$. Now if $b \in U$ there would exist $\varepsilon > 0$ such that $(b - \varepsilon, b + \varepsilon) \subseteq U$. Take ε small enough that $(x, b - (\varepsilon/2)) \subseteq U$. Consequently, $(x, b + \varepsilon) \subseteq U$ contradicting the definition of b.

a) is now obvious since $x \in I_x$ for each $x \in U$.

b) Define $I_x = (a, b)$ and $I_y = (c, d)$, and assume $I_x \cap I_y \neq \emptyset$. Then $a < d$ and $c < b$. Since $c \notin U$ we have $c \notin (a, b)$, and so $c \leqslant a$ (because $c < b$). Since $a \notin U$ we have $a \notin (c, d)$, and so $a \leqslant c$ (because $a < d$). Hence $a = c$, and, similarly, $b = d$. b) follows.

c) For each I_x there is $q_x \in I_x \cap \mathbf{Q}$. If $I_x \neq I_y$ then $q_x \neq q_y$ from b), and we obtain c) since card $\mathbf{Q} = \aleph_0$.

q.e.d.

Proposition 1.2 *Every open set in the plane* $\mathbf{R}^2$ *can be represented as a disjoint union of closed straight line segments.*

A false proof of this result is given in Problem 1.10; a correct proof is given as Solution III of problem E 1434 in Amer. Math. Monthly **68** (1961) 381.

Let $E \subseteq \mathbf{R}$. $x \in \mathbf{R}$ is a point of closure of E if

$$\forall \varepsilon > 0, \exists y \in E \text{ such that } |x - y| < \varepsilon.$$

$\bar{E}$ is the set of points of closure of E, and E is closed if $E = \bar{E}$. If $E \subseteq \mathbf{R}$ is closed then $X \subseteq E$ is dense in E if $\bar{X} = E$. The following facts are easy to check:

$A \subseteq B$ implies $\bar{A} \subseteq \bar{B}$; $\overline{A \cup B} = \bar{A} \cup \bar{B}$; $\bar{\bar{A}} = \bar{A}$; *the finite union of closed sets is closed; the arbitrary intersection of closed sets is closed; the complement of an open set is closed; and the complement of a closed set is open.*

We now define the Cantor set C. This set and its generalizations were crucial motivation for the development of integration theory. Cantor's work in set theory began with his work in trigonometric series; and in this latter research on the problem of unique representation in trigonometric series, Cantor had discussions with his colleague Heine (at Halle) who worked on the same problem (cf. Appendix 3.1.2). After describing C we shall prove the Heine–Borel theorem and observe later that C is an intriguing example related to this result.

C is the set of all $x \in [0, 1]$ with ternary expansion

$$x = .a_1 a_2 \ldots \quad (3), \qquad a_j = 0 \text{ or } 2.$$

Thus $\frac{1}{3} = .0222\ldots$ (3), $\frac{2}{3} = .2000\ldots$ (3) $\in C$.

We observe that $x \notin C$ if $x \in (\frac{1}{3}, \frac{2}{3})$. In fact

$$x = \tfrac{1}{3} + r_1, \qquad 0 < r_1 < \tfrac{1}{3},$$

and so $x = .1000\ldots$ (3) $+ .0a_2 a_3 \ldots$ (3) $= .1a_2 a_3 \ldots$ (3),

where some a_j is non-zero; consequently, $x \notin C$ because of the uniqueness part of Proposition 1.1. In the same way

$$x \notin C \quad \text{if} \quad x \in (k/3^n, (k+1)/3^n), \quad k + 1 < 3^n.$$

This observation leads to the following geometrical description of C. Define

$$\begin{cases} C_1 = C_1^1 \cup C_1^2 = [0, \tfrac{1}{3}] \cup [\tfrac{2}{3}, 1] \\ C_2 = C_2^1 \cup \cdots \cup C_2^4 = [0, \tfrac{1}{9}] \cup [\tfrac{2}{9}, \tfrac{1}{3}] \cup [\tfrac{2}{3}, \tfrac{7}{9}] \cup [\tfrac{8}{9}, 1] \\ C_n = C_n^1 \cup \cdots \cup C_n^{2^n} \\ \quad \vdots \end{cases} \qquad (1.6)$$

Then

$$C = \bigcap_{n=1}^{\infty} C_n, \tag{1.7}$$

and so C is closed.

From this definition it is clear that C contains a countably infinite set with elements of the form $\{k/3^n: k < 3^n, n = 1,...\}$ (although there are integers k such that $k/3^n \notin C$, e.g., $\frac{4}{9} \in (\frac{1}{3}, \frac{2}{3})$). We now prove that card $C > \aleph_0$. Define $f: C \to [0, 1]$ such that if $x = .a_1a_2...$ (3), where a_j is 0 or 2, then $f(x) = .b_1b_2...$ (2), where $b_j = a_j/2$. For example, $f(\frac{1}{3}) = f(.122...$ (3)$) = .0111...$ (2) $= \frac{1}{2}$. Clearly, f maps C onto $[0, 1]$ so that since card $[0, 1] > \aleph_0$, C is uncountable. Thus,

Proposition 1.3 C is a closed uncountable set.

What is the "length" of C? We denote the length of an interval I by mI so that $m[0, 1] = 1$. To construct C we have discarded the following lengths:

first step	$\frac{1}{3}$
second step	$2(\frac{1}{9})$
$\vdots$	
nth step	$2^{n-1}(\frac{1}{3})^n$.

Now

$$1 = \sum_{1}^{\infty} \frac{2^{n-1}}{3^n} \tag{1.8}$$

since the series in (1.8) is

$$\sum_{0}^{\infty} \frac{1}{3}\left(\frac{2}{3}\right)^n = \frac{1}{3}\frac{1}{1-\frac{2}{3}} = 1.$$

Consequently, the natural "length" or "measure" of C is 0.

We generalize the construction of C and define the perfect symmetric set $E \subseteq [0, 1]$ determined by $\{\xi_k: k = 1,...\} \subseteq (0, \frac{1}{2})$: let

$$E_1 = E_1^1 \cup E_1^2,$$

where
$$E_1^1 = [0, \xi_1], \quad E_1^2 = [1 - \xi_1, 1];$$

$$E_2 = E_2^1 \cup ... \cup E_2^4,$$

where
$$E_2^1, E_2^2 \subseteq E_1^1, \ E_2^3, E_2^4 \subseteq E_1^2,$$

and
$$E_2^1 = [0, \xi_1\xi_2], \quad E_2^2 = [\xi_1 - \xi_1\xi_2, \xi_1]$$
$$E_2^3 = [1 - \xi_1, 1 - \xi_1 + \xi_1\xi_2], \quad E_2^4 = [1 - \xi_1\xi_2, 1];$$

continuing this procedure we set

$$E = \bigcap_{k=1}^{\infty} E_k, \quad \text{where} \quad E_k = \bigcup_{j=1}^{2^k} E_k^j.$$

Note that if $\xi_k = \frac{1}{3}$ for each k then $E = C$. In fact, by a proof modelled on that of Proposition 1.3, the statement of Proposition 1.3 remains true for any E. Further note that the length of each E_k^j is given by

$$mE_k^j = \xi_1 \xi_2 ... \xi_k;$$

and so it makes sense to define the measure of E as

$$mE = \lim_{k \to \infty} 2^k \xi_1 ... \xi_k.$$

Consequently, $mC = \lim \left(\frac{2}{3}\right)^k = 0$ which, of course, is the same result that we previously computed for the "length" of C.

Proposition 1.4 For each $\xi \in [0, 1)$ there is $\{\xi_k : k = 1, ...\} \subseteq (0, \frac{1}{2})$ such that $mE = \xi$ for the corresponding perfect symmetric set E.

Proof. Choose $\xi_1 \in (0, \frac{1}{2})$ such that $0 < 2\xi_1 - \xi < 1$. Then take $\xi_2 \in (0, \frac{1}{2})$ such that $0 < 2^2 \xi_1 \xi_2 - \xi < \frac{1}{2}$, and ξ_k such that $0 < 2^k \xi_1 ... \xi_k - \xi < 1/k$; this does it.

To see that the second step (for example) is possible, we know that there is $\alpha \in (0, 1)$ such that $0 < 2\xi_1 \alpha - \xi < \frac{1}{2}$ and therefore we write $\alpha = 2\xi_2$.

q.e.d.

Example 1.1 In order that $mE > 0$ for a perfect symmetric set E we need $\{\xi_k : k = 1, ...\}$ to approach $\frac{1}{2}$ very quickly. For example, take $\xi_k = \frac{1}{2} - 1/2^{k+1}$.

$x \in \mathbf{R}$ is an accumulation point of a set S if $x \in \overline{S \setminus \{x\}}$. S' is the set of accumulation points of S. We say that $x \in S$ is an isolated point of S if $x \notin S'$.

Observe that C has no isolated points. To see this assume $x \in C$ is isolated and let $(x - \delta, x + \delta) \cap C = \{x\}$ for some $\delta > 0$; since $x \in C$ there is $\{C_n^{j_n} : n = 1, ...\}$ for which $x \in \bigcap_n C_n^{j_n}$. Because $mC_n^{j_n} \to 0$ as $n \to \infty$, $C_n^{j_n} \subseteq (x - \delta, x + \delta)$ for all but finitely many n, and this is the desired contradiction.

Also, C is totally disconnected, that is, it contains no open intervals; for if an open interval I were contained in C we'd be able to find an n and j for which

$$(j/3^n, (j + 1)/3^n) \subseteq I,$$

a contradiction.

The same arguments work for any perfect symmetric set E, and thus

Theorem 1.2 Let E be a perfect symmetric set determined by $\{\xi_k : k = 1, ...\} \subseteq (0, \frac{1}{2})$. E is a closed, uncountable, totally disconnected set without isolated points. Also, for each $\xi \in [0, 1)$ there is a perfect symmetric set $E \subseteq [0, 1]$ for which $mE = \xi$.

Our next aim is to prove the Heine–Borel theorem. This is important for several reasons not the least of which is that it serves as the proper motivation to define a compact set.

Theorem 1.3 (Heine–Borel) a) *Let $F \subseteq \mathbf{R}$ be a closed and bounded set. Every open covering $\mathcal{O}$ of F has a finite subcovering.*
b) *Let $\mathcal{C}$ be a collection of closed sets $F \subseteq \mathbf{R}$ such that every finite subcollection of $\mathcal{C}$ has a non-empty intersection. Assume that at least one element of $\mathcal{C}$ is bounded. Then*

$$\bigcap_{F \in \mathcal{C}} F \neq \emptyset.$$

Proof. b) is trivial from a).
a) Let $F = [a, b]$ and define $E = \{x \leqslant b : [a, x]$ can be covered by a finite number of elements of $\mathcal{O}\}$.
$E \neq \emptyset$ since $a \in E$.
Thus, since E is bounded by b it has a least upper bound $c \in [a, b]$. Assume $c < b$.
There is $U \in \mathcal{O}$ such that $c \in (c - \varepsilon, c + \varepsilon) \subseteq U$ for some $\varepsilon > 0$ satisfying $c + \varepsilon \leqslant b$.
$c - \varepsilon$ is not an upper bound for E so we take $x \in E$ for which $x > c - \varepsilon$.
By the definition of E, there exist $U_1, \ldots, U_n \in \mathcal{O}$ such that

$$[a, x] \subseteq \bigcup_1^n U_j,$$

and thus $[a, c + \varepsilon] \subseteq \left(\bigcup_1^n U_j \right) \cup U.$

Consequently, $c + (\varepsilon/2) \in E$, a contradiction; hence, $c = b$.
Next let $F \subseteq \mathbf{R}$ be an arbitrary closed bounded set.
F bounded implies $F \subseteq [a, b]$ for some $a, b \in \mathbf{R}$.
Take the open cover $\mathcal{O} \cup \{F^{\sim}\}$ of $[a, b]$. From the first part there are $U_1, \ldots, U_n \in \mathcal{O}$ such that $F \subseteq [a, b] \subseteq U_1 \cup \ldots U_n \cup F^{\sim}$.
Since $F \cap F^{\sim} = \emptyset$, $F \subseteq U_1 \cup \ldots \cup U_n$.   q.e.d.

1.3 Classical real variable—motivation for the Lebesgue theory

In 1881, Vito Volterra was a student of Ulisse Dini at the Scuola Normale Superiore in Pisa. At this time he published an example of a function on $(0, 1)$, whose derivative exists everywhere, is a bounded function, and is not Riemann integrable [124]. In his thesis [67] (1902), Lebesgue noted that because of Volterra's example one can't consider differentiation and integration as inverse operations for a large enough class of functions; and thus he was motivated to try to find a notion of integration such that integration and differentiation are inverse operations for a much larger class of functions than is usually included by Riemann's theory. (H. J. Smith had published a Volterra example in 1875, but Lebesgue was apparently unaware of Smith's work in 1902.) There were several other important reasons for extending the notion of integration; generally, one could lump all of them together and say that there was a need for an integration theory that included many "natural" results (e.g., the validity of term by term integration of a series of functions) for many functions. We'll see that Lebesgue's theory fills this criterion better than any other that has come along so far.

Now for the above-mentioned example due to Volterra.

Example 1.2 Let $E \subseteq [0, 1]$ be a perfect symmetric set with $mE > 0$. On each contiguous interval (a, b) we temporarily define

$$\varphi(x, a) = (x - a)^2 \sin \frac{1}{x - a}, \qquad x \in [a, b],$$

noting that $\varphi(a, a) = 0$. (The interval (a, b) is contiguous to the closed set $F \subseteq \mathbf{R}$ if $a, b \in F$ and $(a, b) \subseteq F^{\sim}$.) For $x \in (a, b)$,

$$\varphi'(x, a) = 2(x - a) \sin \left(\frac{1}{x - a} \right) - \cos \left(\frac{1}{x - a} \right).$$

Consequently, for all large k and $x = a + (1/k\pi)$,

$$\varphi'(x, a) = -\cos k\pi;$$

and so φ' achieves ± 1 and hence 0, infinitely often. Let $a + y$ be the largest zero of φ' less than or equal to $(a + b)/2$ and define

$$f(x) = \begin{cases} 0, & \text{if } x \in E, \\ \varphi(x, a), & \text{if } a \leqslant x \leqslant a + y, \\ -\varphi(x, b), & \text{if } b - y \leqslant x \leqslant b, \\ \varphi(a + y, a), & \text{if } a + y \leqslant x \leqslant b - y \end{cases}$$

on $[0, 1]$, where (a, b) is a generic contiguous interval of E. It is now straightforward to verify that

a) f' exists on $(0, 1)$ and is 0 on E;
b) f' is discontinuous on E.

To check b), for example, if $x \in E$ then there are contiguous intervals (a_j, b_j) such that $\lim a_j = x = \lim b_j$; thus, since φ' takes on both ± 1 values, we have the required discontinuity.

The state of the art was such at the time of Volterra's example that it was realized he had constructed a function f with a bounded non-Riemann integrable derivative; and in particular, that there was an everywhere differentiable function f for which the (Riemann) fundamental theorem of calculus did not work. The key thing is that f' is discontinuous on a set with strictly positive "length". The technicalities of Volterra's example are based on a simple idea. It was well known that $g(x) = x^2 \sin(1/x)$, $x \neq 0$, and $g(0) = 0$ had a discontinuous derivative at $x = 0$; and Volterra constructed f in terms of these g's and, so, spread the discontinuities of f' to all of E.

Because of Example 1.2 it became important to study set theoretic properties of continuous and differentiable functions, and to take more care in defining notions of small sets.

1.3.1 Continuous functions

A function $f: X \to \mathbf{R}$, where $X \subseteq \mathbf{R}$, is continuous at $x \in X$ if

$$\forall \varepsilon > 0, \exists \delta > 0 \quad \text{such that} \quad \forall y \in X,$$
$$|x - y| < \delta \Rightarrow |f(x) - f(y)| < \varepsilon;$$

a function $f: X \to \mathbf{R}$, where $X \subseteq \mathbf{R}$, is continuous on X if it is continuous at each $x \in X$.

Proposition 1.5 *Let $F \subseteq \mathbf{R}$ be closed and bounded, and assume the function $f: F \to \mathbf{R}$ is continuous. Then f is bounded on F and achieves its maximum and minimum there.*

Proof. a) To show f is bounded.

f continuous on F implies that for each $x \in F$ there is an open interval I_x such that $x \in I_x$ and

$$\forall y \in I_x \cap F, \qquad |f(x) - f(y)| < 1.$$

Since $F \subseteq \bigcup \{I_x : x \in F\}$ we can apply the Heine–Borel theorem, and so there are intervals $I_{x_1}, \ldots, I_{x_n}$ which cover F.

Consequently, $|f(y)| \leqslant 1 + \max \{|f(x_1)|, \ldots, |f(x_n)|\}$ and f is bounded.

b) To show that f achieves its maximum (resp. minimum) on F.

Since f is bounded, $-\infty < \sup \{f(x): x \in F\} = M < \infty$. We must find $x_1 \in F$ such that $f(x_1) = M$. If this is not the case then $f(x) < M$ for each $x \in F$; we obtain a contradiction to this possibility.

From the continuity of f,

$$\forall x \in F, \exists I_x, \text{ an open interval, such that } x \in I_x$$

and $\qquad \forall y \in I_x \cap F, \qquad |f(x) - f(y)| < \tfrac{1}{2}(M - f(x)).$

Consequently,

$$\forall x, \quad \forall y \in F \cap I_x, \qquad f(y) < \tfrac{1}{2}(M + f(x)).$$

Because of the Heine–Borel theorem there is a finite subcover $I_{x_1}, \ldots, I_{x_n}$ (from $\{I_x : x \in F\}$) of F with the property that

$$\forall y \in F, \qquad f(y) < \tfrac{1}{2}(M + A),$$

where $A = \max \{f(x_1), \ldots, f(x_n)\}$.

Therefore, $\frac{1}{2}(M + A)$ is a bound for f on F; and this is a contradiction since $\frac{1}{2}(M + A) < M$.

q.e.d.

We make the following definitions

a) $f\colon X \to \mathbf{R}$ is uniformly continuous on $X \subseteq \mathbf{R}$ if

$$\forall \varepsilon > 0 \quad \exists \delta > 0 \quad \text{such that} \quad \forall x, y \in X,$$
$$|x - y| < \delta \Rightarrow |f(x) - f(y)| < \varepsilon;$$

b) $\{f_n\colon n = 1\}\colon X \to \mathbf{R}$ converges pointwise to f if

$$\forall x \in X, \forall \varepsilon > 0, \exists N \quad \text{such that} \quad \forall n \geqslant N,$$
$$|f_n(x) - f(x)| < \varepsilon;$$

c) $\{f_n\colon n = 1,\ldots\}\colon X \to \mathbf{R}$ converges uniformly on X to f if

$$\forall \varepsilon > 0 \quad \exists N \quad \text{such that} \quad \forall x \in X \quad \text{and} \quad \forall n \geqslant N,$$
$$|f_n(x) - f(x)| < \varepsilon.$$

It is easy to prove:

a) *f is continuous on $X \subseteq \mathbf{R} \Leftrightarrow$ for each open set $U \subseteq \mathbf{R}, f^{-1}(U)$ is open in X;*

b) *if $\{f_n\colon n = 1,\ldots\}\colon X \to \mathbf{R}$ is a sequence of continuous functions which converges uniformly to f on X then f is continuous.*

Example 1.3 Given a perfect symmetric set $E \subseteq [0, 1]$ determined by $\{\xi_k\colon k = 1,\ldots\} \subseteq (0, \frac{1}{2})$. We shall define the Cantor function, C_E, for E. Set $f_1 = \frac{1}{2}$ on $[\xi_1, 1 - \xi_1]$, and extended to $[0, 1]$ so that it is continuous, increases from 0 to 1 (in range values), and is linear on E_1^1 and E_1^2 (e.g. Fig. 1).

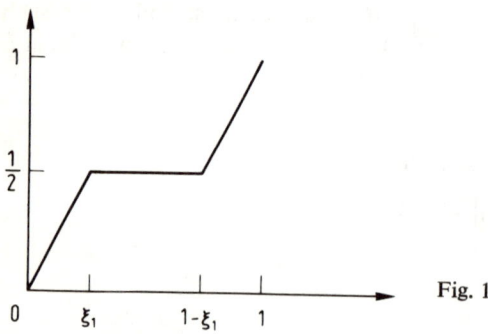

Fig. 1

f_n is defined as the continuous, increasing function with $f_n(0) = 0$ and $f_n(1) = 1$, which is linear on each E_n^j, $j = 1,\ldots, 2^n$, and takes the value $j/2^n$ on the contiguous interval immediately to the right of E_n^j. $\{f_n\colon n = 1,\ldots\}$ converges uniformly to a function f so that f is continuous; it is also easy to check that f increases from 0 to 1 (in range values) and is constant on each interval contiguous to E.

The notion of uniform continuity arose in the following context to define properly the Riemann integral of a continuous function on $[a, b]$.

Proposition 1.6 *Let f be continuous on a closed and bounded set $F \subseteq \mathbf{R}$. Then f is uniformly continuous.*

Proof. For each $\varepsilon > 0$ and $x \in F$ there is $\delta_x > 0$ such that

$$|x - y| < \delta_x \Rightarrow |f(x) - f(y)| < \varepsilon/2.$$

Let $U_x = (x - \frac{1}{2}\delta_x, x + \frac{1}{2}\delta_x)$ so that $\{U_x : x \in F\}$ is an open cover of F. From the Heine–Borel theorem there are sets $U_{x_1}, ..., U_{x_n}$ whose union covers F.

Take $\delta = \frac{1}{2} \min \{\delta_{x_1}, ..., \delta_{x_n}\}$ and let $|z - y| < \delta$.

If $y \in U_{x_i}$ we have

$$|y - x_i| < \tfrac{1}{2}\delta_{x_i} < \delta_{x_i}$$

and thus

$$|z - x_i| \leq |z - y| + |y - x_i| < \delta + \tfrac{1}{2}\delta_{x_i} \leq \delta_{x_i}.$$

Consequently, $|f(z) - f(x_i)|, |f(y) - f(x_i)| < \varepsilon/2$.

q.e.d.

The notion of topology is defined in Appendix I; for now, when you see the word, it is meaningful to translate it as "convergence criterion".

With regard to the characterization of continuous functions in terms of open sets we now define: a function $f: X \to \mathbf{R}$, where $X \subseteq \mathbf{R}$, is open if for every open set $U \subseteq X$, $f(U)$ is open.

It is easy to find open discontinuous functions $f: \mathbf{R} \to \mathbf{R}$ if we make the topology on the domain strictly weaker than on the range. A more interesting example is

Example 1.4 With the usual topology on $\mathbf{R}$ we give an example of an open discontinuous function $f: \mathbf{R} \to \mathbf{R}$. Let f be strictly increasing, continuous, and with range $\mathbf{R}$ on each contiguous interval of the Cantor set C. Set $f = 0$ on C. Thus $f: [0, 1] \to \mathbf{R}$ is certainly not continuous. Without loss of generality we show that $f(I)$ is open for each open interval $I \subseteq [0, 1]$. The result is clear if I is contained in a contiguous interval. If I covers some endpoint c of a "middle third interval" then there is a "middle third interval" $J \subseteq I$ so that $\mathbf{R} = f(J) \subseteq f(I)$ and f is open. The fact that there is such a J follows since I is open and the endpoints of smaller and smaller "middle third intervals" will converge to c.

A countable union of closed sets in $\mathbf{R}$ is an $\mathscr{F}_\sigma$ and a countable intersection of open sets is a $\mathscr{G}_\delta$; thus the complement of an $\mathscr{F}_\sigma$ is a $\mathscr{G}_\delta$ and vice-versa. We can define much more complicated sets. For example, an $\mathscr{F}_{\sigma\delta}$ is the intersection of a countable family of

$\mathscr{F}_\sigma$'s. All of these sets are special cases of Borel sets of which we'll have a lot more to say.

For a given function f, $C(f)$ is its set of points of continuity and $D(f)$ its set of discontinuities.

The oscillation of $f: \mathbf{R} \to \mathbf{R}$ over a closed bounded interval I is

$$\omega(f, I) = \sup_{x \in I} f(x) - \inf_{x \in I} f(x);$$

and the oscillation of $f: \mathbf{R} \to \mathbf{R}$ at x is

$$\omega(f, x) = \inf \{\omega(f, I): I \text{ a closed bounded interval}, x \in \text{int } I\}.$$

Clearly, $\omega(f, x) \geqslant 0$.

Proposition 1.17 *Given a function* $f: \mathbf{R} \to \mathbf{R}$.

a) $x \in C(f) \Leftrightarrow \omega(f, x) = 0$.

b) $C(f)$ *is a* $\mathscr{G}_\delta$ *and* $D(f)$ *is an* $\mathscr{F}_\sigma$.

Proof. a) follows from the definition of continuity.

b) Define

$$E_n = \{x: \inf_I \omega(f, I) \geqslant 1/n\},$$

where I runs through the closed bounded intervals containing x.

It is easy to check that E_n is closed.

Now if $x \in D(f)$, $\omega(f, x) > 0$ and so $x \in E_n$ for some n. Thus

$$D(f) \subseteq \bigcup E_n.$$

Conversely, if $x \in \bigcup E_n$ then x is in some E_n which means that f is discontinuous at x by a) Consequently,

$$\bigcup E_n \subseteq D(f).$$

q.e.d.

Riemann defined the ruler function $r: [0, 1] \to [0, 1]$ as

$$r(x) = \begin{cases} 0, & \text{if } x \in [0, 1] \text{ is irrational}, \\ 1, & \text{if } x = 0, \\ 1/q, & \text{if } x = p/q \in (0, 1], \quad (p, q) = 1. \end{cases}$$

(The notation "$(p, q) = 1$" indicates that p and q are relatively prime.) He proved

Proposition 1.8 *r is continuous at each irrational and discontinuous at each rational.*

Proof. Take any rational p/q satisfying $(p, q) = 1$, and let $0 < \varepsilon < 1/q$.
Since there is an irrational in any interval about p/q there is no interval I about p/q
with the property that

$$\forall x \in I, \qquad |r(x) - r(p/q)| < \varepsilon$$

(because $|r(x) - r(p/q)| = 1/q > \varepsilon$ for irrational $x \in I$). Let $x \in [0, 1]$ be irrational and
let $\varepsilon > 0$. We must find $\delta > 0$ such that

$$\forall y \in (x - \delta, x + \delta), \qquad |r(x) - r(y)| < \varepsilon.$$

Clearly there are only finitely many $p/q \in [0, 1]$, $(p, q) = 1$, for which $1/q \geqslant \varepsilon$. Call
these $x_1, ..., x_n$.
Choose δ so that $x_j \notin (x - \delta, x + \delta) = I$ for any j.
Therefore, for all $y \in I$, $|r(x) - r(y)| - |r(y)| < \varepsilon$.

$$\text{q.e.d.}$$

We generalize r as follows:
Example 1.5 Define $r_y: [0, 1] \to \mathbf{R}$ as

$$r_y(x) = \begin{cases} 0, & \text{if } x \in [0, 1] \text{ is irrational,} \\ 1, & \text{if } x = 0, \\ \gamma_q, & \text{if } x = p/q, \ (p, q) = 1. \end{cases}$$

As in the proof of Proposition 1.8 we see that r_y is continuous at irrational x if $\gamma_q \to 0$
as $q \to \infty$, and discontinuous at each rational x where $\gamma_q \neq 0$. We shall have more to
say about r_y.

Example 1.6 Hardy [50, p. 190] defines $h: \mathbf{R} \to \mathbf{R}$ as

$$h(x) = \begin{cases} x, & \text{if } x \text{ is irrational,} \\ \left(\dfrac{1 + p^2}{1 + q^2}\right)^{1/2}, & \text{if } x = p/q, \ (p, q) = 1. \end{cases}$$

h is discontinuous on $(-\infty, 0) \cup (\mathbf{Q} \cap (0, \infty))$ and continuous on $(\mathbf{R} \setminus \mathbf{Q}) \cap (0, \infty)$.
It is now natural to ask if there is a function $f: \mathbf{R} \to \mathbf{R}$ such that f is continuous on the
rationals and discontinuous on the irrationals. We shall show that the answer is "no";
in light of Proposition 1.7, this reduces to showing that the rationals are not only an
$\mathscr{F}_\sigma$, which is obvious, but are not a $\mathscr{G}_\delta$. To prove this latter property it is convenient to
introduce the notion of a nowhere dense set. $X \subseteq \mathbf{R}$ is nowhere dense if

$$\text{int } \bar{X} = \varnothing.$$

Nowhere dense sets are totally disconnected and the two notions are equivalent (on $\mathbf{R}$)
for closed sets. In particular, any perfect symmetric set is nowhere dense. $X \subseteq \mathbf{R}$ is a

set of first category if it can be written as a countable union of nowhere dense sets. The major result about category is the Baire category theorem which we'll discuss in Appendix I. For now we assume its statement for $\mathbf{R}$:

$$\mathbf{R} \text{ is not of first category.} \tag{1.9}$$

Because of this, $\mathbf{R} \setminus \mathbf{Q}$ is not of first category, for if it were we'd obtain a contradiction to (1.9) since $\mathbf{Q}$ is clearly a set of first category.

Proposition 1.9 a) $\mathbf{Q}$ is not a $\mathcal{G}_\delta$.

b) There is no function $f \colon \mathbf{R} \to \mathbf{R}$, continuous on the rationals and discontinuous on the irrationals.

Proof. b) is clear from Proposition 1.7, the above discussion, and a).

a) If $\mathbf{Q}$ were a $\mathcal{G}_\delta$ then $\mathbf{R} \setminus \mathbf{Q} = \bigcup F_n$, F_n closed.

Since $F_n \subseteq \mathbf{R} \setminus \mathbf{Q}$, it contains no intervals and so is nowhere dense.

Thus $\mathbf{R} \setminus \mathbf{Q}$ is of first category, the required contradiction.

q.e.d.

χ_A will denote the characteristic function of a set A.

Example 1.7 Let $f \colon \mathbf{R} \to \mathbf{R}$ be a non-constant function with the property that every point is a local minimum, i.e.,

$$\forall x \in \mathbf{R} \quad \exists I, \quad \text{an open interval about } x, \text{ such that } \forall y \in I, \quad f(x) \leqslant f(y).$$

It can be shown that $D(f)$ is nowhere dense. The argument proceeds by assuming the result is false and generating a nested sequence of closed intervals in the proper way whose intersection $\{x\}$ is not a local minimum; the details are neither obvious nor impossible. As a sort of converse, take any closed nowhere dense set E and define

$$f = \chi_{\mathbf{R} \setminus E}.$$

Then $D(f) = E$ and f has a local minimum at each $x \in \mathbf{R}$.

1.3.2 Sets of differentiability

Example 1.8 We show that the *ruler function is nowhere differentiable* (of course, r is only continuous on $[0, 1] \setminus \mathbf{Q}$ not on all of $[0, 1]$). Let $0 < x < 1$ be irrational, and note that

$$\frac{r(x + h) - r(x)}{h} = \frac{r(x + h)}{h},$$

where $h \neq 0$. Let $h_i \in \mathbf{R}$ have the property that $x + h_i$ is irrational and $h_i \to 0$; thus

$$\forall i, \quad r(x + h_i)/h_i = 0.$$

Write $x = .a_1a_2...$ (10) and set $h_i = .a_1...a_i - x$. Since x is non-zero and irrational, $a_i \neq 0$ for infinitely many i. Let N be the smallest integer such that $a_N \neq 0$. Now

$$\forall i \geqslant N, \qquad r(x + h_i) = r(.a_1...a_i) \geqslant 10^{-i},$$

by definition of r. Also, $|h_i| = .0...0a_{i+1}... \leqslant 10^{-i}$. Hence

$$\forall i \geqslant N, \qquad |r(x + h_i)/h_i| \geqslant 1.$$

Consequently, $\lim_{h \to 0} r(x + h)/h$ does not exist.

With this example it is natural to inquire if some r_y, for $\gamma_q \to 0$, is differentiable anywhere or even possibly at all irrationals. We shall see in Example 1.9 that there is r_y, for $\gamma_q \to 0$, such that $r'_y(x)$ exists for some (irrational) x, but that there is no $\gamma_q \to 0$ such that $r'_y(x)$ exists for all irrational x. To prove this last assertion we need the following result of M. K. Fort, Jr. (e.g. [13, pp. 126–127], where the second inequality on line 6, p. 127 should be reversed, or [13, 2nd edition]):

> Let $f: [0, 1] \to \mathbf{R}$ be discontinuous on S, $\bar{S} = [0, 1]$; then $\{x: \exists f'(x)\}$
> is a set of first category.   (1.10)

An interesting elaboration of Fort's theorem is found in [9].

Example 1.9 a) Let $\gamma_q = 1/q^4$ and define r_y. A quadratic irrational is the root of a quadratic equation with integer coefficients. It is well known [51, Theorem 188] that if x is a quadratic irrational then for all large q

$$\left| x - \frac{p}{q} \right| > \frac{1}{q^3}.$$

Consequently, for such an $x \in (0, 1)$,

$$\left| \frac{r_y(x) - r_y(p/q)}{x - (p/q)} \right| < \frac{1}{q},$$

and so $r'_y(x) = 0$.

b) If $\gamma_q \to 0$ then r_y is discontinuous on a dense set (the rationals) and so r'_y exists on a set of first category by (1.10); but the irrationals are not of first category by (1.9). Thus there is no r_y, where $\gamma_q \to 0$, for which r'_y exists on all of the irrationals.

Example 1.10 We give examples of functions $f: \mathbf{R} \to \mathbf{R}$ which are differentiable precisely at a single point.

a) Let $f(x) = x^2\chi_\mathbf{Q}(x)$. Clearly f is discontinuous everywhere except at the origin, and $f'(0) = 0$.

b) Assume for the moment that there are everywhere continuous nowhere differentiable functions $g \colon \mathbf{R} \to \mathbf{R}$; in fact, such functions exist as we'll see. Define $f(x) = xg(x)$. Now

$$\frac{f(0 + h) - f(0)}{h} = \frac{hg(h)}{h} = g(h) \to g(0), \qquad \text{as} \quad h \to 0,$$

and so $f'(0) = g(0)$. If $x \neq 0$ then

$$\frac{f(y) - f(x)}{y - x} = x\frac{g(y) - g(x)}{y - x} + g(y);$$

consequently, $f'(x)$ doesn't exist since $g(y) \to g(x)$ and $g'(x)$ doesn't exist.

Example 1.11 Let $f \colon [0, 1] \to \mathbf{R}$ and suppose $f'(x)$ exists and is not zero for each x in an uncountable set S. We observe that $|f|'$ exists on an uncountable set X. In fact, define

$$X = \{x \in S \colon f(x) \neq 0\},$$

and first note (arguing by contradiction) that card $X > \aleph_0$. Also since f has the same sign in some neighborhood of $x \in X$ and since $f'(x)$ exists we have the existence of $|f|'(x)$. For a bit more of a challenge consider the corresponding question for countably infinite sets.

In Example 1.10b we mentioned everywhere continuous nowhere differentiable functions, which we shall denote by *ecnd*. For the sake of clarity, when we say f is not differentiable at x we shall mean that the corresponding difference quotient does not approach a finite limit. In a lecture in 1861, Riemann is supposed to have asserted that

$$R(x) = \sum_{1}^{\infty} \frac{\sin k^2\pi x}{k^2}$$

is *ecnd*. The continuity follows from the uniform convergence of the series. In his attempt to prove Riemann's claim for $R(x)$, Weierstrass proved (although duBois–Reymond first published it in 1874) that

$$K(x) = \sum_{1}^{\infty} a^n \cos b^n\pi x, \qquad \text{where} \quad b \in \mathbf{Z} \text{ is odd}, \, a \in (0, 1),$$

$$\text{and } ab > 1 + \frac{3\pi}{2},$$

is *ecnd*. In 1916, Hardy proved that $R(x)$ is not differentiable at any irrational and some rationals; and he also showed that

$$K(x), \quad b > 1, \quad 0 < a < 1, \quad ab \geqslant 1,$$

is *ecnd*. Naturally enough there have been many subsequent examples of *ecnd* functions.

One of the most popular and elementary is due to van der Waerden (1930) [114, Section 11.23]; we also refer ahead to Problem 4.7. We shall designate van der Waerden's function by $W(x)$ (for future use).

There have been two rather exciting chapters in the business of finding *ecnd* functions. In 1916, G. C. Young brought attention to Cellerier's example of an *ecnd* function,

$$\sum_1^\infty \frac{1}{a^n} \sin a^n \pi x, \qquad a > 1,$$

discovered before 1850; and Bolzano had constructed an *ecnd* function about 1830, a fact not discovered until 1921. It should be noted that Bolzano didn't realize the full potential of his example. The second incident concerning *ecnd* functions is more recent and involves $R(x)$. In 1969, Joseph Gerver, at the time an undergraduate at Columbia University, proved the differentiability of $R(x)$ at certain rational points [41]; he has since written two extensive papers on the matter in the Amer. J. Math., and there have been elaborations and simplifications by others.

Example 1.12 We discuss the fact that

$$H(x) = \sum_1^\infty \frac{1}{2^k} \cos 2^k x$$

is an *ecnd* on **T**. Let $\{n_k : k = 1, ...\}$ be a sequence of positive integers with the property that

$$\inf_k \frac{n_{k+1}}{n_k} > 1.$$

Such sequences were first used by Hadamard to prove that every point on the circle of convergence of a power series with such exponents is a singularity of the function represented; we refer to such a sequence $\{n_k : k = 1, ...\}$ as an Hadamard set. Such sequences and certain generalizations are studied extensively in modern Fourier analysis. The fact that H is *ecnd* follows immediately from

> Given $f(x) = \sum_1^\infty r_k \cos n_k x$, where $\sum r_k < \infty$, $r_k \geqslant 0$, and
> $\{n_k : k = 1, ...\}$ is Hadamard; if f is differentiable at a point
> then $\lim_k n_k r_k = 0$. (1.11)

The proof of (1.11) is found in [59] and is generally simpler than the classical methods used to prove that a function is *ecnd*.

Example 1.13 We now indicate how, for every closed nowhere dense set $E \subset [0, 1]$, there is an *ecnd* f with the properties that

$$f = 0 \quad \text{on} \quad E$$
$$\text{and} \quad f > 0 \quad \text{on} \quad [0, 1] \setminus E$$

(cf. Problem 1.16). We first note that there are problems if we proceed "naturally"; i.e., if we set $f = 0$ on E and define it to be a contraction of W on each contiguous interval (noting that $W(0) = W(1) = 0$) then f is continuous but has points of differentiability. Let $\{(a_j, b_j): j = 1, ...\}$ be the set of contiguous intervals for E and enumerate all of the intervals $(2^{-j}(k - 1), 2^{-j}k)$, $k = 1, 2, ..., 2^j$, $j = 1, 2, ...$, as $\{A_n: n = 1, ...\}$. Next define F_1 to be any (a_j, b_j) in A_1 in case such intervals exist and $\varnothing$ otherwise. Generally, let F_n be any (a_j, b_j) in A_n, where $F_n \neq F_j$ for $j = 1, ..., n - 1$, in case such intervals exist, and $\varnothing$ otherwise. If $x \in F_j = (a_{n_j}, b_{n_j})$ we define

$$f(x) = m(A_j)W((x - a_{n_j})/(b_{n_j} - a_{n_j}));$$

and if $x \in (a_n, b_n)$, a contiguous interval which is not equal to any F_j, $j = 1, ...$, we set

$$f(x) = (b_n - a_n)W((x - a_n)/(b_n - a_n)).$$

Finally set $f = 0$ on E. This function is due to J. Lipiński [73]. The verification that f is ecnd is Problem 1.25.

We now shift our emphasis and consider the situation that f' exists for each $x \in (0, 1)$. We want to find the set of discontinuities of f'.

Example 1.14 We'll find a function $f: (0, 1) \to \mathbf{R}$ such that f' exists on $(0, 1)$ and $D(f') = \mathbf{Q} \cap (0, 1)$. Let $g(x) = x^2 \sin(1/x)$, $g(0) = 0$ so that g' is discontinuous at $x = 0$. If $\{r_n: n = 1, ...\} \subseteq (0, 1)$ is an enumeration of the rationals, then set

$$f(x) = \sum_1^\infty \frac{1}{2^n} g(x - r_n).$$

Clearly, $\sum (1/2^n)g'(x - r_n)$ converges uniformly and so

$$f'(x) = \sum_1^\infty \frac{1}{2^n} g'(x - r_n), \qquad x \in (0, 1).$$

Thus f' is discontinuous at each rational by the definition of g, cf. Volterra's example.

Proposition 1.10 $X = D(f')$ for some function f whose derivative exists at each $x \in [0, 1] \Leftrightarrow X$ is an $\mathscr{F}_\sigma$ of first category.

Proof. ($\Rightarrow$) f' is discontinuous on an $\mathscr{F}_\sigma$ by Proposition 1.7 and it is easy to check that this $\mathscr{F}_\sigma$ is of first category.

($\Leftarrow$) We can write $X = \bigcup F_n$, $F_n \subseteq F_{n+1}$, F_n closed and nowhere dense. As in the Volterra example (Example 1.2) we take f_n differentiable on $[0, 1]$ and with the property that $f'_n = \pm 1$ infinitely often in any neighborhood of any $x \in F_n$.

Consequently (e.g. Example 1.14), $f = \sum f_n/3^n$ is differentiable on $[0, 1]$ and f' is discontinuous precisely on X.

q.e.d.

Remark Because of Proposition 1.10 there is no f: $[0, 1] \rightarrow \mathbf{R}$ whose derivative exists but is discontinuous everywhere. Even more, there is no f: $[0, 1] \rightarrow \mathbf{R}$ whose derivative exists everywhere but is discontinuous on the complement of a countable set.

For extensive surveys of the remarks made in section 1.3.1 and section 1.3.2 we list [13; 23; 111].

Basically we have seen that some of the problems in 19th century analysis led to various methods of characterizing the size of a given set, e.g. countability, category, and "length". As we'll now see, an in-depth study of "length" (i.e. measure) was crucial in getting past the problems of Riemann's integration theory that were focused upon by the likes of Volterra's example. Borel was the mathematician who heralded measure theory in this regard; and Lebesgue was the one who discovered the associated integration theory, which, in a very real way, gets more important theorems than is possible by any other method. Of course, Lebesgue came to his integral after considerably extending Borel's measure, cf. L. W. Cohen's analysis in Scripta Math. **29** (1973) 417–435; and this issue was the major cause for the ensuing disagreement between Borel and Lebesgue, cf. Monna's study in Arch. Hist. Exact Sci. **9** (1972) 57–84 for another aspect of this polemic.

1.4 References for the history of integration theory

The following contains several listings on Fourier series because of the intimate relation in the development of integration theory and Fourier series.

[1] Bliss, G. A.: Integrals of Lebesgue. BAMS **24** (1917) 1–47.

[2] Bourbaki, N.: Intégration (Introduction to chapters I–IV, and the historical remarks for succeeding chapters). Paris 1965

[3] Dauben, J.: The trigonometric background of G. Cantor's theory of sets. Arch. Hist. Ex. Sci. **7** (1971) 181–216

[4] Fréchet, M.: La vie et l'œuvre d'Émile Borel. Monog. L'Enseignement Math. **14** (1965) (in particular, pp. 53–63)

[5] Freudenthal, H.: Did Cauchy plagiarize Bolzano? Arch. Hist. Ex. Sci. **7** (1971) 375–392

[6] Gibson, G. A.: On the history of the Fourier series. Proc. Edinburgh Math. Soc. **11** (1893) 137–166

[7] Grattan-Guinness, I.: The development of the foundations of mathematical analysis from Euler to Riemann. MIT, 1970 (if you read pp. 76–78 then you have a moral obligation to read Freudenthal's article, number 5 above, and Waterhouse's review, number 17 below)

[8] Hawkins, T.: Lebesgue's theory of integration. Madison, Wisc. 1970

[9] Lebesgue, H.: Intégrale, longeur, aire. Ann. Mat. **7** (1902) 231–359

[10, 11] —: Lecons sur l'intégration et la recherche des fonctions primitives (1st and 2nd ed.) Paris 1904 and 1928

[12] Lorch, E. R.: Continuity and Baire functions. Amer. Math. Monthly **78** (1971) 748–762

[13] Paplauskas, A. B.: L'Influence de la théorie des séries trigonométriques sur le développement du calcul intégral. Arch. Int. d'Hist. Sci. **84–85** (1968) 249–260

[14] Pesin, I.: Classical and modern integration theories. New York 1970

[15] Riesz, F.: L'Évolution de la notion d'intégrale depuis Lebesgue. Ann. Inst. Fourier, Grenoble **1** (1949–50) 29–42 (e.g. Chapter A3.2)

[16] Saks, S.: Theory of the integral. New York.

[17] Waterhouse, W. C.: Book Reviews. BAMS **78** (1972) 385–391

Problems for Chapter 1

Some of the more elementary problems in this set are the Problems 1.1, 1.2, 1.3, 1.4, 1.5, 1.10, 1.13, 1.19, 1.21, 1.23, 1.24.

1.1 If $\{\{a\}, \{a, b\}\} = \{\{c\}, \{c, d\}\}$ prove that $a = c$ and $b = d$.

1.2 Given $f: X \to Y$, where f is a function and X and Y are sets.

a) Prove

$$
\begin{aligned}
f(\bigcup A_\alpha) &= \bigcup f(A_\alpha), \\
f^{-1}(\bigcup B_\alpha) &= \bigcup f^{-1}(B_\alpha), \\
f^{-1}(\bigcap B_\alpha) &= \bigcap f^{-1}(B_\alpha), \\
f^{-1}(B^\sim) &= [f^{-1}(B)]^\sim.
\end{aligned}
$$

b) Prove that the following are generally proper inclusions

$$
\begin{aligned}
f(\bigcap A_\alpha) &\subseteq \bigcap f(A_\alpha), \\
f[f^{-1}(B)] &\subseteq B, \\
f^{-1}[f(A)] &\supseteq A.
\end{aligned}
$$

c) Show that if f is surjective then $f[f^{-1}(B)] = B$.

1.3 a) Given a set X and $\mathscr{P}(X)$, the set of all subsets of X. Prove that there is no surjection $f: X \to \mathscr{P}(X)$.

(*Hint.* Set $B = \{x \in X: x \notin f(x)\}$.)

b) Let $S = \{\{a_n: n = 1, \ldots\}: a_n = 0 \text{ or } 1\}$. Prove card $S > \aleph_0$.

(*Hint.* Use the Cantor diagonal process.)

1.4 a) Prove that there is an uncountable closed set of irrationals in [0, 1].

b) Prove that there are uncountable sets of real numbers which do not contain uncountable closed subsets.

(*Hint.* One can show the existence of $X \subseteq \mathbf{R}$ which does not contain any subset Y having the properties: Y is closed and contains no isolated points [53, pp. 201–202].)

1.5 Find a discrete subset of $\mathbf{R}$ with uncountable closure ($X \subseteq \mathbf{R}$ is discrete if for each $x \in X$ there is $I \subseteq \mathbf{R}$, an open interval, such that $I \cap X = \{x\}$).

1.6 a) (Schroeder–Bernstein) Prove that if $f: A \to B$ and $g: B \to A$ are injections, then there is a bijection (i.e. an injection which is also a surjection) $h: A \to B$.

(*Hint.* The basic decomposition required here is perhaps not the first idea that will come to mind; it might be more efficient to read the proof in [63].)

b) Using a), show that a countably infinite set has an uncountable family $\mathscr{F}$ of subsets the intersection of any two of which is finite. Besides using a) there is also the following more geometric solution. Let $P = \{(n, m): n, m \in \mathbf{Z}\} \subseteq \mathbf{R} \times \mathbf{R}$ and take S_φ to be the "strip" in P with angle of inclination φ and width greater than 1. There are uncountably many S_φ (by the uncountability of the φ) and so if we put our given countable set in one to one correspondence with P we obtain the uncountability of $\mathscr{F}$. Clearly $S_\varphi \cap S_\psi$ is finite since this intersection is a bounded parallelogram in $\mathbf{R}^2$.

1.7 Let $\mathscr{F}$ be the field of all functions $f = p/q$ where p and q are polynomials with real coefficients. Define

$$P_a = \{f \subset \mathscr{F}: \exists \varepsilon > 0 \quad \text{for which} \quad f > 0 \quad \text{on} \quad (a, a + \varepsilon)\}$$

and $$P'_a = \{f \in \mathscr{F}: \exists \varepsilon > 0 \quad \text{for which} \quad f > 0 \quad \text{on} \quad (a - \varepsilon, a)\}$$

for each $a \in \mathbf{R}$, and

$$P_\infty = \{f \in \mathscr{F}: \exists N > 0 \quad \text{for which} \quad f > 0 \quad \text{on} \quad (N, \infty)\}$$

and $$P'_\infty = \{f \in \mathscr{F}: \exists N > 0 \quad \text{for which} \quad f > 0 \quad \text{on} \quad (-\infty, -N)\}.$$

Clearly, with each such P, $\mathscr{F}$ is an ordered field with pointwise multiplication and addition (i.e., P is closed under addition and multiplication, and if $f \in \mathscr{F}$ then exactly one of the following is true: $f \in P$, $-f \in P$, or $f = 0$). Prove that the above P's give rise to distinct non-Archimedean ordered fields and that these are the only subsets of $\mathscr{F}$ for which $\mathscr{F}$ is an ordered field. Also prove that if $f \in \mathscr{F}$ is in every P then it can be written as a sum of squares.

1.8 Let $S \subseteq \mathbf{R}^n$ have the property that $d(s_1, s_2)$ is rational for every $s_1, s_2 \in S$ (here d is the usual Euclidean distance). Prove S is countable.

1.9 Let $\mathscr{F}$ be an ordered field with positive elements P. The canonical linear ordering on $\mathscr{F}$ is given by: $f < g$ if $g - f \in P$. Define

$$[f, g] = \{h \in \mathscr{F}: f \leqslant h \leqslant g\}$$
$$(f, g) = \{h \in \mathscr{F}: f < h < g\}.$$

$[f, g]$ is a closed interval and (f, g) is an open interval. A set $\mathscr{S}$ of open intervals covers $[f, g]$ if each $h \in [f, g]$ is contained in at least one member of $\mathscr{S}$. Now assume $\mathscr{F}$ is order-complete (i.e., each non-empty subset of $\mathscr{F}$ which has an upper bound has a supremum). Prove the Heine–Borel theorem: given $[f, g]$ and $\mathscr{S}$ a cover of $[f, g]$; there is a finite subcover.

1.10 Find the error(s) in the following proof of Proposition 1.2:

Every open set $G \subseteq \mathbf{R}^2$ is the disjoint union of open linear sets (e.g. the intersection of G with the set of all horizontal lines in $\mathbf{R}^2$); but it is known that every open linear set is the disjoint union of open straight line segments. Thus the problem is reduced to showing that every open straight line segment can be represented as a disjoint union of closed straight line segments; but this is clear—e.g. $(0, 1) = [\frac{1}{3}, \frac{2}{3}] \cup ([\frac{1}{9}, \frac{2}{9}] \cup [\frac{7}{9}, \frac{8}{9}]) \cup \dots$

1.11 Can you find $\{x_n : n = 1, \dots\} \subseteq \mathbf{R}$, bounded and infinite, such that

$$\forall x \in \mathbf{R}, \qquad \text{card } C \cap \{x_n + x : n = 1, \dots\} < \aleph_0?$$

1.12 Let E_ξ be a perfect symmetric set with $\xi_k = \xi$ for each k.

a) Prove that $E_\xi - E_\xi = \{x - y : x, y \in E_\xi\} \supseteq [0, 1]$ if $\frac{1}{3} \leqslant \xi < \frac{1}{2}$.
b) Prove that $m(E_\xi - E_\xi) = 0$ if $0 < \xi < \frac{1}{3}$.

(*Hint*. First, try the case $0 < \xi < \frac{1}{4}$. H. Steinhaus proved a) for $C = E_{1/3}$ by considering $C \times C \subseteq [0, 1] \times [0, 1]$, and Steinhaus' trick can be used to prove b) for the $E_{1/4}$ case (cf. Problem 3.6b). T. Šalát settled the general case for $\frac{1}{4} < \xi < \frac{1}{3}$.)
In this problem it is first necessary to make a reasonable guess at the definition of "measure 0" or else wait until Chapter 2.

1.13 Prove (1.4).

1.14 a) Prove that $\chi_{\mathbf{R} \setminus \mathbf{Q}}$ is not the pointwise limit of continuous functions. (*Hint*. Assume that a sequence $\{f_n : n = 1, \dots\}$ of continuous functions converges pointwise to $\chi_{\mathbf{R} \setminus \mathbf{Q}}$, and define

$$E_n = \{x : f_n(x) \geqslant \tfrac{1}{2}\}, \qquad F_N = \bigcap_N^\infty E_j;$$

then show that $\mathbf{R} \setminus \mathbf{Q} = \bigcup F_N$. This yields the desired contradiction since $\mathbf{R} \setminus \mathbf{Q}$ is not an $\mathscr{F}_\sigma$.)

b) Let $\{f_n : n = 1, \dots\}$ be a sequence of continuous functions $[0, 1] \to \mathbf{R}$ which converges pointwise to a function f. Prove that $\overline{C(f)} = [0, 1]$. The solution to this is nontrivial; the presentation in [13, pp. 99–101] uses the Baire category theorem three times.

c) Let $\{f_n : n = 1, \dots\}$ be a sequence of continuous functions $[0, 1] \to \mathbf{R}$. Prove that

$$\{x : \exists \lim f_n(x)\}$$

is an $\mathscr{F}_{\sigma\delta}$.

1.15 Let $f : \mathbf{R} \to \mathbf{R}$ be differentiable. Assume that

$$\forall r \in \mathbf{R}, \qquad \{x : f'(x) = r\}$$

is closed. Prove that f' is continuous on $\mathbf{R}$.

1.16 Construct a C^∞-function (i.e. infinitely differentiable) $f : (0, 1) \to \mathbf{R}$ such that $\{x : f(x) = 0\}$ is nowhere dense and uncountable.

1.17 In light of Example 1.7, can the following be proved:

for every nowhere dense $\mathscr{F}_\sigma$, $E \subseteq \mathbf{R}$, there is $f: \mathbf{R} \to \mathbf{R}$ such that $D(f) = E$ and f has a local minimum at each $x \in \mathbf{R}$?

1.18 Given a function $f: [0, 1] \to \mathbf{R}$ with the property that $\overline{C(f)} = [0, 1]$; prove that f is continuous except on a (possibly empty) set of first category (cf. Problem 1.14b and Proposition 1.7).

1.19 Verify the claim of Example 1.2 (Volterra's example) that f' exists on $(0, 1)$ and is 0 on E.

1.20 Find a dense set $S \subseteq \mathbf{C}$ such that for any line segment $L \subseteq \mathbf{C}$, $\overline{S \cap L} \neq L$. (*Hint.* Let $A \subseteq \mathbf{C}$ be the set of algebraic numbers (an algebraic number is the root of a polynomial with integer coefficients. Thus A is countable (cf. Chapter A3.1.2). Define $s_1 = \alpha_1$, $s_2 = \alpha_2$, and

$$\forall n \geqslant 3, \qquad s_n = \alpha_n + \frac{1}{2^n} \exp\{2\pi i/f(n)\}$$

where $f(n)$ is the smallest positive integer for which s_n is not collinear with any two of $s_1, \ldots, s_{n-1}$. Set $S = \{s_n : n = 1, \ldots\}$.)

1.21 Order the rationals $\{r_n : n = 1, \ldots\} \subseteq (0, 1)$ à la Cantor

i.e. $\quad \frac{1}{2}; \frac{1}{3}; \frac{2}{3}, \frac{1}{4}; \frac{1}{5}; \frac{3}{4}, \frac{2}{5}, \frac{1}{6}; \ldots$

and cover r_n by an interval of length $1/2^{n+1}$ and center r_n. The total "length" covered is less than or equal to $\frac{1}{2}$, and so there are real numbers $r \in [0, 1]$ which are not covered by this process. Write down one such r explicitly. (*Hint.* Let $r = 1/\sqrt{2}$ and note that r is a root of $f(x) = 2x^2 - 1$. For any rational $p/q \in (0, 1)$

$$|f(p/q)| = |f(p/q) - f(r)| = 4\xi|r - (p/q)|,$$

$\xi \in (0, 1)$, by the mean value theorem. Since $\sqrt{2}$ is irrational, $|2p^2 - q^2|$ is a positive integer and so

$$|f(p/q)| = |2p^2 - q^2|/q^2 \geqslant 1/q^2.$$

Consequently,

$$|r - (p/q)| \geqslant 1/(4\xi q^2) \geqslant 1/(4q^2). \tag{P1.1}$$

This technique was first used by Liouville to construct transcendental, i.e. non-algebraic, numbers (e.g. [82, pp. 7–8]). We say that $r \in \mathbf{R} \setminus \mathbf{Q}$ is a Liouville number if for every positive integer n there are integers p and q such that $|r - (p/q)| < 1/q^n$, $q > 1$. From properties of the Farey series in terms of the Euler function $\varphi(n)$ (e.g. the proof of Proposition 3.6 for the definition of φ and [51, Theorems 330 and 331]) it can be seen that if p/q is the nth rational then $n + 2 > q^2/4$. This combined with (P1.1) gives the result.)

1.22 Define the function $f: (0, 1) \to \mathbf{R}$ as

$$f(x) = \sum_{0 < \frac{p}{q} \leqslant x} \frac{1}{q^3}, \quad (p, q) = 1.$$

a) Prove that f is increasing on $(0, 1)$, continuous on the irrationals, and discontinuous on the rationals.

b) Since f is increasing, we shall see in Chapter 4 that f' exists for many points. Show that there are points x for which $f'(x)$ does not exist. (*Hint.* Take x to be a Liouville transcendental number, mentioned in Problem 1.21, and verify that

$$\overline{\lim_{u \to x}} \frac{f(u) - f(x)}{u - x} = \infty.)$$

1.23 a) Let $f: [0, 1] \to \mathbf{R}$ be a continuous function, and for each n choose some $x_{k, n} \in [(k - 1)/n, k/n]$, $k = 1, ..., n$. Prove that

$$R \int_0^1 f = \lim_{n \to \infty} \frac{1}{n} \sum_{k=1}^n f(x_{k, n})$$

exists, where the integral denotes the usual Riemann integral.

b) Calculate $\displaystyle \lim_{n \to \infty} \frac{1}{n} \left(\sin \frac{\pi}{n} + \sin \frac{2\pi}{n} + \cdots + \sin \frac{n\pi}{n} \right).$

Remark The partition in a) is "equidistributed". Such evaluations of Riemann integrals play an important role in number theory. This relationship was first developed by H. Weyl in 1916; we refer to [66] for a survey of recent developments on this matter.

1.24 Prove: let $f: [a, b] \to \mathbf{R}$ be continuous; then

a)   there is $F: [a, b] \to \mathbf{R}$ such that $F' = f$ on $[a, b]$,

b) $\displaystyle R \int_a^b f = F(b) - F(a).$

Remark This is the fundamental theorem of calculus for the first 1900 years A.D. We shall generalize it into a deep result (the Vitali–Lebesgue–Radon–Nikodym theorem) which forms a major part of the Lebesgue theory.

1.25 Prove that the function defined in Example 1.13 is *ecnd*.

1.26 Prove that an open interval can not be the countable disjoint union of closed sets.

1.27 Verify that the following function is *ecnd*:

$$f(x) = \sum_{k=1}^{\infty} \frac{1}{2^k} g(2^{2^k} x),$$

where $g: \mathbf{R} \to \mathbf{R}$ has period 4 and is defined by

$$g(x) = \begin{cases} 1 + x, & \text{if} \quad -2 \leqslant x \leqslant 0 \\ 1 - x, & \text{if} \quad 0 \leqslant x \leqslant 2. \end{cases}$$

1.28 Is there an irrational number x such that

$$\{c + x : c \in C\} \subseteq \mathbf{R} \setminus \mathbf{Q}?$$

1.29 Consider

$$\sum_{n=1}^{\infty} \prod_{j=1}^{n} r_j, \qquad r_j \in \mathbf{Q} \cap (0, 1). \tag{P1.2}$$

a) If $\{r_j : j = 1, ...\} = \mathbf{Q} \cap (0, 1)$ is enumerated à la Cantor, can you prove that (P1.2) converges?

b) Find an enumeration of $\mathbf{Q} \cap (0, 1)$ such that (P1.2) diverges.

2 Lebesgue measure and general measure theory

2.1 The theory of measure prior to Lebesgue, and preliminaries

For the historical development of measure theory prior to Lebesgue we refer to Arthur Schoenflies' work on set theory (1900) and [54].

One of the first people to recognize the connection between an integration theory and a precise theory for measuring the "length" of sets was Giuseppe Peano, and his work on these matters is contained in his book "Applicazione geometriche del calcolo infinitesimale" (published in 1887 at Torino). Taking a non-negative, bounded function $f: [a, b] \to \mathbf{R}$, he let E be the point-set bounded by the graph of f and the lines $x = a$, $x = b$, $y = 0$, and then defined

$$c_i(E) = \sup \{a(A): A \subseteq E, A \text{ a polygonal region}\},$$
$$c_e(E) = \inf \{a(A): A \supseteq E, A \text{ a polygonal region}\},$$

where $a(A)$ is the area of A. In these terms he observed that f is (*Riemann*) *integrable if and only if E is "measurable"* (*i.e.* $c_e(E) = c_i(E)$).

When c_e and c_i are equal we write c for their common value, and Peano actually developed the theory of finitely additive set functions (for such c).

Mind you, there was a great deal of activity in the general problem of measuring sets prior to Peano. For example, using the fact that a closed set E can be written as

$$E = (\bigcup I_j)^\sim,$$

where $\{I_j: j = 1, \ldots\}$ is a disjoint family of open intervals, Cantor (and Bendixson) were led to define

$$\text{"}m(C)\text{"} = 1 - \sum m(I_j),$$

whose value, as we showed, is 0.

Five years after Peano's work, Jordan also developed the theory of finitely additive measures, and, although there are no references to Peano, it seems likely that Jordan knew of his work. In any case, mathematically, Jordan carried finitely additive measure theory very far, and the relation between measure and integrability was quite clearly explicated in the Jordan theory.

The next major step in the evolution of the fundamental ideas leading to the present notions of integral and measure was taken by Émile Borel from the time of his doctorate in 1894. To discuss his work we let $X \subseteq \mathbf{R}$ be a set and define an algebra $\mathscr{A} \subseteq \mathscr{P}(X)$ by the properties that

$$\forall A, B \in \mathscr{A}, \qquad A \cup B \in \mathscr{A},$$

and

$$\forall A \in \mathscr{A}, \qquad A^{\sim} \in \mathscr{A}.$$

("$\mathscr{P}(X)$" was defined in Problem 1.3). An algebra $\mathscr{A}$ is a σ-algebra if

$$\forall \{A_n : n = 1, ...\} \subseteq \mathscr{A}, \qquad \bigcup A_n \in \mathscr{A}.$$

It is easy to prove that *if $\mathscr{C} \subseteq \mathscr{P}(X)$ then there is a smallest σ-algebra $\mathscr{A}$ which contains $\mathscr{C}$* (in fact, let $\mathscr{A}$ be the σ-algebra generated by $\mathscr{C}$). In light of the importance of open and closed sets we define the collection $\mathscr{B} = \mathscr{B}(X)$ of Borel sets in $X \subseteq \mathbf{R}$ to be the smallest σ-algebra in $\mathscr{P}(X)$ containing the open sets (in X). If X is any topological space (defined in Appendix I.1) we define the collection $\mathscr{B} = \mathscr{B}(X)$ of Borel sets in the same way; $\mathscr{B}$ is the Borel algebra.

We have defined $\mathscr{F}_\sigma$, $\mathscr{G}_\delta$, and $\mathscr{F}_{\sigma\delta}$ sets. Clearly these are all Borel sets as are

$$\mathscr{G}_{\delta\sigma}, \mathscr{G}_{\delta\sigma\delta}, ..., \mathscr{F}_{\sigma\delta\sigma}, \mathscr{F}_{\sigma\delta\sigma\delta}, \cdots \quad .$$

Example 2.1 There are Borel sets $B \subseteq \mathbf{R}$ which are not of the form $\mathscr{F}_n$ or $\mathscr{G}_m$, where, for example,

$$\mathscr{F}_n = \mathscr{F}_{\underbrace{\sigma\delta\sigma\cdots}_{n}} .$$

In fact, set

$$B = \bigcup_{n=1}^{\infty} A_n$$

where A_n is an $\mathscr{F}_n$ but not an $\mathscr{F}_{n-1}$ (e.g. [53, p. 182]).

Example 2.2 There are Borel sets $B_1, B_2 \in \mathscr{B}(\mathbf{R})$ such that $B_1 - B_2 = \{x - y : x \in B_1, y \in B_2\} \notin \mathscr{B}(\mathbf{R})$ [90].

Besides first proving the Heine–Borel theorem in the form of Theorem 1.3, Borel gave a reasonable definition of measure 0 and stressed countable additivity for his measures. In fact, by dividing $[0, 1]$ into equal parts Jordan had reached the conclusion that "$m(\mathbf{Q} \cap [0, 1])$" $= 1$. Borel, on the other hand, attached to each $r_n \in \mathbf{Q} \cap [0, 1]$ the segment of length ε/n^2. Consequently, he concluded that "$m(\mathbf{Q} \cap [0, 1])$" $< \varepsilon \sum 1/n^2$ for each ε; and so "$m(\mathbf{Q} \cap [0, 1])$" $= 0$. This example led Borel into his study of measure.

We shall see how countable additivity and integration are related very soon. The countable additivity of Borel, as opposed to the finite additivity of Peano–Jordan, was crucial in Lebesgue's theory to attain many fundamental results.

Let $\mathscr{A}$ be a σ-algebra (on **R**). μ is a countably additive measure (or, just simply, a measure) if

$$\mu: \mathscr{A} \to \mathbf{R}^+ \cup \{\infty\}$$

has the property that

$$\mu(\bigcup A_n) = \sum \mu A_n,$$

where $\{A_n: n = 1, ...\} \subseteq \mathscr{A}$ is a disjoint family.

Remark 1 It is not à priori clear that a non-trivial measure exists. Lebesgue solved this existence problem in a very strong way by constructing a σ-algebra $\mathscr{M}(\mathbf{R}) \supseteq \mathscr{B}(\mathbf{R})$ and a measure m on $\mathscr{M}(\mathbf{R})$ with the reasonable properties that

$$\forall x \in \mathbf{R} \quad \text{and} \quad \forall A \in \mathscr{M}(\mathbf{R}), \qquad m(x + A) = mA, \tag{2.1}$$

and

$$\forall I \subseteq \mathbf{R}, \quad I \text{ an interval, } mI \text{ is the length of } I; \tag{2.2}$$

m is Lebesgue measure, and conditions (2.1) and (2.2) are peculiar to m.

Condition (2.1) is the translation invariance of Lebesgue measure. It turns out that there are many other measures on **R**, and we'll study spaces of measures later. In Example 2.8 we'll show that it is impossible to define a measure on the σ-algebra $\mathscr{P}(\mathbf{R})$ which is both translation invariant on $\mathscr{P}(\mathbf{R})$ and satisfies (2.2). We shall make explicit use of the axiom of choice in this result.

Remark 2 The continuum hypothesis is:

$$\forall A \subseteq \mathbf{R}, \qquad \text{card } A > \aleph_0 \Rightarrow \text{card } A = \text{card } \mathbf{R}.$$

Explicitly using the continuum hypothesis, Banach and Kuratowski [7] and Ulam [115] proved that there is no nontrivial measure on all of $\mathscr{P}(\mathbf{R})$ which satisfies (2.2) [109, pp. 107–109]. In fact, the result is that there are no non-trivial continuous measures (such measures are defined in Chapters 4 and 5) defined on $\mathscr{P}(\mathbf{R})$. These are stronger conclusions than what we'll prove in Example 2.8. In 1950 Kakutani, Kodaira, and Oxtoby proved a positive result in this area by extending Lebesgue measure to a large σ-algebra $\mathscr{A} \supsetneq \mathscr{M}(\mathbf{R})$ while preserving (2.1) on $\mathscr{A}$. It is, of course, easy to define translation invariant measures on $\mathscr{P}(\mathbf{R})$ which don't have property (2.2) (e.g. Example 2.9).

Remark 3. With regard to the continuum hypothesis, it can be verified that if $F \subseteq \mathbf{R}$ is closed and uncountable then

$$\text{card } F = \text{card } \mathbf{R}.$$

This follows because:

a) *If* card $F > \aleph_0$ *then* $F = P \cup D$ *where* D *is countable and* P *is a closed set without isolated points;*

b) *If $P \subseteq \mathbf{R}$ is a non-empty closed set without isolated points then* card $P =$ card $\mathbf{R}$.

a) is the Cantor–Bendixson theorem. b) is proved using Cantor's theorem which asserts that *if* $\{F_n: n = 1, ...\} \subseteq \mathscr{P}(\mathbf{R})$ *is a nested decreasing sequence of closed sets whose diameters tend to* 0 *then* $\bigcap F_n$ *is a single point*; we'll prove Cantor's result in the proof of the Baire category theorem (Theorem I.6). Compare Cantor's theorem with Theorem 1.3b.

As we've indicated, if I is an interval then mI will designate its length. For any $A \subseteq \mathbf{R}$ the Lebesgue outer measure of A is

$$m^*A = \inf \{\sum_n mI_n : A \subseteq \bigcup_n I_n \text{ and } \{I_n: n = 1, ...\} \text{ is a countable family of}$$

open intervals}.

Clearly,

a) *if* $A \subseteq B$ *then* $m^*A \leqslant m^*B$,

b) $m^*\emptyset = 0$.

Proposition 2.1 *Let $I \subseteq \mathbf{R}$ be an interval. Then $m^*I = mI$.*

Proof. Let $I = [a, b]$, $-\infty < a \leqslant b < \infty$.
For each $\varepsilon > 0$, $I \subseteq (a - \varepsilon, b + \varepsilon)$, and so

$$m^*I \leqslant b - a + 2\varepsilon = m(a - \varepsilon, b + \varepsilon). \tag{2.3}$$

Since (2.3) is true for each $\varepsilon > 0$,

$$m^*I \leqslant b - a.$$

To prove $m^*I \geqslant b - a$, we'll show that for any sequence $\{I_n: n = 1, ...\}$ of open intervals,

$$I \subseteq \bigcup I_n \Rightarrow \sum mI_n \geqslant b - a. \tag{2.4}$$

By the Heine–Borel theorem each such sequence $\{I_n: n = 1, ...\}$ (for which $I \subseteq \bigcup I_n$) has a finite subcollection $I_{n_1}, ..., I_{n_k}$ covering I, and, in particular

$$\sum_1^\infty mI_n \geqslant \sum_{j=1}^k mI_{n_j}.$$

Consequently we need only prove (2.4) for finite covers $\{J_1, ..., J_k\}$. Since $a \in \bigcup_{j=1}^k J_j$, we have $a \in J_1 = (a_1, b_1)$ (we take J_1 without loss of generality).
Thus, if $b_1 \geqslant b$,

$$\sum_{j=1}^k mJ_j \geqslant b_1 - a_1 > b - a,$$

in which case we are finished.

If $a < b_1 < b$, then since $b_1 \in I$ there is $J_2 = (a_2, b_2)$ in our finite subcollection such that $a_2 < b_1 < b_2$. Again, if $b_2 \geqslant b$ we are finished. If $b_2 < b$ we choose $J_3 = (a_3, b_3)$ from our finite collection, etc.

Since the collection is finite, this process ends with $J_n = (a_n, b_n)$, $n \leqslant k$; and by hypothesis and construction we have $a_n < b < b_n$.

Hence,

$$\sum_{j=1}^{k} mJ_j \geqslant \sum_{j=1}^{n} (b_j - a_j)$$

$$= b_n - (a_n - b_{n-1}) - (a_{n-1} - b_{n-2}) - \cdots - (a_2 - b_1) - a_1. \qquad (2.5)$$

Now $a_j < b_{j-1}$ (recall $a_2 < b_1 < b_2$, etc.), and so, from (2.5),

$$\sum_{j=1}^{k} mJ_j > b_n - a_1 > b - a;$$

this gives (2.4).

For the case of a bounded interval I, take any $\varepsilon > 0$ and choose a closed interval $J \subseteq I$ for which

$$mJ > mI - \varepsilon.$$

Hence,

$$mI - \varepsilon < mJ = m^*J \leqslant m^*I \leqslant m^*\bar{I} = m\bar{I} = mI;$$

that is, for each $\varepsilon > 0$

$$mI - \varepsilon < m^*I \leqslant mI,$$

and so $mI = m^*I$.

If I is an infinite interval then for each $r > 0$ there is a closed interval $J \subseteq I$ such that $mJ = r$.

Therefore, $m^*I \geqslant m^*J = mJ = r$, and, so, "letting $r \to \infty$", we have $m^*I = +\infty$.

q.e.d.

Proposition 2.2 Let $\{A_n : n = 1, ...\} \subseteq \mathscr{P}(\mathbf{R})$. Then

$$m^* \bigcup A_n \leqslant \sum m^*A_n. \qquad (2.6)$$

Proof. If $m^*A_n = \infty$ for some n we are finished.

Therefore assume $m^*A_n < \infty$ for each n.

Given $\varepsilon > 0$ and n, there is a sequence $\{I_{n,i} : n = 1, ..., i = 1, ...\}$ of open intervals such that

$$A_n \subseteq \bigcup_i I_{n,i} \qquad \text{and} \qquad \sum_i mI_{n,i} < m^*A_n + 2^{-n}\varepsilon,$$

by the definition of m^*.

Now, card $\{I_{n,i}: i, n\} \leqslant \aleph_0$ and $\bigcup_n A_n \subseteq \bigcup_{n,i} I_{n,i}$.

Therefore,

$$m^*(\bigcup_{n,i} A_n) \leqslant \sum_{n,i} mI_{n,i} = \sum_n \sum_i mI_{n,i} <$$

$$\sum_n m^*A_n + \varepsilon \sum_n \frac{1}{2^n} = \sum_n m^*A_n + \varepsilon.$$

Since this is true for each ε we have (2.6).

q.e.d.

Taking A_n to be a one-point set (in Proposition 2.2) we have

a) *if A is countable then* $m^*A = 0$.

a) combined with Proposition 2.1 gives an ingenious, though complicated, proof that
b) card $[0, 1] > \aleph_0$.

The following result is straightforward to prove and the proof is left as an exercise (Problem 2.1).

Proposition 2.3 *Given* $A \subseteq \mathbf{R}$.

a) *For each* $\varepsilon > 0$ *there is an open set* $U \subseteq \mathbf{R}$ *such that*

$$A \subseteq U \quad \text{and} \quad m^*U \leqslant m^*A + \varepsilon.$$

b) *There is a* $\mathscr{G}_\delta$ *set G such that*

$$A \subseteq G \quad \text{and} \quad m^*A = m^*G.$$

2.2 The existence of Lebesgue measure

We use the Caratheodory approach (1914) to define Lebesgue measure. $A \subseteq \mathbf{R}$ is (Lebesgue) measurable if

$$\forall E \subseteq \mathbf{R}, \quad m^*E = m^*(A \cap E) + m^*(E \cap A^\sim).$$

Thus, A is measurable if "no matter how you cut it" (with any E), m^* is nicely additive just as we hoped it would be.

It is clear that

a) *if* $m^*A = 0$ *then A is measurable;*
b) *if* A_1, A_2 *are measurable then* $A_1 \cup A_2$ *is measurable.*

From b) we have

Proposition 2.4 *The family* $\mathscr{M}(\mathbf{R})$ *of measurable sets is an algebra.*

More important is the following

Theorem 2.1

a) $\mathscr{M}(\mathbf{R})$ *is a σ-algebra.*
b) $\mathscr{B}(\mathbf{R}) \subseteq \mathscr{M}(\mathbf{R})$.

Proof. a.i) We first show that if $E \subseteq \mathbf{R}$ and $\{A_1, ..., A_n\} \subseteq \mathscr{M}(\mathbf{R})$ is a disjoint family then

$$m^*\left(E \cap \left[\bigcup_{j=1}^{n} A_j\right]\right) = \sum_{j=1}^{n} m^*(E \cap A_j). \tag{2.7}$$

We'll prove (2.7) by induction noting that it is obvious for $n = 1$. Assume (2.7) is true for $A_1, ..., A_{n-1}$ and take $A_1, ..., A_n$.
Then

$$E \cap \left[\bigcup_{j=1}^{n} A_j\right] \cap A_n = E \cap A_n \tag{2.8}$$

and

$$E \cap \left[\bigcup_{j=1}^{n} A_j\right] \cap A_n^{\sim} = E \cap \left[\bigcup_{j=1}^{n-1} A_j\right] \tag{2.9}$$

((2.9) follows since $\left(\bigcup_{j=1}^{n} A_j\right) \cap A_n^{\sim} = \bigcup_{j=1}^{n} (A_j \cap A_n^{\sim}) = \left(\bigcup_{j=1}^{n-1} (A_j \cap A_n^{\sim})\right) \cup (A_n \cap A_n^{\sim})$
$= \bigcup_{j=1}^{n-1} A_j$, where we've used the fact that $\{A_1, ..., A_n\}$ is disjoint in the last step).
Thus, $A_n \in \mathscr{M}(\mathbf{R})$ implies, from (2.8) and (2.9), that

$$m^*\left(E \cap \left[\bigcup_{j=1}^{n} A_j\right]\right) = m^*(E \cap A_n) + m^*\left(E \cap \left[\bigcup_{j=1}^{n-1} A_j\right]\right)$$

$$= m^*(E \cap A_n) + \sum_{j=1}^{n-1} m^*(E \cap A_j)$$

by the induction hypothesis. Consequently, (2.7) holds.
a.ii) We now prove that $\mathscr{M}(\mathbf{R})$ is a σ-algebra.

Hence, given $\{B_j : j = 1, ...\} \subseteq \mathscr{M}(\mathbf{R})$ we must show that $A = \bigcup_{1}^{\infty} B_j \in \mathscr{M}(\mathbf{R})$. Since $\mathscr{M}(\mathbf{R})$ is an algebra, there is a disjoint family $\{A_j : j = 1, ...\} \subseteq \mathscr{M}(\mathbf{R})$ such that

$$A = \bigcup B_j = \bigcup A_j;$$

in fact, let $A_1 = B_1$, $A_2 = B_2 \setminus B_1$, $A_3 = B_3 \setminus (B_1 \cup B_2)$, ...
Let $C_n = \bigcup_{j=1}^{n} A_j$ so that $C_n \in \mathscr{M}(\mathbf{R})$ (again using the fact that $\mathscr{M}(\mathbf{R})$ is an algebra).

Next note that $A^\sim \subseteq C_n^\sim$ because

$$C_n = \bigcup_{j=1}^{n} A_j \subseteq A.$$

Taking any $E \subseteq \mathbf{R}$ we calculate

$$m^*E = m^*(E \cap C_n) + m^*(E \cap C_n^\sim) \geqslant m^*(E \cap C_n) + m^*(E \cap A^\sim).$$
(2.10)

By a.i) and the disjointness of $\{A_j : j = 1, \dots\}$

$$m^*(E \cap C_n) = m^*\left(E \cap \left(\bigcup_{j=1}^{n} A_j\right)\right) = \sum_{j=1}^{n} m^*(E \cap A_j).$$
(2.11)

Combining (2.10) and (2.11) gives

$$m^*E \geqslant \sum_{j=1}^{n} m^*(E \cap A_j) + m^*(E \cap A^\sim),$$

and since the left-hand side is independent of n,

$$m^*E \geqslant \sum_{j=1}^{\infty} m^*(E \cap A_j) + m^*(E \cap A^\sim).$$

Thus, by the subadditivity of m^* (Proposition 2.2),

$$m^*E \geqslant m^*\left(E \cap \left(\bigcup_{1}^{\infty} A_j\right)\right) + m^*(E \cap A^\sim).$$
(2.12)

The opposite inequality to (2.12) is always true and so

$$A \in \mathcal{M}(\mathbf{R}).$$

b.i) We first show that $(a, \infty) \in \mathcal{M}(\mathbf{R})$. Take $E \subseteq \mathbf{R}$ and set

$$E_1 = E \cap (a, \infty), \qquad E_2 = E \cap (-\infty, a];$$

it is sufficient to prove that $\varepsilon + m^*E \geqslant m^*E_1 + m^*E_2$ for each $\varepsilon > 0$, and without loss of generality (obviously) we take $m^*E < \infty$. Fix $\varepsilon > 0$.

Since $m^*E < \infty$ there is a sequence $\{I_n : n = 1, \dots\}$ of open intervals covering E such that

$$\sum mI_n \leqslant \varepsilon + m^*E.$$

$I'_n = I_n \cap (a, \infty)$ and $I''_n = I_n \cap (-\infty, a]$ are intervals (possibly empty), and

$$mI_n = mI'_n + mI''_n = m^*I'_n + m^*I''_n$$

by Proposition 2.1.

Clearly, $E_1 \subseteq \bigcup I'_n$ and $E_2 \subseteq \bigcup I''_n$; consequently,

$$m^*E_1 \leqslant m^*(\bigcup I'_n) \leqslant \sum m^*I'_n,$$
$$m^*E_2 \leqslant m^*(\bigcup I''_n) \leqslant \sum m^*I''_n$$

from Proposition 2.2. Therefore

$$m^*E_1 + m^*E_2 \leqslant \sum (m^*I'_n + m^*I''_n) = \sum mI_n \leqslant \varepsilon + m^*E.$$

b.ii) $\mathscr{B}(\mathbf{R})$ is the smallest σ-algebra containing each (a, ∞); but we've just proved that $\mathscr{M}(\mathbf{R})$ is a σ-algebra containing each (a, ∞).

q.e.d.

For each $A \in \mathscr{M}(\mathbf{R})$, $mA = m^*A$ is the Lebesgue measure of A. Clearly, $A \in \mathscr{M}(\mathbf{R})$ has zero measure if and only if

$$\forall \varepsilon > 0 \quad \exists\{I_n: I_n \text{ an open interval and } n = 1, \ldots\} \text{ such that}$$

$$A \subseteq \bigcup I_n \quad \text{and} \quad \sum mI_n < \varepsilon.$$

N. Wiener (AMS semicentennial) has made an historical case for the definition of measure 0 based on considerations from statistical mechanics.

The following result is interesting in light of the ruler function (e.g. Chapter 1.3.1); we have only outlined its proof (cf. Problem 2.5).

Proposition 2.5 *There is a function $f: [0, 1] \to \mathbf{R}$ and a set $D \subseteq [0, 1]$ which is dense in $[0, 1]$, such that $D \in \mathscr{M}(\mathbf{R})$, f is continuous on D, f is discontinuous on $[0, 1] \setminus D$, and $mD = 0$.*

Proof. Let $\{r_n: n = 1, \ldots\}$ be an enumeration of $\mathbf{Q} \cap [0, 1]$, denote the open interval about r_j of radius $1/(k2^j)$ by I_{jk}, and set

$$D = \bigcap_{k=3}^{\infty} U_k, \quad \text{where } U_k = \bigcup_{j=1}^{\infty} I_{jk}.$$

D is not of first category by an application of the Baire category theorem, and $mD = 0$ (e.g. Theorem 2.2c below). Define $f = \sum_{k \geqslant 3} f_k$, where

$$f_k(x) = \begin{cases} 0, & \text{if } x \in U_k \\ 1/2^k, & \text{if } x \in [0, 1] \setminus U_k. \end{cases}$$

q.e.d.

See Problem 2.2 for the following

Theorem 2.2 *Let $\{A_n: n = 1, \ldots\} \subseteq \mathscr{M}(\mathbf{R})$.*
a) $m(\bigcup A_n) \leqslant \sum mA_n$.
b) *If $\{A_n: n = 1, \ldots\}$ is a disjoint family then $m(\bigcup A_n) = \sum mA_n$.*
c) *If $A_1 \supseteq A_2 \supseteq \ldots \supseteq A_n \supseteq \ldots$ and $mA_1 < \infty$, then $m\left(\bigcap_1^{\infty} A_n\right) = \lim mA_n$.*

We'll prove Theorem 2.2c in a more general setting in Theorem 2.4.

Example 2.3 Because of Theorem 2.2c, if E is a perfect symmetric set then the definition of mE which we gave in Chapter 1 is precisely the Lebesgue measure of E.

Example 2.4 Let $A \subseteq \mathbf{R}$ have the property:

$\exists q \in (0, 1)$ such that $\forall (a, b) \subseteq \mathbf{R}$ $\exists \{I_n : I_n$ an open interval and $n = 1, ...\}$ for which $A \cap (a, b) \subseteq \bigcup I_n$ and $\sum mI_n < q(b - a)$.

We'll prove that $mA = 0$. It is sufficient to prove that $m(A \cap (a, b)) = 0$ since $mA = m(A \cap (\bigcup J_i)) \leqslant \sum m(A \cap J_i)$, where $\{J_i : i = 1, ...\}$ is a cover of $\mathbf{R}$ by open intervals. Now cover $A \cap (a, b)$ by intervals (a_n, b_n) having the property that $\sum (b_n - a_n) \leqslant q(b - a)$. Then cover each $A \cap (a_n, b_n)$ by a countable collection of intervals having total length $q(b_n - a_n)$. Thus we've covered A by open intervals of total length L where

$$L \leqslant q(b_1 - a_1) + q(b_2 - a_2) + ... = q((b_1 - a_1) + ...) \leqslant q^2(b - a).$$

Repeating this process we obtain the inequality $mA \leqslant q^n(b - a)$, and so $mA = 0$ since $0 < q < 1$.

Let $A \in \mathcal{M}(\mathbf{R})$. Suppose $G \subseteq A$ is Lebesgue measurable so that $A \setminus G \in \mathcal{M}(\mathbf{R})$. Then

$$A = G \cup (A \setminus G).$$

Because of this and Theorem 2.3 below we conclude the following important fact:

$$\forall A \in \mathcal{M}(\mathbf{R}) \quad \exists B \in \mathcal{B}(\mathbf{R}) \quad \text{and} \quad \exists E \in \mathcal{M}(\mathbf{R}), \quad mE = 0, \quad \text{such that}$$
$$A = B \cup E, B \cap E = \varnothing, \text{ and } mA = mB$$

i.e., *every Lebesgue measurable set is a Borel set up to a set of measure zero.*

With regard to Proposition 2.3 we have

Theorem 2.3 *The following are equivalent:*

a) $A \in \mathcal{M}(\mathbf{R})$.
b) $\forall \varepsilon > 0$, $\exists U \supseteq A$, *open, such that* $m^*(U \setminus A) < \varepsilon$.
c) $\forall \varepsilon > 0$, $\exists F \subseteq A$, *closed, such that* $m^*(A \setminus F) < \varepsilon$.
d) $\exists G$, *a* $\mathcal{G}_\delta$ *set, such that* $A \subseteq G$ *and* $m^*(G \setminus A) = 0$.
e) $\exists F$, *an* $\mathcal{F}_\sigma$ *set, such that* $F \subseteq A$ *and* $m^*(A \setminus F) = 0$.

Example 2.5 In Problem 1.4a we wanted to find a closed uncountable subset of the irrationals. This is most easily done by taking an open interval of radius $\varepsilon/2^n$ about the nth rational and then looking at the complement of the union of these intervals. Theorem 2.3c gives another proof since $m^*([0, 1] \cap (\mathbf{R} \setminus \mathbf{Q})) = 1$ (i.e. take $A = [0, 1] \cap (\mathbf{R} \setminus \mathbf{Q})$ and use the facts that its outer measure is positive and that it is measurable).

With regard to Problem 1.4b, if $X \in \mathcal{M}(\mathbf{R})$ and $mX > 0$, then we can find a closed uncountable set $F \subseteq X$ by the same reason (i.e. by using Theorem 2.3c) since $m(X \setminus F) = mX - mF$.

Example 2.6 Let A be the set of those $x \in [0, 1]$ whose decimal expansion consists of not more than 9 distinct digits. Then card $A > \aleph_0$ and $mA = 0$. To prove this let $S_j \subseteq A$ consist of those elements whose decimal expansions don't contain j. Thus $\bigcup_0^9 S_j = A$. Now S_j is equivalent (bijectively) to $[0, 1]$ when we view $x \in [0, 1]$ as having an expansion to the base 9. Thus card $S_j > \aleph_0$ and so card $A > \aleph_0$. The proof that $mA = 0$ reduces to showing that for each j, $mS_j = 0$. Fix j and consider the ten intervals $\left[\dfrac{k}{10}, \dfrac{k+1}{10}\right)$, $k = 0, ..., 9$. Then throw away the $(j + 1)$st interval and so dispense with all decimals whose first term is j; note that the length of this interval is 1/10. Next we divide each of the remaining 9 intervals into 10 intervals, each of length $1/10^2$. From each of these divisions we throw away the $(j + 1)$st interval and thus we have removed all decimals whose second term is j; at this stage, then, we dispense with lengths totaling $9/10^2$. Continuing this process we end up with precisely S_j; on the other hand we have thrown away, altogether, lengths totalling

$$\frac{1}{10} + \frac{9}{10^2} + \frac{9^2}{10^3} + \cdots = \frac{1}{10}\left(1\Big/\left[1 - \frac{9}{10}\right]\right) = 1,$$

and so $mS_j = 0$.

We've defined Lebesgue measure on $\mathbf{R}$; it can also be defined on n-dimensional Euclidean space, $\mathbf{R}^n$. We shall not give the details here. Systematic treatments from various points of view are given in the books on Lebesgue integration by Asplund and Bungart, Dixmier (course at the Sorbonne), Fleming (pp. 136–205), Gurevich and Shilov, Rudin [94, pp. 49–52] and Williamson. Naturally, we can define Lebesgue outer measure on $\mathbf{R}^n$ in a manner analogous to our treatment for $\mathbf{R}$, using n-dimensional "cubes" instead of intervals; and the construction of Lebesgue measure m^n on the σ-algebra $\mathcal{M}^n$ of Lebesgue measurable sets proceeds analogously. There are, however, complications; for instance

Example 2.7 The measure of the unit ball

$$B_n = \{x \in \mathbf{R}^n : |x| \leqslant 1\}$$

is

$$m^n B_n = \pi^{n/2}/[(n/2)\Gamma(n/2)], \qquad n = 1, 2, ...,$$

where

$$\Gamma(x) = \int_0^\infty t^{x-1} e^{-t} \, dt, \quad x > 0.$$

We've indicated that $\mathcal{M}(\mathbf{R}) \setminus \mathcal{B}(\mathbf{R}) \neq \varnothing$ (Example 2.1) and $\mathcal{P}(\mathbf{R}) \setminus \mathcal{M}(\mathbf{R}) \neq \varnothing$. We now prove the latter fact (Example 2.8) as we promised in Section 2.1.

Remark In 1904, Lebesgue [68] posed the problem of measure: does there exist a non-trivial countably additive set function on $\mathcal{P}(\mathbf{R})$ which is translation invariant on $\mathcal{P}(\mathbf{R})$ and satisfies (2.2). Vitali settled this question in the negative in 1905 [120] (cf. Remark 2 after Example 2.2). The example we now give is due to Sierpiński [107].

Example 2.8 For $x, y \in \mathbf{R}$ we write $x \sim y$ if $x - y \in \mathbf{Q}$. Clearly, "$\sim$" is an equivalence relation. We now make explicit use of the axiom of choice and define A to be a subset of $(0, 1)$ which contains exactly one point from each equivalence class (for each $x \in \mathbf{R}$, an equivalence class is of the form $A_x = x + \mathbf{Q}$). We'll prove that $A \notin \mathcal{M}(\mathbf{R})$. Clearly,

$$\text{if } x \in (0, 1) \text{ then } \exists r \in \mathbf{Q} \cap (-1, 1) \text{ such that } x \in A + r;$$

also, by the definition of A we argue by contradiction to prove that

$$\text{if } r, s \in \mathbf{Q}, r \neq s, \text{ then } (A + r) \cap (A + s) = \varnothing.$$

Assume $A \in \mathcal{M}(\mathbf{R})$ and define

$$S = \bigcup \{r + A : r \in \mathbf{Q} \cap (-1, 1)\}.$$

From the translation invariance and the fact that $(A + r) \cap (A + s) = \varnothing$ for $r \neq s$ (where $r, s \in \mathbf{Q}$), we have that $S \in \mathcal{M}(\mathbf{R})$, and

$$mS = \infty \quad \text{or} \quad mS = 0.$$

Since $A \subseteq (0, 1)$, $S \subseteq (-1, 2)$ and so $mS \leqslant 3$. Hence $mS = 0$. On the other hand, $(0, 1) \subseteq S$; and so $mS \geqslant 1$, the desired contradiction. Therefore $A \notin \mathcal{M}(\mathbf{R})$, i.e., $\mathcal{M}(\mathbf{R})$ can't be all of $\mathcal{P}(\mathbf{R})$ if m is to satisfy (2.2) as well as being translation invariant.

The following result is true in any measure space (see section 2.3) having non-measurable sets, and indicates how difficult it is to approximate non-measurable sets with measurable ones.

Proposition 2.6 *Let $E \subseteq \mathbf{R}$ be non-measurable. There is $\varepsilon > 0$ such that if $E \subseteq A$ and $E^{\sim} \subseteq B$, where A and B are measurable, then $m(A \cap B) \geqslant \varepsilon$.*

Proof. Assume the result is false. Then for each n there are sets $G_n, D_n \in \mathcal{M}(\mathbf{R})$ such that

$$E \subseteq G_n, \quad E^{\sim} \subseteq D_n \quad \text{and} \quad m(G_n \cap D_n) < 1/n.$$

$G = \bigcap G_n$ and $D = \bigcap D_n$ are Lebesgue measurable, $E \subseteq G$, $E^{\sim} \subseteq D$, and $m(G \cap D) = 0$.

Since $G \in \mathcal{M}(\mathbf{R})$, $m^*S = m^*(S \cap G) + m^*(S \cap G^{\sim})$ for each $S \subseteq \mathbf{R}$.

Now, $D \subseteq (G \cap D) \cup G^{\sim}$, and so

$$S \cap D \subseteq (G \cap S \cap D) \cup (G^{\sim} \cap S).$$

This implies that

$$m^*(S \cap D) \leqslant m^*(G^\sim \cap S) + m^*((G \cap S) \cap D)$$
$$\leqslant m^*(G^\sim \cap S) + m^*(G \cap D) = m^*(G^\sim \cap S).$$

Consequently, $m^*(S \cap D) + m^*(S \cap G) \leqslant m^*(G^\sim \cap S) + m^*(G \cap S) = m^*S$.
Observe that

$$E \subseteq G \Rightarrow m^*(E \cap S) \leqslant m^*(S \cap G),$$

and $\qquad E^\sim \subseteq D \Rightarrow m^*(E^\sim \cap S) \leqslant m^*(D \cap S).$

Thus, for all $S \subseteq \mathbf{R}$,

$$m^*(E \cap S) + m^*(E^\sim \cap S) \leqslant m^*S,$$

and so $E \in \mathcal{M}(\mathbf{R})$, a contradiction.

$$\text{q.e.d.}$$

We close this section with two set theoretic remarks related to Example 2.8.

Remark 1 In 1964, R. M. Solovay proved that the Zermelo–Fraenkel set theory, ZF (which does not contain the axiom of choice, AC) is consistent with

LM: every $A \subseteq \mathbf{R}$ is Lebesgue measurable.

The main paper in which his work appears is [112]. His theorem does not say that

if AC fails then LM $\qquad\qquad\qquad (2.13)$

(i.e. that AC is necessary to find $A \notin \mathcal{M}(\mathbf{R})$); in fact, in P. J. Cohen's model for establishing certain results concerning the independence of the axiom of choice, AC fails and there are non-measurable sets. His (Solovay's) result does say that LM is consistent with but independent of the failure of AC. If we wanted to prove that the existence of a non-measurable set implies the axiom of choice then we'd have to show that in every model in which AC fails every set is Lebesgue measurable. Solovay has given one such model in which this occurs. He also proves that ZF is consistent with the property that every uncountable set $X \subseteq \mathbf{R}$ contains an uncountable closed set (e.g. Problem 1.4b, Example 2.5, and Problem 2.6). Related to statement (2.13) he shows that even with AC we can't produce a definable (from a countable sequence of ordinals) set $A \notin \mathcal{M}(\mathbf{R})$.

Remark 2 In Example 2.8 the countable subadditivity of m was crucial to establish the existence of an element $A \in \mathcal{P}(\mathbf{R}) \setminus \mathcal{M}(\mathbf{R})$. We shall now see that for the special space $\mathbf{R}^3$ we obtain the much more general result:

there is no non-trivial finitely-additive set function (e.g. section 2.3)

$$\mu: \mathcal{P}(\mathbf{R}^3) \to \mathbf{R}^+ \cup \{+\infty\}$$

such that $\mu B < \infty$ for bounded sets $B \subseteq \mathbf{R}^3$ and for which   (2.14)

$$\mu A = \mu B$$

if A and B are isometric (see Appendix I.1).

(2.14) is proved using the axiom of choice in the form of Hausdorff's paradox: *let S be the surface of the unit sphere in $\mathbf{R}^3$; then there is a countable set $D \subseteq S$ and there are disjoint subsets A, B, $C \subseteq S \setminus D$ such that*

$$S \setminus D = A \cup B \cup C \quad and \quad A \simeq B \simeq C \simeq B \cup C,$$

where "$\simeq$" designates congruence (i.e. a surjective isometry). We refer to [110] for a clear proof of the Hausdorff paradox. To prove (2.14) from the Hausdorff paradox, assume such a μ exists and define $v: \mathscr{P}(S) \to \mathbf{R}^+ \cup \{+\infty\}$ as follows: if $X \subseteq S$, $v(X)$ is μ of the union of all radii of the unit sphere with endpoints in X. Then v satisfies the same conditions as μ and we obtain a contradiction to the hypothesis of finite additivity because of Hausdorff's result. The Banach–Tarski paradox is also a corollary of the Hausdorff paradox. We'll settle for a statement of the Banach–Tarski result and again refer to [110] for details: *let A, $B \subseteq \mathbf{R}^3$ be disjoint solid spheres having the same radius; there exist $C_1,...,C_{41} \subseteq A$ and $D_1,...,D_{41} \subseteq A \cup B$ such that $A = \bigcup C_j$, $A \cup B = \bigcup D_j$, $C_j \simeq D_j$ for each j, and*

$$C_i \cap C_j = D_i \cap D_j = \emptyset, \quad 1 \leqslant i < j \leqslant 41.$$

As a final remark on these matters, Banach has shown that on $\mathbf{R}$ and $\mathbf{R}^2$ there are non-negative finitely additive set functions on $\mathscr{P}(\mathbf{R})$ and $\mathscr{P}(\mathbf{R}^2)$ which are finite for bounded sets, translation invariant, and which equal Lebesgue measure for Lebesgue measurable sets. It is consistent with ZF (i.e., there is a model of set theory) to extend Banach's result to the countably additive case.

2.3 General measure theory

As we shall later see, countable additivity is crucial for proving general and good theorems about taking limits under integral signs, switching iterated limits, etc. Thus, as an efficiency move, we now proceed directly to this notion.

Let X be a set and $\mathscr{A}$ a σ-algebra in $\mathscr{P}(X)$ (i.e. $\emptyset \in \mathscr{A}$ and $\mathscr{A}$ is closed with respect to taking complements and countable unions). $(X, \mathscr{A})$ is a measurable space and $A \in \mathscr{A}$ is measurable with respect to $\mathscr{A}$. A measure (resp. finitely additive measure) μ on $(X, \mathscr{A})$ is a function $\mathscr{A} \to \mathbf{R}^+ \cup \{+\infty\}$ such that

$$\mu(\emptyset) = 0$$

and

$$\mu(\bigcup A_n) = \sum \mu A_n,$$

where $\{A_n : n = 1, \ldots\} \subseteq \mathscr{A}$ is a countable (resp. finite) disjoint family. A **measure space** $(X, \mathscr{A}, \mu)$ is a measurable space $(X, \mathscr{A})$ and a measure μ. We frequently use the terminology, **countably** (resp. **finitely**) **additive set function**, to designate a measure (resp. finitely additive measure). If X is a topological space (see Appendix I.1) and $\mathscr{A}$ contains the Borel sets then a measure μ on $(X, \mathscr{A})$ is a **Borel measure** (cf. Appendix III.2).

Johann Radon was the first to define a general measure space (in 1913). The idea was certainly in the air because of **Vitali's** and **Lebesgue's** work to extend the fundamental theorem of calculus to $\mathbf{R}^n$; in fact, in such a setting it became important to consider set functions, etc. (cf. Chapter A5.1). The key result of **Radon's** paper was the first form of the now crucial Radon–Nikodym theorem which is the natural generalization of the fundamental theorem of calculus.

Example 2.9 We give some examples of measure spaces.

a) The Lebesgue measure space $(\mathbf{R}, \mathscr{M}(\mathbf{R}), m)$.
b) For any set X and any σ-algebra $\mathscr{A} \subseteq \mathscr{P}(X)$, fix x and define

$$\forall A \in \mathscr{A}, \qquad \delta_x A = \begin{cases} 0, & \text{if } x \notin A \\ 1, & \text{if } x \in A. \end{cases}$$

$(X, \mathscr{A}, \delta_x)$ is a measure space. δ_x is the **Dirac measure** at x.

c) $(\mathbf{R}, \mathscr{B}(\mathbf{R}), m)$ is a measure space.
d) For any set X and any σ-algebra $\mathscr{A} \subseteq \mathscr{P}(X)$, define

$$\forall A \in \mathscr{A}, \qquad cA = \begin{cases} \text{card } A, & \text{if } \text{card } A < \aleph_0 \\ \infty, & \text{if } \text{card } A \geqslant \aleph_0. \end{cases}$$

$(X, \mathscr{A}, c)$ is a measure space and c is called **counting measure**. For $\mathscr{A} = \mathscr{P}(\mathbf{R})$, c is translation invariant (i.e. c satisfies (2.1) on $\mathscr{P}(\mathbf{R})$); compare this with Example 2.8.

Theorem 2.4b is the result we mentioned after Theorem 2.2.

Theorem 2.4 *Given the measure space* $(X, \mathscr{A}, \mu)$.

a) *If* $A, B \in \mathscr{A}, A \subseteq B$, *then* $\mu A \leqslant \mu B$.
b) *If* $\{A_n : n = 1, \ldots\} \subseteq \mathscr{A}$ *satisfies the conditions that* $\mu A_1 < \infty$ *and* $A_j \subseteq A_{j-1}$ *for each* $j \geqslant 2$, *then* $\mu(\bigcap A_j) = \lim \mu A_n$.
c) *For each sequence* $\{A_n : n = 1, \ldots\} \subseteq \mathscr{A}, \mu(\bigcup A_j) \leqslant \sum \mu A_j$.

Proof. a) $B = A \cup (B \setminus A)$, and so $\mu B = \mu A + \mu(B \setminus A)$. Since $\mu(B \setminus A) \geqslant 0$ we have $\mu A + \mu(B \setminus A) \geqslant \mu A$.
b) Let $A = \bigcap A_n$ so that A_1 is the disjoint union (e.g. Fig. 2)

$$A_1 = A \cup \left(\bigcup_1^\infty (A_j \setminus A_{j+1}) \right).$$

Consequently,

$$\mu A_1 = \mu A + \sum \mu(A_j \setminus A_{j+1}).$$

Also $A_j = A_{j+1} \cup (A_j \setminus A_{j+1})$ is a disjoint union, from which we conclude that

$$\mu A_j - \mu A_{j+1} = \mu(A_j \setminus A_{j+1})$$

since all the terms, μA_k, are finite.

By our hypothesis on A_1, $\sum \mu(A_j \setminus A_{j+1})$ converges and $\mu A < \infty$; thus

$$\mu A_1 = \mu A + \sum_{1,}^{\infty} (\mu A_j - \mu A_{j+1}) = \mu A + \lim_n \sum_{1}^{n} (\mu A_j - \mu A_{j+1})$$

$$= \mu A + \mu A_1 - \lim \mu A_{n+1}.$$

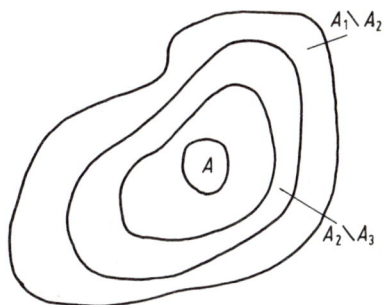

Fig. 2

c) Set $G_n = A_n \setminus \bigcup_1^{n-1} A_j$. Thus $\{G_n : n = 1, ...\}$ is a disjoint family and $\bigcup G_n = \bigcup A_n$.

Also $G_n \subseteq A_n$, and so

$$\mu(\bigcup A_n) = \mu(\bigcup G_n) = \sum \mu G_n \leqslant \sum \mu A_n.$$

q.e.d.

Let $(X, \mathscr{A}, \mu)$ be a measure space. X or μ is finite (or bounded) if $\mu X < \infty$; and X or μ is σ-finite if there is a sequence $\{A_n : n = 1, ...\} \subseteq \mathscr{A}$ such that $\mu A_n < \infty$ for each n and $\bigcup A_n = X$. Clearly, we can always take $\{A_n : n = 1, ...\}$ to be a disjoint family. A measure space $(X, \mathscr{A}, \mu)$ is complete if

$$\forall A \in \mathscr{A}, \text{ for which } \mu A = 0, \text{ and } \forall B \subseteq A \text{ we have } B \in \mathscr{A}$$

(and thus $\mu B = 0$). It follows from the definition of Lebesgue measure that $(\mathbf{R}, \mathscr{M}(\mathbf{R}), m)$ is complete. In Example 2.11 we'll "construct" a set $L \in \mathscr{M}(\mathbf{R}) \setminus \mathscr{B}(\mathbf{R})$ such that $L \subseteq C$; and, thus, $(\mathbf{R}, \mathscr{B}(\mathbf{R}), m)$ is not complete measure space.

Note that the complete measure space $(\mathbf{R}, \mathscr{M}(\mathbf{R}), m)$ is formed by "adding" all of the sets A having Lebesgue measure $mA = 0$ to $\mathscr{B}(\mathbf{R})$. This phenomenon is general in the following sense:

Theorem 2.5 *Let $(X, \mathscr{A}, \mu)$ be a measure space. Then there is a measure space $(X, \mathscr{A}_0, \mu_0)$ such that*

a) $\mathscr{A} \subseteq \mathscr{A}_0$,

b) $\mu_0 = \mu$ *on* $\mathscr{A}$,

c) $A \in \mathscr{A}_0 \Leftrightarrow A = B \cup E$, *where* $B \in \mathscr{A}$ *and* $E \subseteq D$, *for some* $D \in \mathscr{A}$ *which satisfies* $\mu D = 0$,

d) *if* $A \in \mathscr{A}_0$, $\mu_0 A = 0$, *and* $S \subseteq A$ *then* $S \in \mathscr{A}_0$ *and* $\mu_0 S = 0$.

Proof. i) We first show that $\mathscr{A}_0$, defined by c), is a σ-algebra. Let $A_n = B_n \cup E_n$, where $B_n \in \mathscr{A}$ and $E_n \subseteq D_n$, for some $D_n \in \mathscr{A}$ which satisfies $\mu D_n = 0$. $\bigcup A_n = \bigcup (B_n \cup E_n) = (\bigcup B_n) \cup (\bigcup E_n)$, and $\bigcup B_n \in \mathscr{A}$ since $\mathscr{A}$ is a σ-algebra. Clearly $\bigcup E_n \subseteq \bigcup D_n$ and $\bigcup D_n \in \mathscr{A}$ since $D_n \in \mathscr{A}$; also $\mu(\bigcup D_n) \leqslant \sum \mu D_n = 0$ by Theorem 2.4c.

Consequently, $\mathscr{A}_0$ is closed under countable unions. Obviously, $\emptyset \in \mathscr{A}_0$.

Finally we must show that $A^\sim \in \mathscr{A}_0$ if $A \in \mathscr{A}_0$.

Note that $A^\sim = B^\sim \cap E^\sim = (B^\sim \cap D^\sim) \cup (D \cap E^\sim \cap B^\sim)$, $B^\sim \cap D^\sim \in \mathscr{A}$, $D \cap E^\sim \cap B^\sim \subseteq D \in \mathscr{A}$, and $\mu D = 0$. Thus $A^\sim \in \mathscr{A}_0$ by the definition of $\mathscr{A}_0$.

ii) If $A = B \cup E$, with notation as in i), we define $\mu_0 A = \mu B$.

We must check that μ_0 is well-defined. Letting $A = B_1 \cup E_1 = B_2 \cup E_2$ it is sufficient to prove that $\mu B_1 = \mu B_2$. Clearly, $B_1 \subseteq B_2 \cup E_2 \subseteq B_2 \cup D_2$, and so $\mu B_1 \leqslant \mu B_2 + \mu D = \mu B_2$. Similarly, we compute $\mu B_2 \leqslant \mu B_1$.

iii) Next we show that μ_0 is a measure. From ii), $\mu_0 \geqslant 0$. Now consider $\bigcup A_n$, where $A_n = B_n \cup E_n$ is decomposed as in c) and $\{A_n : n = 1, \ldots\}$ is a disjoint collection.

$$\mu_0(\bigcup A_n) = \mu_0((\bigcup B_n) \cup (\bigcup E_n)) = \mu(\bigcup B_n) = \sum \mu B_n = \sum \mu_0 A_n.$$

iv) (and d). Let $\mu_0 A = 0$ and take $S \subseteq A$. Assume $A = B \cup E$ is decomposed as in c). $\mu B = 0$ since $\mu_0 A = 0$. We write $A = \emptyset \cup (B \cup E)$, noting that

$$B \cup E \subseteq B \cup D \in \mathscr{A} \quad \text{and} \quad \mu(B \cup D) \leqslant \mu B + \mu D = 0.$$

Then $S = \emptyset \cup S$ and $S \subseteq A = B \cup E \subseteq B \cup D \in \mathscr{A}$ where $\mu(B \cup D) = 0$. Thus, $S \in \mathscr{A}_0$ and $\mu_0 S = \mu \emptyset = 0$.

q.e.d.

$(X, \mathscr{A}_0, \mu_0)$ is the *complete measure space corresponding to* $(X, \mathscr{A}, \mu)$.

Theorem 2.6 a) *Let* $\{(X_\alpha, \mathscr{A}_\alpha, \mu_\alpha)\}$ *be a collection of measure spaces. Define the triple* $(X, \mathscr{A}, \mu)$ *as* $X = \bigcup X_\alpha$,

$$\mathscr{A} = \{A \subseteq X : \forall \alpha, A \cap X_\alpha \in \mathscr{A}_\alpha\}, \quad \text{and} \quad \forall A \in \mathscr{A}, \quad \mu A = \sum \mu_\alpha(A \cap X_\alpha).$$

Then $(X, \mathscr{A}, \mu)$ is a measure space; and it is σ-finite if and only if all but a countable number of the μ_α are zero and the remainder are σ-finite.

b) *Given the measure space $(X, \mathscr{A}, \mu)$ and $Y \in \mathscr{A}$.*

Set

$$\mathscr{A}_Y = \{A \in \mathscr{A} : A \subseteq Y\}.$$

Define $\mu_Y A = \mu A$ for $A \in \mathscr{A}_Y$. Then $(Y, \mathscr{A}_Y, \mu_Y)$ is a measure space.

In light of Theorem 2.6b we'll make the following notational convention. If we wish to consider Lebesgue measure m restricted to the Lebesgue measurable sets contained in a fixed set $X \subseteq \mathbf{R}$, we'll write $(X, \mathscr{M}, m)$ for the corresponding measure space. A similar remark applies to a measure μ defined on the Borel sets of a topological space X, in which case we will write $(X, \mathscr{B}, \mu)$.

Let $\mathbf{R}^* = \mathbf{R} \cup \{\pm\infty\}$. By convention, $\infty + \infty = \infty$, $-\infty - \infty = -\infty$, $\infty \cdot (\pm\infty) = \pm\infty$, $-\infty \cdot (\pm\infty) = \mp\infty$, and $\mp\infty \pm \infty$ is undefined.

Proposition 2.7 *Given the measure space $(X, \mathscr{A}, \mu)$ and the function $f: X \to \mathbf{R}^*$. The following are equivalent:*

a) $\forall \alpha \in \mathbf{R}, \{x : f(x) > \alpha\} \in \mathscr{A}$,

b) $\forall \alpha \in \mathbf{R}, \{x : f(x) \geq \alpha\} \in \mathscr{A}$,

c) $\forall \alpha \in \mathbf{R}, \{x : f(x) < \alpha\} \in \mathscr{A}$,

d) $\forall \alpha \in \mathbf{R}, \{x : f(x) \leq \alpha\} \in \mathscr{A}$,

e) $\forall U \subseteq \mathbf{R}$, open, $f^{-1}(U) \in \mathscr{A}$ and

$\qquad f^{-1}(\pm\infty) \in \mathscr{A}$.

Proof a) $\Rightarrow$ d). Clearly, $\{x : f(x) \leq \alpha\} = X \setminus \{x : f(x) > \alpha\}$ so that since $\{x : f(x) > \alpha\} \in \mathscr{A}$ and $\mathscr{A}$ is an algebra, we have $\{x : f(x) \leq \alpha\} \in \mathscr{A}$.

Similarly, $d) \Rightarrow a)$ and $b) \Leftrightarrow e)$.

a) $\Rightarrow$ b) $\{x : f(x) \geq \alpha\} = \bigcap \{x : f(x) > \alpha - (1/n)\}$ so that we have the required implication since $\mathscr{A}$ is a σ-algebra.

Similarly, b) $\Rightarrow$ a), and thus a) through d) are equivalent.

Assume a)–d) and let $U = \bigcup I_j$, where each I_j is an open interval. $f^{-1}(U) = \bigcup f^{-1}(I_j)$ and $f^{-1}(I_j) \in \mathscr{A}$ because of a)–d). Thus $f^{-1}(U) \in \mathscr{A}$ since $\mathscr{A}$ is a σ-algebra. Conversely, suppose we assume e) and take $U = (\alpha, \infty]$ (or $U = (\alpha, \infty)$); then $f^{-1}(U) \in \mathscr{A}$ and we have a).

$\qquad\qquad\qquad\qquad\qquad\qquad\qquad\qquad\qquad\qquad\qquad\qquad$ q.e.d.

An extended real-valued function f (i.e. a function with range contained in $\mathbf{R}^*$) defined on a measure space $(X, \mathscr{A}, \mu)$ is measurable (more precisely, $\mathscr{A}$-measurable or μ-measurable) if any one of the above conditions (in Proposition 2.7) holds. If $\mathscr{A}$ is $\mathscr{M}$ or $\mathscr{B}$ the corresponding measurable function is Lebesgue or Borel measurable, respectively.

Note that when X is a topological space (defined in Appendix I.1), a real-valued function f on X is continuous if $f^{-1}(U)$ is open for all open sets in $\mathbf{R}$. Whenever X is a topological space and $(X, \mathscr{A}, \mu)$ is a measure space we shall assume that $\mathscr{B}(X) \subseteq \mathscr{A}$; hence continuous functions $f: (X, \mathscr{A}, \mu) \to \mathbf{R}$ are $\mathscr{A}$-measurable. Of course, every continuous function $f: X \to \mathbf{R}$ is $\mathscr{B}(X)$-measurable.

Also, $A \in \mathscr{A}$ if and only if χ_A is a measurable function.

Proposition 2.8 *Let f and g be real-valued measurable functions on a measure space $(X, \mathscr{A}, \mu)$. Then $f \pm g, fg, f + c$, and cf are measurable, where $c \in \mathbf{R}$.*

Proof. We first indicate the proof that $f + g$ is measurable. Let $S = \{x : f(x) + g(x) < \alpha\}$. Thus if $x \in S$ there is $r \in \mathbf{Q}$ such that $f(x) < r < \alpha - g(x)$. Hence,

$$S = \bigcup_{r \in \mathbf{Q}} (\{x : f(x) < r\} \cap \{x : g(x) < \alpha - r\}).$$

Since f and g are measurable and the union is countable, $f + g$ is measurable.

To show that fg is measurable note that we need only prove f^2 is measurable since $fg = \frac{1}{2}[(f + g)^2 - f^2 - g^2]$.

$\{x : f^2(x) > \alpha\} = \{x : f(x) > \sqrt{\alpha}\} \cup \{x : f(x) < -\sqrt{\alpha}\} \in \mathscr{A}$ implies f^2 is measurable.

q.e.d.

Example 2.10 We'll prove that there is a perfect symmetric set $E \subseteq [0, 1]$ with a subset $N \notin \mathscr{M}$. From Example 2.8, choose $S \notin \mathscr{M}$, where $S \subseteq [0, 1]$; and let $\{E_n : n = 1, \ldots\}$ be a sequence of nowhere dense sets such that

$$\forall n, \qquad 1 \geqslant mE_n \geqslant 1 - (1/n).$$

If $S \cap E_n \in \mathscr{M}$ for each n then

$$\bigcup(S \cap E_n) \in \mathscr{M}.$$

Let $[0, 1] \cap (\bigcup E_n)^\sim = F \subseteq [0, 1]$. Then $mF = 0$ and

$$S = S \cap (F \cup (\bigcup E_n)).$$

Now $S \cap F \subseteq F$ implies $m(S \cap F) = 0$ and so $(S \cap F) \cup (\bigcup (S \cap E_n)) \in \mathscr{M}$, i.e. $S \in \mathscr{M}$, the desired contradiction.

Example 2.11 We'll find a set $L \in \mathscr{M} \setminus \mathscr{B}$ such that $L \subseteq C$. Since $C \in \mathscr{B}$ and $mC = 0$ we conclude that $(\mathbf{R}, \mathscr{B}, m)$ is not a complete measure space.

a) Take the set E of Example 2.10 and put the contiguous intervals of C in one-to-one correspondence with those of E by the "almost linear" map depicted in Fig. 3. Thus g is defined on $[0, 1] \setminus E$, it is increasing, and it maps $[0, 1] \setminus E$ onto $[0, 1] \setminus C$. By the monotonicity, $g(x \pm 0)$ exist for all $x \in [0, 1]$; and, since C is nowhere dense, g can be extended to a continuous increasing surjection $g: [0, 1] \to [0, 1]$. Thus g is a Borel measurable function. Take $N \subseteq E$ as in Example 2.10 and define $L = g(N)$. Thus $L \subseteq C$ and, since $mC = 0$, we conclude that $L \in \mathscr{M}$ and $mL = 0$. Also $g^{-1}L = N$ because g is injective.

b) Finally, observe that $L \notin \mathscr{B}$ for, by a routine property of Borel measurable functions $f, f^{-1}\mathscr{B} \subseteq \mathscr{B}$ (cf. Problem 2.15); and this would imply that N is a Borel set if in fact L were a Borel set.

Given a measure space $(X, \mathscr{A}, \mu)$ and a statement $S(x)$ about a point $x \in X$; for example, for a given function $f: X \to \mathbf{R}$, $S(x)$ could be the statement, $f(x) > 0$. A statement $S(x)$ is valid almost everywhere if there is a set $N \in \mathscr{A}$, for which $\mu N = 0$, such that

$$\forall x \in X \setminus N, \qquad S(x) \text{ is true;}$$

in this case we write S, a.e. or S, μ-a.e. Thus, for two functions f and g, $f = g$, μ-a.e., if $\mu\{x: f(x) \neq g(x)\} = 0$. We shall always assume that our measurable functions are finite a.e.

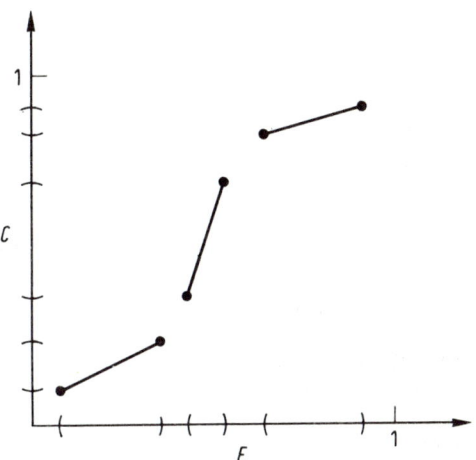

Fig. 3

Proposition 2.9 *Let $(X, \mathscr{A}, \mu)$ be a complete measure space. If f is measurable and $f = g$ μ-a.e. then g is measurable.*

Proof. Let $E = \{x: f(x) \neq g(x)\}$; and so $\mu E = 0$. Observe that

$$\{x: g(x) > \alpha\} = (\{x: f(x) > \alpha\} \cup \{x \in E: g(x) > \alpha\}) \setminus \{x \in E: g(x) \leqslant \alpha\}.$$

Since $(X, \mathscr{A}, \mu)$ is complete and $\mu E = 0$,

$$\mu\{x \in E: g(x) > \alpha\} = 0 = \mu\{x \in E: g(x) \leqslant \alpha\};$$

thus $\{x: g(x) > \alpha\} \in \mathscr{A}$ because f is measurable.

$$\text{q.e.d.}$$

Example 2.12 Let $L \in \mathcal{M} \setminus \mathcal{B}$ have Lebesgue measure $mL = 0$, and let $G \supseteq L$ be a $\mathcal{G}_\delta$ set for which $mG = mL$. Define $f = \chi_{\mathbf{R}}$ and

$$g(x) = \begin{cases} 1, & \text{if} \quad x \in \mathbf{R} \setminus G, \\ \tfrac{1}{2}, & \text{if} \quad x \in G \setminus L, \\ 0, & \text{if} \quad x \in L. \end{cases}$$

Then, since $G \in \mathcal{B}$, $f = g$, m-a.e., in the measure space $(\mathbf{R}, \mathcal{B}, m)$; but $g^{-1}(0) \notin \mathcal{B}$ and so g is not measurable on $(\mathbf{R}, \mathcal{B}, m)$.

Remark 1 Recall that if $A \in \mathcal{M}$ then $A = B \cup E$, where $B \in \mathcal{B}$ and $mE = 0$. On the other hand there are measure spaces $(\mathbf{R}, \mathcal{A}, \mu)$, for which $\mathcal{B} \subseteq \mathcal{A}$, where this decomposition is not true.

Remark 2 W. Sierpiński [104] has given an example of a set $A \notin \mathcal{M}(\mathbf{R}^2)$ which has at most two points on each straight line. Using this set A one can find a function $f: \mathbf{R} \times \mathbf{R} \to \mathbf{R}$ whose projections are Borel measurable functions but which itself is not Lebesgue measurable. There are positive results in the opposite direction, dating from Lebesgue in 1905, which are important for the finer study of "Fubini theorems" (the basic statement of Fubini's theorem is given in Appendix II).

2.4 Approximation theorems for measurable functions

Recall that our measurable functions are finite *a.e.*

Proposition 2.10 *Given a sequence* $\{f_n: n = 1, \dots\}$ *of* $\mathbf{R}^*$-*valued measurable functions on the measure space* $(X, \mathcal{A}, \mu)$.

a) *For each n, $g = \sup\{f_1, \dots, f_n\}$ and $h = \inf\{f_1, \dots, f_n\}$ are measurable functions.*

b) *$g = \sup_n f_n$ and $h = \inf_n f_n$ are measurable functions if they are finite μ-a.e.*

c) *$g = \overline{\lim} f_n$ and $h = \underline{\lim} f_n$ are measurable functions if they are finite μ-a.e.*

Proof. We prove one case for each part.

a) $\{x: g(x) > \alpha\} = \bigcup_{j=1}^{n} \{x: f_j(x) > \alpha\} \in \mathcal{A}$, and so g is measurable.

b) A routine argument shows that

$$\{x: g(x) > \alpha\} = \bigcup_{1}^{\infty} \{x: f_j(x) > \alpha\} \in \mathcal{A}.$$

c) $\overline{\lim} f_n = \inf_n \sup_{k \geqslant n} f_k$ so that g is measurable because of part b).

$$\text{q.e.d.}$$

Proposition 2.11 *Let $\{f_n: n = 1,...\}$ be a sequence of measurable functions on the measure space $(X, \mathscr{A}, \mu)$, and assume that $\lim f_n(x) = f(x)$ exists pointwise for each $x \in X$. Then f is a measurable function.*

Example 2.13 Naturally we would like to have Proposition 2.11 be true with the weaker hypothesis that $f_n \to f$, μ-a.e. Unfortunately such is not the case generally unless $(X, \mathscr{A}, \mu)$ is complete (e.g. Theorem 2.7). As a counterexample let $(X, \mathscr{A}, \mu) = (\mathbf{R}, \mathscr{B}, m)$, set $f_n = 0$ on $\mathbf{R}$, and $f = \chi_L$, where $L \in \mathscr{M} \setminus \mathscr{B}$ and $mL = 0$. Then f is not Borel measurable whereas $f_n \to f$, m-a.e., in $(\mathbf{R}, \mathscr{B}, m)$.

Theorem 2.7 *Let $(X, \mathscr{A}, \mu)$ be a complete measure space and let $\{f_n: n = 1,...\}$ be a sequence of measurable functions each defined μ-a.e. on X. If f is defined μ-a.e. and $\lim f_n = f$ pointwise μ-a.e. then f is measurable.*

For a given measure space $(X, \mathscr{A}, \mu)$, a function $f: X \to \mathbf{R}$ is simple if it can be written in the form

$$f = \sum_1^n a_j \chi_{A_j}, \quad \text{where } a_j \in \mathbf{R} \text{ and } A_j \in \mathscr{A}.$$

In particular, simple functions are measurable. We can approximate measurable functions by simple functions as follows:

Theorem 2.8 *Let f be a measurable function on $(X, \mathscr{A}, \mu)$. There is a sequence $\{f_n: n = 1,...\}$ of simple functions such that*

a) *$\forall j$ and $\forall x \in X, |f_j(x)| \leqslant |f_{j+1}(x)|$*

and

b) *$\forall x \in X, \lim\limits_j f_j(x) = f(x)$.*

If f is bounded the convergence in b) is uniform.

Proof. We'll do the details for the key part of the proof. Assume $f \geqslant 0$. For each $n = 1,...$ and $1 \leqslant k \leqslant n2^n$ define

$$A_{n,k} = \left\{ x \in X: \frac{k-1}{2^n} \leqslant f(x) < \frac{k}{2^n} \right\} \quad \text{and} \quad B_n = \{x \in X: f(x) \geqslant n\}.$$

Set $\quad f_n(x) = \sum\limits_{k=1}^{n2^n} \frac{k-1}{2^n} \chi_{A_{n,k}}(x) + n\chi_{B_n}(x).$

Clearly, f_n is a simple function for which $|f_j| \leqslant j$, and for each x

$$0 \leqslant f_j(x) \leqslant f_{j+1}(x) \leqslant f(x).$$

If $x \in \tilde{B}_n$, $|f(x) - f_n(x)| < 1/2^n$.

If $x \in B_n$ and $f(x) = \infty$ then $f_n(x) = n \to \infty$; for $x \in B_n$ and $f(x)$ finite we see that $x \in B_m^\sim$, eventually.

Consequently, we have b).   q.e.d.

The idea of the above proof is to make a finer and finer grid of $X \times [0, \infty]$ and then to draw the appropriate simple function on the grid.

Corollary 2.8.1 *Given a measure space $(X, \mathscr{A}, \mu)$ and a function $f: X \to \mathbf{R}^*$ which is finite μ-a.e.*

a) *f is measurable $\Leftrightarrow f$ is the pointwise limit (everywhere) of simple functions increasing in absolute value to $|f|$.*

b) *Assume f is bounded. f is measurable $\Leftrightarrow f$ is the uniform limit of simple functions increasing in absolute value to $|f|$.*

In 1911, D. T. Egoroff proved the following result for the case of interval.

Theorem 2.9 (Egoroff) *Given a finite measure space $(X, \mathscr{A}, \mu)$ and a family $\{f, f_n: n = 1, ...\}$ of measurable functions. Assume that $f_n \to f$, μ-a.e. Then*

$$\forall \delta > 0 \quad \exists A \in \mathscr{A} \text{ such that } \mu A < \delta \text{ and } \lim f_n = f \text{ uniformly on } A^\sim.$$

Proof. Using the convenience (tradition) of tradition (convenience) we give the proof in the following two parts:

a) Given $\varepsilon > 0$ and $\delta > 0$, we'll find a set $A_{\varepsilon\delta} \in \mathscr{A}$ and an integer N such that $\mu A_{\varepsilon\delta} < \delta$

and $\quad\quad \forall x \notin A_{\varepsilon\delta}, \forall n \geqslant N \quad |f_n(x) - f(x)| < \varepsilon.$

Without loss of generality assume $f_n \to f$ everywhere and set

$$A_m = \{x: \exists n \geqslant m \text{ such that } |f_n(x) - f(x)| \geqslant \varepsilon\}.$$

Clearly $A_m \supseteq A_{m+1}$ and for $x \in X$ there is an m for which $x \notin A_m$. Therefore $\bigcap A_m = \emptyset$ and since $\mu A_j < \infty$ we have $\lim \mu A_m = 0$. Consequently choose $A_{\varepsilon\delta} = A_N$ where $\mu A_m < \delta$ if $m \geqslant N$.

b) Given $\delta > 0$ and m, we use a) to find $B_m \in \mathscr{A}$ and N_m such that $\mu B_m < \delta/2^m$ and

$$\forall x \notin B_m, \forall n \geqslant N_m, \quad |f_n(x) - f(x)| < \frac{1}{m}.$$

Set $A = \bigcup B_m$.   q.e.d.

Example 2.14 The hypothesis that $\mu X < \infty$ is necessary in Theorem 2.9. Take $f_n = \chi_{[n, n+1]}$ on $(\mathbf{R}, \mathscr{M}, m)$. Clearly, $f_n \to f = 0$ pointwise, whereas if we let $mA < 1$ for some $A \in \mathscr{A}$ then

$$\forall n \quad \exists x_n \in [n, n+1] \cap A^\sim \text{ such that } |f(x_n) - f_n(x_n)| = 1.$$

We define

$$\|f\|_\infty = \inf\{M: \mu\{x: |f(x)| > M\} = 0\},$$

where $(X, \mathscr{A}, \mu)$ is a measure space and f is a measurable function on X (cf. Appendix I.2); notationally, we (sometimes) write

$$\|f\|_\infty = \operatorname*{ess\,sup}_{x \in X} |f(x)|.$$

It is trivial to check that

Proposition 2.12 *Given a sequence* $\{f, f_n: n = 1, \ldots\}$ *of measurable functions on the measure space* $(X, \mathscr{A}, \mu)$. $\lim\limits_{n} \|f_n - f\|_\infty = 0$ *if and only if there is a set* $A \in \mathscr{A}$, *such that* $\mu A = 0$ *and* $f_n \to f$ *uniformly on* $A^\sim$.

We noted earlier that when X is both a topological space and a measure space then continuous functions $f: X \to \mathbf{R}$ are also measurable. We close this section by looking more closely at the relation between continuous and measurable functions, especially in light of the results of the first part of this section.

Recall that $\chi_{\mathbf{R} \setminus \mathbf{Q}}$ is not the pointwise limit of continuous functions (e.g. Problem 1.14), although, from Theorem 2.8, every measurable function is the pointwise limit of simple functions.

On the other hand, we have Dirichlet's example

Example 2.15 Set

$$g_m(x) = \lim_{n} [1 - (\cos m!\pi x)^{2n}].$$

Clearly,
$$g_m(x) = \begin{cases} 0, & \text{if} \quad m!x \in \mathbf{Z} \\ 1, & \text{if} \quad m!x \notin \mathbf{Z}; \end{cases}$$

and so,

$$\forall x \in \mathbf{R}, \quad \lim_{m} g_m(x) = \chi_{\mathbf{R} \setminus \mathbf{Q}}(x).$$

Remark In Problem 1.14b we gave necessary conditions in order that a sequence of continuous functions converges pointwise to a function on $[a, b]$. In fact there are the following necessary and sufficient conditions (e.g. [13, pp. 99–102]): *a sequence* $\{f_n: n = 1, \ldots\}$ *of continuous functions* $[a, b] \to \mathbf{R}$ *converges pointwise to a function* f *if and only if for every closed set* $P \subseteq [a, b]$ *without isolated points*

$$\overline{C(f) \cap P} = P.$$

Observe, from Theorem 2.9, say, that if a topological space X has enough continuous functions defined on it, e.g. if $X = [a, b]$, then we have: *let* $f: X \to \mathbf{R}^*$ *be measurable,* $\mu X < \infty$, *and let* $\varepsilon > 0$; *then there is a continuous function* $h: X \to \mathbf{R}$ *such that* $\mu\{x: |f(x) - h(x)| \geq \varepsilon\} < \varepsilon$. A much more powerful result is the Vitali–Lusin theorem (commonly called Lusin's theorem). We'll give Vitali's original proof from 1905 [119] for two reasons: first, for historical reasons, and, second, since it is an efficient and intuitive proof. Lusin's proof appeared in 1912 [75]; and there are standard proofs due to Sierpiński [105] and L. W. Cohen [27]. All the proofs that I've seen have a similar conceptual flavor. Such a theorem, relating topological and measure theoretic notions, was certainly thought of quite early, and Vitali suggests that Borel and Lebesgue might have known of it before him; in any case Vitali published the first proof. Naturally the setting for Vitali was ($[a, b]$, $\mathcal{M}$, m). I'll state the result in its most convenient present day setting and this requires some definitions.

A Hausdorff topological space X is locally compact if each point has a neighborhood basis of compact sets (cf. Appendix I.1). Such spaces are auspicious in measure theory since they guarantee the existence of non-trivial continuous functions, and we do want to integrate continuous functions. If the notion of local compactness is not in your toolkit yet you may think of X as an interval for the time being.

Given a locally compact space X and a measure space $(X, \mathcal{A}, \mu)$ for which $\mathcal{B}(X) \subseteq \mathcal{A}$. μ is regular if

$$\forall F \subseteq X, \text{ compact, } \mu F < \infty,$$

$$\forall A \in \mathcal{A}, \mu A = \inf \{\mu U : A \subseteq U \text{ and } U \text{ is open}\},$$

$$\forall U \subseteq X, \text{ open or } \mu U < \infty, \mu U = \sup \{\mu F : F \subseteq U \text{ and } F \text{ is compact}\}.$$

Theorem 2.10 (*Vitali–Lusin*) *Given a complete measure space* $(X, \mathcal{A}, \mu)$, *where* μ *is regular and* X *is a locally compact space. Choose* $A \in \mathcal{A}$ *for which* $\mu A < \infty$ *and take a measurable function* $f : X \to \mathbf{R}^*$ *which vanishes on* $A^\sim$. *Then for each* $\varepsilon > 0$ *there is a continuous function* $g : X \to \mathbf{R}$, *which vanishes outside of a compact set, such that*

$$\mu\{x : f(x) \neq g(x)\} < \varepsilon.$$

Proof. Without loss of generality assume that $\mu X < \infty$ and $A = X$. We first prove:

$$\forall \varepsilon, \delta > 0 \quad \exists F \subseteq X, \text{ compact, such that } \mu F > \mu X - \varepsilon \tag{2.15}$$

and

$$\forall x \in F \quad \exists U_x, \text{ an open neighborhood of } x, \text{ such that}$$

$$\sup_{y \in U_x \cap F} f(y) - \inf_{y \in U_x \cap F} f(y) \leq \delta \tag{2.16}$$

(cf. the notation "$\omega(f, I)$").

To do this set

$$A_{n, \delta} = \{x : n\delta \leq f(x) < (n + 1)\delta\},$$

and note that

$$\mu X = \sum_{-\infty}^{\infty} \mu_{n, \delta}, \quad \text{where } \mu_{n, \delta} = \mu A_{n, \delta}.$$

Choose $n_1, ..., n_k$ such that

$$\mu X - \sum_{j=1}^{k} \mu_{n_j, \delta} < \varepsilon/2,$$

and let $r_j < \mu_{n_j, \delta}, j = 1, ..., k$ have the property that $\sum_{1}^{k} r_j < \varepsilon/2$.

From the regularity of μ there are compact sets $F_j \subseteq A_{n_j, \delta}, j = 1, ..., k$, for which $\mu F_j > \mu_{n_j, \delta} - r_j$.

Consequently, $F = \bigcup_{1}^{k} F_j$ is compact and

$$\mu F = \sum \mu F_j > \sum (\mu_{n_j, \delta} - r_j) > \mu X - \varepsilon.$$

(2.16) follows from the definition of $A_{n, \delta}$.

We now use (2.15) and (2.16) countably often to obtain the result.

Given $\varepsilon > 0$. Choose $\varepsilon_j, \delta_j > 0$ such that $\sum \varepsilon_j < \varepsilon$ and $\lim \delta_j = 0$. At the first step let F_1 be compact, with measure $\mu F_1 > \mu X - \varepsilon_1$, and let it also satisfy the oscillation condition (2.16) with $\delta = \delta_1$.

Choose a compact set $F_2 \subseteq F_1$ such that $\mu F_2 > \mu F_1 - \varepsilon_2$ and such that the oscillation condition (2.16) is satisfied for $\delta = \delta_2$. Thus $\mu(\cap F_j) > \mu X - \varepsilon$ and f is continuous on $\cap F_j$. The fact that we can extend f from the compact set $\cap F_j$ to a continuous function g on X follows from Urysohn's lemma (Appendix I.1).

q.e.d.

Bourbaki [18, Chapitre IV.5.1] uses the Vitali–Lusin criterion as the definition of measurable function. In fact, we have the following corollary to Theorem 2.10

Corollary 2.10.1 *Given a complete measure space $(X, \mathscr{A}, \mu)$ where X is locally compact and μ is regular. A function $f: X \to \mathbf{R}^*$, which is finite μ-a.e., is measurable if and only if for every compact set $K \subseteq X$ there is $A \in \mathscr{A}$, such that $A \subseteq K$ and $\mu A = 0$, and there is a disjoint family of compact sets $K_n \subseteq K \setminus A, n = 1, ...,$ such that $K = A \cup (\bigcup K_n)$, for which f restricted to each K_n is continuous.*

A locally compact space is σ-compact if it is the countable union of compact sets.

Corollary 2.10.2 *Given a complete measure space $(X, \mathscr{A}, \mu)$, where X is σ-compact and μ is regular. A function $f: X \to \mathbf{R}^*$, which is finite μ-a.e., is measurable if and only if there is a sequence of continuous functions $f_n: X \to \mathbf{R}$ such that $f_n \to f, \mu$-a.e.*

Proof. The sufficient conditions follow from Theorem 2.7. For the necessary conditions we use Theorem 2.10, and for each n choose a continuous function f_n such that

$$\mu A_n = \mu\{x : f(x) \neq f_n(x)\} < 1/2^n. \tag{2.17}$$

Setting

$$A = \bigcap_{k=1}^{\infty} \bigcup_{n=k}^{\infty} A_n,$$

we have $\mu A = 0$ since $\mu\left(\bigcup_{m=k}^{\infty} A_m\right) \leqslant 1/2^{k-1}$.

Also, by definition, A is precisely the set of points x which are in infinitely many A_k. Consequently, if $x \notin A$ then x is in at most finitely many A_k and so $f_n(x) = f(x)$ for all large n.

Then $f_n \to f$, μ-a.e.

q.e.d.

Problems for Chapter 2

Some of the more elementary problems in this set are the Problems 2.1, 2.2, 2.3, 2.5, 2.8, 2.13, 2.14, 2.15, 2.17, 2.18.

2.1 Prove Proposition 2.3.

2.2 a) Prove Theorem 2.2.

b) Give an example to show that the hypothesis is necessary in Theorem 2.2c.

c) Find $\{A_j : j = 1, \ldots\} \subseteq \mathscr{P}([0, 1])$, pairwise disjoint, such that

$$m^*(\bigcup A_j) < \sum m^* A_j.$$

d) Find $\{A_j : j = 1, \ldots\} \subseteq \mathscr{P}([0, 1])$ such that $A_j \supseteq A_{j+1}$ and

$$m^*(\bigcap A_j) < \lim m^* A_j$$

(obviously, $m^* A_j < \infty$ for each j).

2.3 Define $\mathscr{A} \subseteq \mathscr{P}(\mathbf{R})$ by the property that either card A or card $(\mathbf{R} \setminus A)$ is countable for $A \in \mathscr{A}$. The set function $\mu : \mathscr{A} \to \mathbf{R}$ is given by

$$\mu A = \begin{cases} 0, & \text{if } \operatorname{card} A \leqslant \aleph_0 \\ \infty, & \text{if } \operatorname{card} \mathbf{R}\setminus A \leqslant \aleph_0. \end{cases}$$

Prove that $(\mathbf{R}, \mathscr{A}, \mu)$ is a measure space.

2.4 For each $X \subseteq \mathbf{R}$, define

$$X_d = \{x \in (0, 1) : \exists m \in \mathbf{Z} \text{ such that } mx \in X\}.$$

Prove that

$$\forall N > 0 \text{ and } \forall \varepsilon > 0 \quad \exists X \subseteq \mathbf{R} \text{ such that } mX > N \text{ and } mX_d < \varepsilon.$$

(*Hint.* Begin with the following special case. Take

$$X = \bigcup_{k=1}^{[n/\delta]} \left(\frac{n}{k}, \frac{n+\delta}{k} \right),$$

where $[n/\delta]$ is the greatest integer less than or equal to n/δ. Show that

$$\forall n, \qquad mX = \sum_{k=1}^{[n/\delta]} \delta/k > \delta \log (n/\delta)$$

and that for large n

$$mX_d = \delta + \sum_{k=j+1}^{[n/\delta]} \delta/k < \delta + 2\delta \log (1/\delta).$$

Consequently, "$mX \to \infty$" and "$mX_d \to 0$".)

2.5 a) Find $X \subseteq [0, 1]$ such that $mX = 0$ and X is not of first category. (*Hint.* Consider a perfect symmetric set with perfect symmetric sets in each of its contiguous intervals, etc.; make the measures add up to 1 and look at the complementary set.)

b) For each $\varepsilon > 0$ construct an open set $U \subseteq [0, 1]$ such that $\bar{U} = [0, 1]$ and $mU = \varepsilon$. (*Hint.* Take intervals of length $\varepsilon/2^k$ about the rationals.)

c) Find $A, B \subseteq [0, 1]$ such that $A, B \in \mathcal{M}$, $A \cap B = \emptyset$,

$$mA = 0 \text{ and } A \text{ is not of first category,}$$

and

$$mB = 1 \text{ and } B \text{ is of first category.}$$

(*Hint.* Check Proposition 2.5. Can you find a simpler example?)

2.6 a) Prove that every $B \in \mathcal{B}(\mathbf{R})$ which is not of first category contains a closed uncountable subset.

b) Is a) true for uncountable Borel sets of first category?

c) Prove that the F. Bernstein example, which is what we reference in Problem 1.4b, contains uncountable sets of measure 0.

2.7 Let m^2 be Lebesgue measure on $\mathbf{R} \times \mathbf{R}$.

a) A **packing problem**. Let U be the open unit disc in $\mathbf{R} \times \mathbf{R}$ and let $\{U_n : n = 1, ...\}$ be a sequence of open discs in $\mathbf{R} \times \mathbf{R}$ such that

i) $\bar{U}_n \subseteq U$ for each n,
ii) $\{\bar{U}_n : n = 1, ...\}$ is pairwise disjoint,
iii) $\sum r_n < \infty$, where r_n is the radius of U_n.

Define $X = U \setminus \bigcup_{1}^{\infty} U_n$ and prove $mX > 0$ (cf. [126]).

b) Show that every set of positive Lebesgue measure in $\mathbf{R} \times \mathbf{R}$ contains the vertices of an equilateral triangle.

c) Is b) true for other polygons in $\mathbf{R} \times \mathbf{R}$?

2.8 a) Prove that there are no numbers ε and δ, such that $0 < \varepsilon \leqslant \delta < 1$, which have the following property: if $\{A_n : n = 1, ...\} \subseteq \mathscr{P}([0, 1])$ is any sequence of Lebesgue measurable sets each satisfying $mA_n \geqslant \delta$ then there is a set A of measure ε for which $A \subseteq A_n$ for infinitely many n. (*Hint.* Let A_n be the set of $x \in (0, 1)$ such that the nth digit in its decimal expansion is non-zero.)

b) Given $\{A_n : n = 1, ...\} \subseteq \mathscr{M}$, $A_n \subseteq [0, 1]$, and assume that 1 is a cluster point of $\{mA_n : n = 1, ...\}$. Prove that there is a subsequence $\{n_k : k = 1, ...\}$ for which

$$m\left(\bigcap_k A_{n_k}\right) > 0. \tag{P2.1}$$

(*Hint.* $\sum_k (1 - mA_{n_k}) < 1$.)

Remark If we begin by taking an uncountable collection then (P2.1) is always possible. For further results on this type of problem we refer to [42; 43].

2.9 Let $f : \mathbf{R} \to \mathbf{R}$ be a Lebesgue measurable function and let $G : \mathbf{R} \times \mathbf{R} \to \mathbf{R}$ be continuous. Show that if

$$\forall x, y \in \mathbf{R}, \qquad |f(x + y)| \leqslant G(f(x), f(y)),$$

then f is bounded on bounded sets. (*Hint.* $h(x) = |f(x)| + |f(-x)|$ is measurable and so there is a set $X \subseteq \mathbf{R}$ such that $mX > 0$ and h is bounded on X; thus for all $x, y \in X$, $f(x - y)$ is bounded. Since $X - X$ is a neighborhood U of 0 (a useful result with an ingenious proof due to Steinhaus, which we'll give in Problem 3.6b) f is bounded on U. By applying induction to the hypothesis we see that $f(nz)$ is bounded, for $z \in U$; and so f is bounded on any bounded set.)

2.10 a) Given $A \subseteq [0, 1]$, and assume $m^*A > 0$. Show that there is a non-measurable set $E \subseteq A$. Can you show that for any $\alpha \in (0, 1)$ there is a non-measurable set $E \subseteq [0, 1]$ for which $m^*E = \alpha$? For purposes of comparison note Example 2.10.

b) For $Y \subseteq \mathbf{R}$ let $\tau_\alpha Y = \{y + \alpha : y \in Y\}$.

i) If $S = \bigcup_{-\infty}^{\infty} [2k, 2k + 1)$, prove that $\tau_\alpha S = S^\sim$ when α is an odd integer.

ii) Let $D = \{m/3^n : m, n \in \mathbf{Z}, n \geqslant 0\}$. If we define $x \sim y$ by the condition that $x - y \in D$ then "$\sim$" gives rise to an equivalence relation; we then define X to be the set consisting of exactly one point from each equivalence class. Now, for each $x \in X$, set

$$A_x = \left\{x + \frac{m}{3^n} : m, n \in \mathbf{Z}, m \text{ even}, n \geqslant 0\right\},$$

and let

$$S = \bigcup_{x \in X} A_x.$$

Prove that $\tau_\alpha S = S^\sim$ when $\alpha = 1/3^n$.

c) Define $\tau S = \inf \{\alpha \in \mathbf{R}, \alpha > 0: \tau_\alpha S = S^\sim\}$; thus τS is 1 and 0 for bi) and bii), respectively. Prove that if $\tau S = 0$ then $S \notin \mathcal{M}(\mathbf{R})$.

d) $H \subseteq \mathbf{R}$ is an Hamel basis if

$$\forall x \in R, \exists \{r_\alpha\} \subseteq \mathbf{Q} \text{ and } \exists \{h_\alpha\} \subseteq H, \text{ such that } x = \sum r_\alpha h_\alpha,$$

where the sum is finite and the representation is unique. Using Zorn's lemma, which is an equivalent form of the axiom of choice, it is easy to prove that Hamel bases exist using the following argument: let $\mathscr{F}$ be the family of all subsets $S \subseteq \mathbf{R}$ which are linearly independent over $\mathbf{Q}$; then there is a maximal element $H \in \mathscr{F}$. Prove that if H is an Hamel basis and $H \in \mathcal{M}$ then $mH = 0$. Thus, if H is an Hamel basis and $m^*H > 0$, we conclude that $H \notin \mathcal{M}$.

e) The existence of an Hamel basis with $m^*H > 0$ is a corollary of the following fact:

there is an Hamel basis H_B which intersects every closed uncountable

set in $\mathbf{R}$. (P2.2)

(P2.2) was first proved by Burstin in 1916; a simpler proof of this fact is due to Abian [1]. Assuming (P2.2) prove that $m^*H_B > 0$.

f) Let $H = \{x\} \cup \{x_\alpha: \alpha \in A, \text{ an index set}\} \subseteq \mathbf{R}$ be an Hamel basis and define X to be the set of elements $y \in (0, 1)$ whose Hamel expansions do not use x. Prove that $X \notin \mathcal{M}$ (e.g. [2; 103]).

Remark In [109], Sierpiński defines a set $X \subseteq \mathbf{R}$ to have the property S if

$$\forall A \subseteq \mathbf{R}, mA = 0, \quad \text{we have} \quad \operatorname{card} A \cap X \leqslant \aleph_0.$$

Thus X has the property S if and only if every uncountable subset of X is non-measurable. He then proves that uncountable sets with the property S exist if the continuum hypothesis is assumed (cf. [32] for a recent development).

2.11 Prove that there is a continuous function $f: [0, 1] \to \mathbf{R}$ such that $f(x) \in \mathbf{Q}$, m-a.e., and f is not constant on any interval. (*Hint.* At the first step form a sequence $\{f_{n,1}: n = 1, ...\}$ of continuous functions such that $f_{n,1} \to f_1$ uniformly and f_1 has rational values on a set of Lebesgue measure $\frac{1}{2}$.)

2.12 For each $x \in (0, 1]$ let $f(x) = \sum a_j/j$ when $x = .a_1...$ (2) (where we take the expansion of x to have an infinite number of 1's in the case that x has the form $.a_1...a_n$ (2)). Are $f^{-1}(\mathbf{R}^+)$ and $f^{-1}(\infty)$ Lebesgue measurable sets?

2.13 Let f be a decreasing and bounded real-valued function on $[0, 1]$. Show that there is a sequence $\{f_n: n = 1, ...\}$ of continuous decreasing functions such that $f_n \to f$, m-a.e.

2.14 Prove Theorem 2.7. (*Hint.* Define $A \in \mathcal{A}$ with null complement in terms of $\{f_n : n = 1, ...\}$ and $\{f\}$, set $g_n = f_n \chi_A$, $g = f \chi_A$, and use Proposition 2.9.)

2.15 Let $f: [a, b] \to \mathbf{R}$ be a Lebesgue measurable function.
Prove that

$$\forall B \in \mathcal{B}(R), \qquad f^{-1}(B) \in \mathcal{M}.$$

(*Hint.* Let $\mathscr{C} = \{A \subseteq \mathbf{R}: f^{-1} A \in \mathcal{M}\}$ and prove that $\mathscr{C}$ is a σ-algebra.)

2.16 Find an everywhere discontinuous function $f: \mathbf{R} \to \mathbf{R}$ such that f is measurable on $(\mathbf{R}, \mathcal{M}, m)$ but not measurable on $(\mathbf{R}, \mathcal{B}, m)$. (*Hint.* Use the idea of Example 2.12.)

2.17 Let the functions $f, g: (0, 1) \to \mathbf{R}$ have the following property:

$$\forall \alpha \in \mathbf{R}, \qquad m\{x: f(x) \geqslant \alpha\} = m\{x: g(x) \geqslant \alpha\}. \tag{P2.3}$$

Prove that if f and g are continuous on the left for all $x \in (0,1)$ as well as being decreasing, then $f = g$ on $(0, 1)$. (*Hint.* Assume $f(x_0) > g(x_0)$, let $\varepsilon = f(x_0) - g(x_0)$, and find $\delta > 0$ such that

$$m\{x: f(x) \geqslant f(x_0)\} = x_0$$
$$m\{x: g(x) \geqslant f(x_0)\} = x_0 - \delta.)$$

2.18 Find a sequence $\{f_n : n = 1, ...\}$ of functions $[0, 1] \to \mathbf{R}$ such that

$$\forall x \in [0, 1], \qquad \lim f_n(x) = 0,$$

whereas for each $[a, b] \subseteq [0, 1]$, $\{f_n : n = 1, ...\}$ does not converge uniformly on $[a, b]$.

2.19 With respect to Problem 1.12 we'd like to know if for any fixed n there is a set $X \subseteq \mathbf{R}$ with $mX = 0$ such that

$$\forall d_1, ..., d_n > 0 \quad \exists x_0, ..., x_n \in X \quad \text{for which}$$
$$d_j = x_j - x_{j-1}, \quad j = 1, ..., n.$$

2.20 Let $\alpha_n \to 0$. Find a bounded Lebesgue measurable function $f: [0, 1] \to \mathbf{R}$ such that $f(x - \alpha_n)$ does not tend to $f(x)$, m-a.e. (*Hint.* Let $f = \chi_E$, where E is a perfect symmetric set of positive Lebesgue measure.)

2.21 Let $\mathscr{L}$ be a collection of line segments in $\mathbf{R}^n$ and let $E(\mathscr{L})$ be the set of all endpoints of the members of $\mathscr{L}$. Any set $X \subseteq \mathbf{R}^n$ of two or more points has the form $E(\mathscr{L})$ for some $\mathscr{L}$. $X \subseteq \mathbf{R}^n$ is an endset if $X = E(\mathscr{L})$ for a disjoint collection $\mathscr{L}$. Prove the following results.

a) In $\mathbf{R}^1$, if S is an endset then card $S \leqslant \aleph_0$.

b) In $\mathbf{R}^2$, if S is a closed bounded endset then $m^2 S = 0$.

Remark Part b) does not extend to $\mathbf{R}^n$ for $n \geqslant 4$; and the result is not known for $\mathbf{R}^3$.

2.22 a) Given $A \in \mathcal{M}$, where $A \subseteq [0, 1]$ and $0 < mA < 1$. Prove that

$$\inf m(A \cap I)/mI = 0 \qquad \text{and} \qquad \sup m(A \cap I)/mI = 1,$$

where the inf and sup are taken over all non-trivial proper subintervals I of $[0, 1]$.
b) Find $A \in \mathcal{M}$, where $A \subseteq [0, 1]$, such that for each nontrivial proper subinterval $I \subseteq [0, 1]$,

$$m(A \cap I) > 0 \qquad \text{and} \qquad m(A^\sim \cap I) > 0.$$

Remark In the Bernstein example, mentioned in Problems 1.4b and 2.6c, $\mathbf{R}$ is decomposed as a disjoint union $A \cup B$ where neither A nor B contains uncountable closed sets F but where for each such F, $A \cap F, B \cap F \neq \emptyset$. It is easy to check that neither A nor B is measurable (in fact, assume A is measurable, approximate the measure from within by compact sets, and obtain a contradiction to the decomposition properties). Compare this situation with Problem 2.22b.

2.23 An homeomorphism $h: [0, 1] \to [0, 1]$ is a continuous bijection whose inverse is also continuous. Note that the Borel sets are invariant under homeomorphisms $h: [0, 1] \to [0, 1]$; and that such is not the case for the Lebesgue measurable sets (Example 2.11). This is why in probability theory the class of "probabilizable" sets are the Borel sets and not the Lebesgue measurable sets.

a) Let $E \subseteq [0, 1]$ be a closed nowhere dense set. Show that there is an homeomorphism $h: [0, 1] \to [0, 1]$ such that $m(hE) = 0$. (*Hint:* Define

$$h(x) = m([0, x] \cap ([0, 1] \setminus E))/m([0, 1] \setminus E).)$$

b) Let $E \subseteq [0, 1]$ be a set of first category. Show that there is an homeomorphism $h: [0, 1] \to [0, 1]$ such that $m(hE) = 0$. (*Hint.* Let $\mathcal{H}$ be the family of homeomorphisms $[0, 1] \to [0, 1]$ with metric $d(f, g) = \sup_x |f(x) - g(x)|$. If $E = \bigcup E_n$, where E_n is nowhere dense, define

$$A_{n, k} = \{h \in \mathcal{H} : m\bar{E}_n < 1/k\}.$$

Prove that $A_{n, k}$ is open in $\mathcal{H}$, and, setting $A = \bigcap_{n, k} A_{n, k}$, note that $m(hE) = 0$ for each $h \in A$.)

Remark Actually it is possible to find an uncountable set $F \subseteq [0, 1]$ such that

$$\forall h \in \mathcal{H}, \qquad m(hF) = 0. \tag{P2.4}$$

This result makes explicit use of the continuum hypothesis and was first proved by A. S. Besicovitch [11]. For each $h \in \mathcal{H}$ there is a corresponding Hausdorff measure μ_h (we define this notion in the next exercise) and Besicovitch actually proved $\mu_h(F) = 0$ for each $h \in \mathcal{H}$; (P2.4) follows trivially from this. Relative to our remark about Borel sets and homeomorphisms at the beginning of this exercise, we know that $F \in \mathcal{M} \setminus \mathcal{B}$.

Besicovitch's solution to (P 2.4) is essentially equivalent to his solution of a famous conjecture made by Borel (1919). In order to state the conjecture we say that a set $F \subseteq [0, 1]$ has the property C if

$$\forall \{a_n : a_n > 0, n = 1, \ldots\} \quad \exists \{I_n = (c_n, d_n) : d_n - c_n = a_n, n = 1, \ldots\}$$

such that

$$F \subseteq \bigcup I_n.$$

Borel's conjecture was that every set with the property C had to be countable. In order to construct his counterexample (to the Borel conjecture), Besicovitch defined the notion of a *concentrated set F in the neighborhood of a given countable set H* by the property that

$$\operatorname{card} (F \setminus (U \cap F)) \leqslant \aleph_0$$

for each open set U containing H. Using the continuum hypothesis, Besicovitch was able to construct such sets F for which $\operatorname{card} F > \aleph_0$. These concentrated sets provide the solution to Borel's conjecture as well as to (P2.4) (e.g. Problem 2.24c).

2.24 Let $\mathcal{H}$ be as in Problem 2.23. If $A \subseteq [0, 1]$, set $dA = \sup \{|x - y| : x, y \in A\}$. For each $h \in \mathcal{H}$ and $\varepsilon > 0$ define

$$\mu_h^\varepsilon A = \inf \left\{ \sum_{j=1}^\infty h(dA_j) : A = \bigcup A_j, dA_j < \varepsilon \right\}$$

and $\mu_h A = \sup_{\varepsilon > 0} \mu_h^\varepsilon A$.

for each $A \subseteq [0, 1]$. If $h(t) = t^p$ we write $\mu_h^\varepsilon = \mu_p^\varepsilon$ and $\mu_h = \mu_p$. Observe that if $p < q$ and $\mu_p A < \infty$ then $\mu_q A = 0$; also $\mu_1 = m^*$. μ_h is a Hausdorff measure and, in fact, it is an outer measure on $\mathcal{P}([0, 1])$. By the canonical Caratheodory method each μ_h is a measure on a σ-algebra $\mathcal{A}_h$. Further $\mathcal{B}[0, 1] \subseteq \mathcal{A}_h$ and μ_h is regular. The Hausdorff dimension of $A \subseteq [0, 1]$ is

$$H\text{-dim } A = \sup \{p : \mu_p A > 0\}.$$

Some standard references on Hausdorff measure are [52; 58; 60, Chapter 2; 91].

a) Compute H-dim $C = \log 2/\log 3$. (*Hint.* For each n divide C into 2^n parts P_k, where $dP_k = (\frac{1}{3})^n$; then for $\varepsilon = (\frac{1}{3})^n$ and $p = \log 2/\log 3$ compute $\mu_p^\varepsilon C$.)

b) Prove that $\mu_h A = 0$ for all $h \in \mathcal{H}$ if and only if for every strictly decreasing sequence $\{a_n : n = 1, \ldots\}$ tending to 0 there is a sequence of intervals I_n such that $A \subseteq \bigcup I_n$ and $mI_n \leqslant a_n$.

c) If F is a concentrated set in a neighborhood of a countable set H prove that $\mu_h F = 0$ for all $h \in \mathcal{H}$.

Remark In these problems the purpose is to analyze sets of measure zero for any sort of thickness it might have. A profound study in this area has been made by Borel [17, Chapter 4] and there are still many answers (and questions) to be found. Before moving on to some known results in this field (in Problem 2.25) we note that $\mu_p L = 0$ for each $p > 0$, where $L \subseteq [0, 1]$ is the set of Liouville numbers. Compare this with Besicovitch's result. Borel was interested in such matters.

2.25 a) Prove that if $E \subseteq [0, 1]$ and $mE = 0$ then

$$\exists\{a_n: a_n > 0, \sum a_n < \infty\}$$

such that

$$\forall \varepsilon > 0 \quad \exists\{I_n = (c_n, d_n): n = 1, \ldots\}$$

with the properties that

$$E \subseteq \bigcup I_n \quad \text{and} \quad mI_n \leqslant \varepsilon a_n.$$

(*Hint.* Let $\{I_{jk}: j = 1, \ldots, k = 1, \ldots\}$ be a covering of E by open intervals such that $\sum_k mI_{jk} < (\frac{1}{4})^j$. Enumerate $\{I_{jk}: j, k\}$ as $\{J_n: n = 1, \ldots\}$ and let I_n be an interval with the same center as J_n such that $2^j mJ_n = mI_n$, where J_n corresponds to some I_{jk}. Let $a_n = mI_n$.)

b) Given $E \subseteq [0, 1]$ with Hausdorff dimension H-dim $E > 0$. Prove that if

$$\forall p > 0, \quad \sum_{n=1}^{\infty} a_n^p \text{ converges,}$$

then $\{a_n: n = 1, \ldots\}$ can not be associated with E in the sense of part a). Thus there is $\varepsilon > 0$ so that for any cover of C by open intervals $\{I_n: n = 1, \ldots\}$, there is n for which $mI_n > \varepsilon/2^n$.

c) Let $0 < r < \log_2 3 - 1$ and set $a_n = 1/n^{\log_2 3 - r}$. Prove that $\{a_n: n = 1, \ldots\}$ is associated with C in the sense of part a).

d) In light of part b) it is natural to ask if for any convergent series $\sum a_n (a_n > 0)$ there is a set $E \subseteq [0, 1]$, $mE = 0$, such that $\{a_n: n = 1, \ldots\}$ is not associated with E in the sense of part a). The answer is that such an E always exists and the proof depends on a result of A. Dvoretzky (1948) [91, pp. 68 ff]; the details are due to E. Boardman [14].

3 The Lebesgue integral

3.1 Motivation

An excellent description of the motivation to develop the notion of the Lebesgue integral has been given by Lebesgue himself in an article, "Development of the integral concept" (1926), which appears in [71].

We begin by recalling the definition of the Riemann integral for (certain) bounded functions $f: [a, b] \to \mathbf{R}$. For any partition

$$P: a = x_0 < x_1 < \ldots < x_n = b$$

of $[a, b]$ consider the numbers

$$S_P = \sum_1^n M_i (x_i - x_{i-1}) \quad \text{and} \quad s_P = \sum_1^n m_i(x_i - x_{i-1}), \tag{3.1}$$

where

$$M_i = \sup \{f(x): x_{i-1} < x \leqslant x_i\} \quad \text{and} \quad m_i = \inf \{f(x): x_{i-1} < x \leqslant x_i\}.$$

Define

$$R\overline{\int_a^b} f = \inf_P S_P \quad \text{and} \quad R\underline{\int_a^b} f = \sup_P s_P. \tag{3.2}$$

Clearly, $R\overline{\int} \geqslant R\underline{\int}$ and we say that f is Riemann integrable if $R\overline{\int} = R\underline{\int}$; in this case, the Riemann integral of f over $[a, b]$ is

$$R\int_a^b f = R\underline{\int_a^b} f = R\overline{\int_a^b} f.$$

Note that

$$R\overline{\int_a^b} f = \inf \left\{ \int_a^b \psi : \psi \geqslant f, \psi = \sum_1^n c_j \chi_{(x_{j-1}, x_j)} \right\},$$

where $x_0 < x_1 < ... < x_n$ is a partition of $[a, b]$ and

$$\int_a^b \psi = \sum_1^n c_j(x_j - x_{j-1}).$$

Example 3.1 Define the function $f: [a, b] \to \mathbf{R}$ as

$$f(x) = \begin{cases} 0, & \text{if } x \in [a, b] \text{ is irrational,} \\ 1, & \text{if } x \in [a, b] \text{ is rational.} \end{cases}$$

Clearly,

$$R\overline{\int_a^b} f = b - a \qquad \text{and} \qquad R\underline{\int_a^b} f = 0$$

so that f is not Riemann integrable.

Lebesgue's observation goes something like this. The numbers $R\underline{\int}$ and $R\overline{\int}$ will be close if somehow there is a lot of continuity in each interval of each partition. Example 3.1 shows that this will not happen if f has many discontinuities. Lebesgue's goal was to collect approximately equal values of f. He proceeded in the following way. Let $f: [a, b] \to \mathbf{R}$ be a bounded Lebesgue measurable function, and consider the partition

$$Q: \alpha = y_0 < y_1 < ... < y_n = \beta,$$

where

$$\alpha = \inf \{f(x): x \in [a, b]\} \qquad \text{and} \qquad \beta = \sup \{f(x): x \in [a, b]\}.$$

If $A_j = \{x: y_j \leqslant f(x) < y_{j+1}\}$, $j = 0, ..., n - 1$, $A_n = \{x: f(x) = \beta\}$, and $y_{n+1} = \beta$ we define

$$S_Q = \sum_{j=0}^n y_{j+1} m A_j, \qquad s_Q = \sum_{j=0}^n y_j m A_j,$$

and

$$\overline{\int_a^b} f = \inf_Q S_Q, \qquad \underline{\int_a^b} f = \sup_Q s_Q.$$

A major initial result is

Theorem 3.1 *Let $f: [a, b] \to \mathbf{R}$ be a bounded Lebesgue measurable function. Then*

$$\underline{\int_a^b} f = \overline{\int_a^b} f,$$

and the common value is denoted by $\int_a^b f.$

Proof. Given $\varepsilon > 0$, we divide the interval (α, β) into two partitions, Q_1 and Q_2, such that for any y_i, $i \geqslant 1$, in Q_1 or Q_2, we have

$$y_i - y_{i-1} < (\varepsilon/2)(b - a).$$

Then

$$S_{Q_j} - s_{Q_j} = \sum_0^n (y_{i+1} - y_i)mA_i < \frac{\varepsilon}{2(b - a)} \sum_0^n mA_i = \frac{\varepsilon}{2}.$$

Let Q be the partition formed by the points in both Q_1 and Q_2. Note that $S_{Q_j} \geqslant S_Q$ and $s_Q \geqslant s_{Q_j}$; for example, if the partition Q_1 is

$$Q_1 : \alpha = y_0 < \ldots < y_n = \beta$$

and $y_{j-1} < y' < y_j$ then

$$y_j m B_j + y' m C_j \leqslant y_j(mB_j + mC_j) = y_j mA_j,$$

where

$$B_j = \{x : y' \leqslant f(x) < y_j\} \qquad \text{and} \qquad C_j = \{x : y_{j-1} < f(x) \leqslant y'\}.$$

Thus the interval $(s_Q, S_Q) \subseteq (s_{Q_1}, S_{Q_1}) \cap (s_{Q_2}, S_{Q_2})$. Consequently, s_{Q_j} and S_{Q_j}, for $j = 1, 2$, are points in an interval of length less than ε. Hence, by the Cauchy criterion, $\inf\limits_Q S_Q = \sup\limits_Q s_Q$.

q.e.d.

Remark 1 The clever technical trick of partitioning the "$f(x)$-axis" is not the conceptual key; the idea lies in trying to collect approximately equal values of $f(x)$ in the delicate manner made possible by measure theory.

Remark 2 If g is a simple function

$$g = \sum_1^k a_j \chi_{A_j},$$

defined on $([a, b], \mathcal{M}, m)$, it is natural to define the integral, $\int_a^b g$, as

$$\int_a^b g = \sum_1^k a_j mA_j$$

(cf. the Remark at the very beginning of section 3.2). In light of Theorem 2.8, Theorem 3.1 can be restated as follows: *if $f : [a, b] \to \mathbf{R}$ is a bounded Lebesgue measurable function then there is a sequence $\{f_n : n = 1, \ldots\}$ of simple functions such that $f_n \to f$ pointwise (in fact, the convergence is uniform and $\|f_n\|_\infty \leqslant \|f\|_\infty$) and*

$$\int_a^b f_n \to \int_a^b f. \tag{3.3}$$

In this form it is interesting to compare this result with the Lebesgue dominated convergence theorem (section 3.3).

Because of Theorem 3.1 we define the Lebesgue integral of a bounded Lebesgue measurable function $f: [a, b] \to \mathbf{R}$ to be $\int_a^b f$.

Example 3.2 a) Take f as in Example 3.1. Clearly f is Lebesgue measurable. For the partition $R: 0 = y_0 < y_1 < \ldots < y_n = 1$ we have

$$S_R = y_1(b - a).$$

Therefore, $\inf_Q S_Q = \int_a^b f = 0$.

b) Let $\{r_n : n = 1, \ldots\}$ be an ordering of $\mathbf{Q} \cap [a, b]$ and define the function $f_m : [a, b] \to \mathbf{R}$ as

$$f_m(x) = \begin{cases} 1, & \text{if } x = r_1, \ldots, r_m, \\ 0, & \text{if otherwise.} \end{cases}$$

Clearly $f_m \geqslant 0$ and $\{f_m : m = 1, \ldots\}$ increases pointwise to the function f defined in Example 3.1. Also,

$$\forall m, \qquad R \int_a^b f_m = 0. \tag{3.4}$$

Even though (3.4) is true, $R \int_a^b f$ does not exist as we showed in Example 3.1. One of the beauties of Lebesgue's theory is that (on a finite interval $[a, b]$, say) if $g_n \geqslant 0$ is measurable and the sequence $\{g_n : n = 1, \ldots\}$ increases pointwise to a bounded function g, then $\int g$ exists and equals $\lim \int g_n$.

Remark Example 3.2 is not a good example of a function which is Lebesgue integrable but not Riemann integrable. In fact f is 0 m-a.e. and so there is a Riemann integrable function, viz., g identically 0, in the same equivalence class as f (noting that "a.e." defines an equivalence relation). In Example 3.11 we shall give examples of functions g whose Lebesgue integral exists but for which there is no Riemann integrable function h with the property that $g = h$, m-a.e.

3.2 The Lebesgue integral

Let $(X, \mathscr{A}, \mu)$ be a measure space. We define the integral of a simple function

$$f = \sum_1^n a_j \chi_{A_j}, \qquad A_j \in \mathscr{A}, a_j \in \mathbf{R}, \tag{3.5}$$

to be

$$\int f \, d\mu = \sum a_j \mu A_j$$

if $A_j = \{x : f(x) = a_j\}$ and $\mu A_j < \infty$. For each such f we write

$$\int_A f \, d\mu = \int \chi_A \, f \, d\mu, \qquad A \in \mathscr{A}.$$

Remark We can write a simple function f in many ways. Our canonical criterion will be (3.5) with the property that $A_j = \{x : f(x) = a_j\}$. On the other hand, the operation "$\int$" is a linear map on the vector space of simple functions which vanish outside of a set of finite measure (and so we do not have to worry if $A_j \cap A_k \neq \emptyset$). This fact can be proved in the following way. Write a given simple function $f = \sum b_j \chi_{B_j}$ canonically as $\sum a_j \chi_{A_j}$; define $\int f \, d\mu$ as $\sum a_j \mu A_j$ and check that this is well-defined; finally, calculate that $\sum a_j \mu A_j = \sum b_j \mu B_j$. All the details are routine.

Theorem 3.1 can be generalized in a straightforward way to the following general context.

Theorem 3.2 *Given a complete finite space $(X, \mathscr{A}, \mu)$, and let $f \colon X \to \mathbf{R}$ be a bounded function. f is μ-measurable $\Leftrightarrow$*

$$\inf \left\{ \int h \, d\mu : f \leqslant h \text{ and } h \text{ is simple} \right\} = \sup \left\{ \int g \, d\mu : f \geqslant g \text{ and } g \text{ is simple} \right\}.$$

$$(3.6)$$

Note that we do not use the completeness hypothesis in Theorem 3.2 to prove (3.6).

Because of Theorem 3.2 we define the μ-integral of a μ-measurable bounded function $f \colon X \to \mathbf{R}$, $\mu X < \infty$, as

$$\int f \, d\mu = \inf \left\{ \int h \, d\mu : f \leqslant h \text{ and } h \text{ is simple} \right\}.$$

$$(3.7)$$

Example 3.3 Given $([0, 1], \mathscr{M}, m)$. For this example we shall refer to a finite disjoint collection $\{A_j : j = 1, \ldots, n\} \subseteq \mathscr{M}$, whose union is $[0, 1]$, as a "partition" P; we set $|P| = \sup_j mA_j$. Recalling the definition of the Riemann integral one might be tempted to define the integral of $f \colon [0, 1] \to \mathbf{R}$ as

$$\int f = \lim_{|P| \to 0} \sum f(x_j) m A_j,$$

where $x_j \in A_j$. On the other hand we see immediately that this limit is less likely to exist than the ordinary Riemann integral since we'll be considering many more partitions (of the same type as the Riemann case) than for $R \int$. In fact $\int f = J$ exists if and only if there is a sequence $\{x_j : j = 1, \ldots\} \subseteq [0, 1]$ such that $\{x : f(x) \neq J\} = \{x_j :$

$f(x_j) \to J$. We provide the details in one direction, the other direction being of about the same difficulty. Assume (without loss of generality) that $J = 0$, and suppose that $\{x: f(x) \neq 0\} = \{x_j: f(x_j) \to 0\}$. Thus, $\sup |f(x)| = M < \infty$, and, for each $\varepsilon > 0$, $2|f| \geq \varepsilon$ for at most a finite subset $\{x_1, ..., x_r\}$ of $\{x_j: j = 1, ...\}$. Now consider a partition P' defined in the following way. Make sure that

$$2rM|P'| < \varepsilon,$$

and choose an $x_j \in A_j$, for $j = 1, ..., r$. Then

$$\left| \sum_1^n f(x_j) m A_j \right| \leq \frac{\varepsilon}{2} + \sum_{r+1}^n |f(x_j)| m A_j < \varepsilon.$$

Epsilon more epsilonics entails that

$$\left| \sum_1 f(x_j) m A_j \right| < \varepsilon$$

for each "partition" P for which $|P| \leq |P'|$; and so $\int f = 0$.

If $f \geq 0$ is a μ-measurable function on the measure space $(X, \mathscr{A}, \mu)$, we define

$$\int f \, d\mu = \sup_{h \leq f} \int h \, d\mu, \tag{3.8}$$

where h is a bounded μ-measurable function and $\mu\{x: h(x) \neq 0\} < \infty$. f is μ-integrable with integral $\int f d\mu$ if $\int f d\mu < \infty$. For $X \subseteq \mathbf{R}$, $\mathscr{A} = \mathscr{M}$, and $\mu = m$, we write

$$\int f \, dm = \int f(x) \, dx = \int f \, dx = \int f.$$

This definition of a μ-integrable function (for $f \geq 0$) is reasonable in light of Theorem 3.2. On the other hand there is a slight possible pathology involved if there are not enough such functions h on the given measure space $(X, \mathscr{A}, \mu)$. This point arises explicitly in Theorem 3.6 and we shall tacitly assume that the measure spaces with which we deal do not have this deficiency. Also, with this definition of a μ-integrable function, the innocent looking linearity in Theorem 3.3a (and a consequence of this linearity in Theorem 3.4a) really depends on the results in section 3.3; since no logical problems evolve and since aesthetically it (the linearity) should be mentioned now we give it here.

Next, let $f: X \to \mathbf{R}^*$ be a μ-measurable function on the measure space $(X, \mathscr{A}, \mu)$, and set

$$f^+(x) = \max \{f(x), 0\} \quad \text{and} \quad f^-(x) = \max \{(-f)(x), 0\}.$$

Since f is μ-measurable, both f^+ and f^- are μ-measurable and

$$f = f^+ - f^-, \quad |f| = f^+ + f^-.$$

f is μ-integrable if f^+ and f^- are μ-integrable and we define the μ-integral of f as

$$\int f \, d\mu = \int_X f \, d\mu = \int f^+ \, d\mu - \int f^- \, d\mu. \qquad (3.9)$$

Similarly, we define $\int f \, d\mu$ for $f \colon X \to \mathbf{C}$.

Instead of considering the space of μ-integrable functions it is more advantageous to employ the space, $L_\mu^1(X)$, each of whose elements is the collection of μ-integrable functions which are equal μ-a.e. There are two reasons that this is done: first, the operation of integration assigns the same value to two functions which are equal μ-a.e.; and, second, the natural norm topology on $L_\mu^1(X)$ is Hausdorff, a notion which essentially dispenses with the topological crises inherent in the lives of identical twins. Standard conventions and precise definitions are given in Appendix I.2.1 and Appendix I.2.2, and these two sections should now be read. From this point on, we shall assume the notation introduced there. On the other hand there is no problem in computation if we deal with integrable functions instead of the corresponding elements in $L_\mu^1(X)$.

Theorem 3.3 *Let $(X, \mathscr{A}, \mu)$ be a measure space.*

a) *$L_\mu^1(X)$ is a vector space, and*

$$\int (\alpha f + \beta g) \, d\mu = \alpha \int f \, d\mu + \beta \int g \, d\mu,$$

where $f, g \in L_\mu^1(X)$ and α and β are scalars.

b) *If $f, g \in L_\mu^1(X)$ and $f \leqslant g$, μ-a.e., then*

$$\int f \, d\mu \leqslant \int g \, d\mu.$$

c) *Let $A \in \mathscr{A}$ and $B \in \mathscr{A}$ be disjoint, and choose $f \in L_\mu^1(X)$. Then*

$$\int_{A \cup B} f \, d\mu = \int_A f \, d\mu + \int_B f \, d\mu.$$

Naturally, $L_\mu^1(X)$ is a real (resp., complex) vector space if the elements of $L_\mu^1(X)$ are **R**-valued (resp., **C**-valued).

Theorem 3.4 *Let $(X, \mathscr{A}, \mu)$ be a measure space.*

a) *$f \in L_\mu^1(X) \Leftrightarrow |f| \in L_\mu^1(X)$ and f is μ-measurable.*

b) *$\left| \int f \, d\mu \right| \leqslant \int |f| \, d\mu.$*

c) *$\left| \int f \, d\mu \right| = \int |f| \, d\mu \Leftrightarrow$ there is $c \in \mathbf{C}$ such that $|c| = 1$ and $cf \geqslant 0$, μ-a.e.*

d) *Let f be a μ-measurable function which is non-negative μ-a.e.* $\int f \, d\mu = 0 \Leftrightarrow f = 0$
μ-a.e.

Proof. a) follows from Theorem 3.3 and the decomposition of f in terms of f^+
and f^-.

b) There is $c \in \mathbf{C}$, for which $|c| = 1$, such that

$$\left| \int f \, c \, d\mu \right| = \int f c \, d\mu.$$

Thus, by the definition of the integral,

$$\left| \int f c \, d\mu \right| = \int f c \, d\mu = \int \operatorname{Re}(cf) \, d\mu + i \int \operatorname{Im}(cf) \, d\mu$$
$$= \int \operatorname{Re}(cf) \, d\mu \leqslant \int |cf| \, d\mu,$$

where the inequality follows since $|cf| - \operatorname{Re}(cf) \geqslant 0$. (Re a and Im a are the real
and imaginary parts of $a \in \mathbf{C}$).

c) Because of b) we have the desired equality if and only if $\operatorname{Re}(cf) = |cf|$, μ-a.e., and
this occurs if and only if $cf \geqslant 0$, μ-a.e.

d) Define the sets $A_n = \{x : f(x) > 1/n\}$ and $A = \{x : f(x) > 0\}$, and let $\int f \, d\mu = 0$.
Thus, $A = \bigcup A_n$, and we need only prove that for each n, $\mu A_n = 0$. If $\mu A_n > 0$ we set
$g_n = (1/n)\chi_{A_n}$. Hence, $g_n \leqslant f$ and $\int g_n \, d\mu = (1/n)\mu A_n$. Thus $\int f \, d\mu > 0$, the required
contradiction.

$$\text{q.e.d.}$$

Example 3.4 It is necessary to assume that f is μ-measurable in Theorem 3.4a. In fact,
if we consider the measure space $([0, 1], \mathscr{M}, m)$ and choose $A \notin \mathscr{M}$, then the function

$$f(x) = \begin{cases} 1, & \text{if } x \in A \\ -1, & \text{if } x \notin A \end{cases}$$

is not m-measurable whereas $|f|$ is m-measurable.

Remark As we shall see, we can find functions $f : [a, b] \to \mathbf{R}$ such that f' exists everywhere but $f' \notin L_m^1[a, b]$. There are general integration theories with the property that
the fundamental theorem of calculus holds whenever f' exists everywhere on $[a, b]$
(these are due to Perron and Denjoy). Such theories are important, but, as of now,
they have not achieved the general success of Lebesgue's theory; one of the reasons
for this is precisely because they do not have the property of Theorem 3.4a. We refer
to [83] (translated from the Russian) for a modern and historical approach to the
Perron–Denjoy theories. Classically, there are some introductory remarks in [114,
Chapter 11], and more full-fledged treatments in [64] and [97, Chapters 6–8]. Denjoy
has written extensively on the subject.

Example 3.5 Let $f_0 > 0$ on $(0, 1)$ be an element of $L_m^1(0, 1)$ and set

$$f_{n+1}(x) = \left(\int_0^x f_n(t)\, dt \right)^{1/2}.$$

It can be shown that $\lim_{n \to \infty} f_n(x) = x/2$.

Example 3.6 a) Recall that the derivative of the Volterra function f (Example 1.2) vanishes on E, and for x just to the right of a (in a contiguous interval (a, b)) it has the value

$$2(x - a) \sin \left(\frac{1}{x - a} \right) - \cos \left(\frac{1}{x - a} \right).$$

From the symmetric definition of f on (a, b) we therefore see that f' is bounded and Lebesgue measurable. Thus $f' \in L_m^1([0, 1])$. Since f' is not continuous m-a.e. we can conclude from Theorem 3.13 that $R \int f'$ does not exist.

b) Let $f(x) = x^2 \sin (1/x^2)$ for $x \in (0, 1]$, and set $f(0) = 0$. Then

$$f'(x) = \begin{cases} 0, & \text{if } x = 0 \\ 2x \sin\dfrac{1}{x^2} - \dfrac{2}{x} \cos \dfrac{1}{x^2}, & \text{if } x \in (0, 1]. \end{cases}$$

Note that for $x > 0$

$$|f'(x)| \geq \frac{2}{x} \left| \cos \frac{1}{x^2} \right| - 2x \left| \sin \frac{1}{x^2} \right| \geq \frac{2}{x} \left| \cos \frac{1}{x^2} \right| - 2x.$$

Define

$$I_n = [[(2n + \tfrac{1}{3})\pi]^{-1/2}, [(2n - \tfrac{1}{3})\pi]^{-1/2}].$$

Observe that

$$\forall x \in I_n, \qquad \left| \cos \frac{1}{x^2} \right| \geq \frac{1}{2};$$

in fact, for $x = [(2n + \tfrac{1}{3})\pi]^{-1/2}$,

$$\left| \cos \frac{1}{x^2} \right| = \left| \cos \left(2n\pi + \frac{\pi}{3} \right) \right| = \left| \cos \frac{\pi}{3} \right|.$$

Thus, for any $x \in I_n$, $|f'(x)| \geq (1/x) - 2x$ and so

$$\int_{I_n} |f'| \geq \int_{I_n} \left(\frac{1}{x} - 2x \right) = \frac{1}{2} \log \frac{2n + (\tfrac{1}{3})}{2n - (\tfrac{1}{3})} - \frac{2}{3} \frac{1}{\pi(2n + (\tfrac{1}{3}))(2n - (\tfrac{1}{3}))}.$$

The fact that the sequence $\{I_n : n = 1, ...\}$ is a disjoint family follows since $2n + 2 - \frac{1}{3} > 2n + \frac{1}{3}$ implies $[(2n + 2 - \frac{1}{3})\pi]^{-1/2} < [(2n + \frac{1}{3})\pi]^{-1/2}$. Consequently, if we let $a_N = \sum_1^N 1/[(2n + \frac{1}{3})(2n - \frac{1}{3})]$ we have

$$\int_0^1 |f'| \geq \sum_1^N \int_{I_n} |f'| \geq \frac{1}{2} \sum_1^N \log \frac{2n + (\frac{1}{3})}{2n - (\frac{1}{3})} - \frac{2a_N}{3\pi}$$

$$\geq \frac{1}{2} \sum_1^N \log \left(1 + \frac{1}{6n}\right) - \frac{2a_N}{3\pi}.$$

Because the sequence $\{a_N : N = 1, ...\}$ is convergent and $\sum_1^N \log\left(1 + \frac{1}{6n}\right)$ is divergent

(noting that $\sum \frac{1}{6n}$ and, hence, $\Pi\left(1 + \frac{1}{6n}\right)$ diverge) we conclude that

$$f' \notin L_m^1[0, 1].$$

c) Consider the function f of part b). From the fundamental theorem of calculus for Riemann integration (e.g. Problem 1.24) we compute

$$R\int_\varepsilon^1 f' = f(1) - f(\varepsilon) = \sin 1 - \varepsilon^2 \sin \frac{1}{\varepsilon^2}$$

for each $\varepsilon > 0$, and so

$$\lim_{\varepsilon \to 0} R \int_\varepsilon^1 f' = \sin 1.$$

We'll see in section 3.4 that $f' \in L_m^1[\varepsilon, 1]$ since it is bounded and Riemann integrable on $[\varepsilon, 1]$; consequently

$$\lim_{\varepsilon \to 0} \int_\varepsilon^1 f' = \sin 1.$$

d) We again consider the function f from part b). We now observe that even though f' exists everywhere on $[0, 1]$, f is not a function of bounded variation. In fact,

$$f(1/\sqrt{k\pi}) = 0 \quad \text{and} \quad f\left(1/\sqrt{k\pi + \frac{\pi}{2}}\right) = (-1)^k / \left(k\pi + \frac{\pi}{2}\right),$$

and so the variation of f is larger than

$$\sum \left| 0 - \frac{(-1)^k}{k\pi + (\pi/2)} \right| = \infty.$$

As we'll see later, if g' exists everywhere on $[a, b]$ and $g' \in L_m^1[a, b]$ then not only is g a function of bounded variation, but it is also absolutely continuous.

3.3 The Lebesgue dominated convergence theorem

In his thesis (cf. Chapter 1.3) Lebesgue notes that his dominated convergence theorem is a generalization, with simplification in proof, of a theorem by W. F. Osgood [81]. Osgood's result is Proposition 3.1 for the special case of a continuous function f, and Proposition 3.1 was originally proved by C. Arzelà [3]. We'll state Arzelà's result before making additional remarks.

A sequence $\{f_n: n = 1, ...\}$ of functions $[a\ b] \to \mathbf{R}$ converges boundedly to a function $f: [a, b] \to \mathbf{R}$ if $\{f_n: n = 1, ...\}$ converges pointwise to f and

$$\sup_n \|f_n\|_\infty < \infty.$$

Clearly, if $\{f_n: n = 1, ...\}$ is a sequence of bounded functions on $[a, b]$ and if $f_n \to f$ uniformly on $[a, b]$ then $f_n \to f$ boundedly.

Proposition 3.1 *Given a sequence* $\{f, f_n: n = 1, ...\}$ *of Riemann integrable functions* $[a, b] \to \mathbf{R}$, *and assume that* $f_n \to f$ *boundedly. Then*

$$\lim_{n \to \infty} R\int_a^b f_n = R\int_a^b f.$$

Starting with the definition of the Riemann integral it is non-trivial to prove Proposition 3.1, whereas, starting from the axioms of measure theory it is not difficult to prove the corresponding and more general Lebesgue dominated convergence theorem (Theorem 3.5). The reason for this is not so mysterious. Arzelà's result depends on a countable additivity property and, starting with Riemann's definition of integral, the route to proving such a property requires some effort; on the other hand, in Lebesgue's theory the countable additivity is essentially built into the preliminaries. The history of the elementary (that is, without Lebesgue's theory) proofs of Proposition 3.1 has recently been recorded by W. A. J. Luxemburg [77]; he also gives another elementary proof of his own which is basically a corrected version of an old proof due to Hausdorff (1927). The problem of "taking limits under the integral"—which, of course, has many forms—is one of the absolutely fundamental issues in analysis; consequently, new proofs of such results as Proposition 3.1 are valuable for providing insights on the matter of "switching limits".

Arzelà's original proof depended on a complicated lemma which is, in fact, an easy corollary of our Problem 2.8b. It was in this lemma that he derived the "countable additivity property"—mentioned in the previous paragraph—that was necessary for his theorem. Mind you, Problem 2.8b is straightforward to prove when one begins with the countable additivity of Lebesgue measure. Using Problem 2.8b, we now give Arzelà's proof, properly streamlined, of Proposition 3.1:

Proof. (Proposition 3.1) Without loss of generality assume that $[a, b] = [0, 1], f = 0$, and $f_n(x) \in [-1, 1]$ for all x and n.

If the result is false, then $\overline{\lim_{n}} \; R \int_0^1 f_n$ or $\underline{\lim_{n}} \; R \int_0^1 f_n$ is not zero. Assume $\underline{\lim_{n}} \; R \int_0^1 f_n = r > 0$.

Define $A_n = \{x : f_n(x) \geqslant r/2\}$ so that $\underline{\lim_{n}} \; mA_n > 0$.

By Problem 2.8b we see that there is a point $y \in [0, 1]$ such that $f_n(y) \geqslant r/2$ for infinitely many n.

This contradicts the hypothesis that $f_n(y) \to 0$.

q.e.d.

Observe that we assumed f to be Riemann integrable in Proposition 3.1. The corresponding assumption will not have to be made in Theorem 3.5.

Theorem 3.5 *Let* $(X, \mathscr{A}, \mu)$ *be a finite measure space and let* $\{f_n : n = 1, \ldots\}$ *be a sequence of measurable functions* $X \to \mathbf{R}^*$ *for which*

$$\sup_n \|f_n\|_\infty = M < \infty.$$

If $f_n \to f$ *pointwise on* X *then* $f \in L^1_\mu(X), \|f\|_\infty \leqslant M$, *and*

$$\lim \int |f_n - f| \, d\mu = 0;$$

in particular,

$$\lim \int f_n \, d\mu = \int f \, d\mu. \qquad (3.10)$$

Proof. Clearly f is measurable by Proposition 2.11, and $\|f\|_\infty \leqslant M$; thus $f \in L^1_\mu(X)$. Take $\varepsilon > 0$ and choose N and $A \in \mathscr{A}$, by Egoroff's theorem, such that $\mu A < \varepsilon/4M$ and

$$\forall n \geqslant N, \forall x \notin A, \qquad |f_n(x) - f(x)| < \frac{\varepsilon}{2\mu X}.$$

Thus, for all $n \geqslant N$

$$\int |f_n - f| \, d\mu = \int_A + \int_{X \backslash A} |f_n - f| \, d\mu \leqslant \frac{2M\varepsilon}{4M} + \mu(X \backslash A)\frac{\varepsilon}{2\mu X} \leqslant \varepsilon.$$

q.e.d.

Example 3.7 a) Obviously we can find a sequence $\{f_n: n = 1, \ldots\}$ of simple functions on $([0, 1], \mathcal{M}, m)$ such that

i) $\qquad\qquad f_n \geqslant 0$

ii) $\qquad\qquad \int f_n(x)\, dx = 1$

iii) $\qquad\qquad f_n \to 0$ pointwise;

iii') $\qquad\qquad f_n \nrightarrow 0$ pointwise

iv) $\qquad\qquad \varlimsup_{n \to \infty} \|f_n\|_\infty = \infty.$

In fact, take $f_n = n(n+1)\chi_{[1/(n+1),\, 1/n]}$ and $f_n = 2n\chi_{[1/2 - 1/n,\, 1/2 + 1/n]}$, respectively.

b) Observe that ii) and iii) in part a) cannot hold along with the condition that $\sup_n \|f_n\|_\infty < \infty$; for if we had iii) and the norm boundedness we would contradict ii) by Theorem 3.5.

c) Define $f_n(x) = x^n$ on $[0, 1]$. Then $f_n \to f = 0$ pointwise on $[0, 1)$ but the convergence is not uniform. On the other hand, $|f_n| \leqslant 1$, $\int_0^1 x^n\, dx = 1/(n+1)$, and $\int_0^1 f\, dx = 0.$

d) Set

$$f_n(x) = \begin{cases} 0, & \text{if } x \leqslant 0 \text{ and } x \geqslant 1/n \\ n^2, & \text{if } 0 < x < 1/n. \end{cases}$$

Then $f_n \to f = 0$ pointwise on $\mathbf{R}$ but once again the convergence is not uniform. $\int f_n\, dx = n \to \infty$ and $\int f\, dx = 0.$

e) Set

$$f_n(x) = \begin{cases} 1/n, & \text{if } |x| \leqslant n^2 \\ 0, & \text{if } |x| > n^2. \end{cases}$$

Then $f_n \to f = 0$ uniformly on $\mathbf{R}$, $\int f_n\, dx = 2n \to \infty$, and $\int f\, dx = 0.$

We now give Fatou's lemma (Theorem 3.6) which Fatou, a friend of Lebesgue, published in his famous thesis [38]. In order to motivate this result we first establish some notation and make some remarks on Fourier series.

$f \in L_m^1(\mathbf{T})$ if f is a 2π-periodic $\mathbf{R}$ or $\mathbf{C}$-valued m-measurable function defined on $\mathbf{R}$ such that for some (and therefore for all) $r \in \mathbf{R}$,

$$\int_r^{r+2\pi} |f| < \infty;$$

the associated measure space is designated by $(\mathbf{T}, \mathcal{M}, m)$ and the $L_m^1(\mathbf{T})$ norm of

$f \in L_m^1(\mathbf{T})$ is

$$\|f\|_1 = \frac{1}{2\pi} \int_0^{2\pi} |f|.$$

We also define $f \in L_m^\infty(\mathbf{T})$ with norm $\|f\|_\infty$ if $f \in L_m^1(\mathbf{T})$ and

$$\|f\|_\infty = \operatorname*{ess\,sup}_{x \in \mathbf{R}} |f(x)| < \infty.$$

If $f \in L_m^1(\mathbf{T})$ the Fourier coefficients of f are defined by

$$\hat{f}(n) = c_n = \frac{1}{2\pi} \int_0^{2\pi} f(x) \, e^{-inx} \, dx, \qquad n \in \mathbf{Z},$$

and the Fourier series of f is

$$f \sim \sum c_n e^{inx}.$$

Setting

$$u(r, x) = \sum c_n r^n e^{inx}, \qquad r \in (0, 1),$$

Fatou proved

$$\lim_{r \to 1} u(r, x) = f(x), \quad m - a.e. \tag{3.11}$$

This result is quite important in function theory, and in order to prove it Fatou used Theorem 3.6. Combining (3.11) and Theorem 3.6 he also obtained Parseval's equality

$$\frac{1}{2\pi} \int_0^{2\pi} |f(x)|^2 \, dx = \sum |c_n|^2$$

for $f \in L_m^2(\mathbf{T}) = \{f : f^2 \in L_m^1(\mathbf{T})\}$, a fact which Lebesgue initially proved only for $f \in L_\mu^\infty(\mathbf{T})$. Shortly thereafter, F. Riesz used Fatou's theory to prove that $L_m^2(\mathbf{T})$ is a complete metric space with metric ρ defined by

$$\rho(f, g) = \left(\frac{1}{2\pi} \int |f(x) - g(x)|^2 \, dx \right)^{1/2}$$

(e.g. Appendix I.2).

Theorem 3.6 *Let* $\{f_n : n = 1, \ldots\}$ *be a sequence of measurable functions* $X \to \mathbf{R}^*$ *defined on the measure space* $(X, \mathscr{A}, \mu)$. *Assume that* $\{f_n : n = 1, \ldots\}$ *is bounded below by some* $g \in L_\mu^1(X)$, $f_n \to f$, μ-*a.e., and* f *is* μ-*measurable. Then*

$$\int f \, d\mu \leqslant \underline{\lim} \int f_n \, d\mu.$$

Proof. First note that we make no claims about integrability either in the hypothesis or conclusion since our integrals may be unbounded.

Without loss of generality assume that g is identically 0, that $f_n \to f$ pointwise everywhere, and that $\underline{\lim} \int f_n \, d\mu < \infty$. Take a bounded measurable function h such that $0 \leqslant h \leqslant f$ and $h \neq 0$ on a set Y of finite measure (cf. the remark on pathology after Example 3.3).

Define $h_n(x) = \min \{h(x), f_n(x)\}$. Thus $0 \leqslant h_n \leqslant h$ and $h_n = 0$ on $Y^\sim$.

Since $f_n \to f$ pointwise and $h \leqslant f$, we have $h_n \to h$ by the definition of h_n. Consequently we apply Theorem 3.5 to obtain

$$\int_Y h \, d\mu = \int h \, d\mu = \lim \int_Y h_n \, d\mu. \tag{3.12}$$

Now $f_n \geqslant h_n$ implies $\int f_n \, d\mu \geqslant \int_Y f_n \, d\mu \geqslant \int_Y h_n \, d\mu$ and so

$$\underline{\lim} \int_Y f_n \, d\mu \geqslant \underline{\lim} \int_Y h_n \, d\mu = \lim \int_Y h_n \, d\mu.$$

This combined with (3.12) yields the result from the definition of $\int f \, d\mu$.

q.e.d.

The following result allows us to prove Theorem 3.3a. This is important since we use linearity in Theorem 3.8.

Theorem 3.7 (*Lebesgue*) *Let* $(X, \mathcal{A}, \mu)$ *be a measure space and let* $\{f_n : n = 1, ...\}$ *be a sequence of μ-measurable $\mathbf{R}^*$-valued functions defined on X. Assume that* $\{f_n : n = 1, ...\}$ *is bounded below by some $g \in L^1_\mu(X)$ and that* $\{f_n : n = 1, ...\}$ *converges μ-a.e. to a μ-measurable function f. If*

$$\forall n, \qquad f_n \leqslant f, \quad \mu\text{-a.e.}$$

then

$$\int f \, d\mu = \lim \int f_n \, d\mu.$$

Proof. Since $f_n \leqslant f$, μ-a.e., we have $\int f_n \, d\mu \leqslant \int f \, d\mu$.

By Theorem 3.6, $\int f \, d\mu \leqslant \underline{\lim} \int f_n \, d\mu \leqslant \overline{\lim} \int f_n \, d\mu \leqslant \int f \, d\mu$, and we are done.

q.e.d.

If f_n increases to f then the conditions of Theorem 3.7 hold, and it was in this form that the result was proved by B. Levi in 1906. The fundamental criterion for taking limits under the integral sign is

Theorem 3.8 (*Lebesgue*) *Let* $(X, \mathscr{A}, \mu)$ *be a measure space and let* $\{f_n: n = 1, ...\}$ *be a sequence of μ-measurable functions each of which is $\mathbf{C}$-valued μ-a.e. Assume that $f_n \rightarrow f$, μ-a.e., that f is μ-measurable, and that there is an element $g \in L^1_\mu(X)$ such that*

$$\forall n, \qquad |f_n| \leqslant g, \quad \mu\text{-a.e.}$$

Then $f \in L^1_\mu(X)$ *and* $\int f \, d\mu = \lim \int f_n \, d\mu.$

Proof. $f \in L^1_\mu(X)$ since $|f| \leqslant g$, μ-a.e. Without loss of generality take f_n real-valued μ-a.e. and so

$$g - f_n \geqslant 0, \quad \mu\text{-a.e.}$$

Thus from Theorem 3.6

$$\int (g - f) \, d\mu \leqslant \underline{\lim} \int (g - f_n) \, d\mu;$$

consequently from properties of the $\underline{\lim}$ and $\overline{\lim}$ and since $f, f_n \in L^1_\mu(X)$,

$$\int g \, d\mu - \int f \, d\mu \leqslant \int g \, d\mu - \overline{\lim} \int f_n \, d\mu.$$

This yields

$$\int f \, d\mu \geqslant \overline{\lim} \int f_n \, d\mu.$$

For the opposite direction we consider $g + f_n$ and compute

$$\int f \, d\mu \leqslant \underline{\lim} \int f_n \, d\mu.$$

This does it.

q.e.d.

One means of generalizing Theorem 3.8 is

Theorem 3.9 *Let* $(X, \mathscr{A})$ *be a measurable space and let* $\{\mu_n: n = 1, ...\}$ *be a sequence of measures on $\mathscr{A}$ such that*

$$\forall A \in \mathscr{A}, \qquad \lim \mu_n A = \mu A, \tag{3.13}$$

where μ is a measure on $\mathscr{A}$. Assume that the sequence, $\{g, g_n: n = 1, ...\} \subseteq L^1_\mu(X)$, satisfies the conditions that $g_n \rightarrow g$ pointwise and

$$\lim_n \int g_n \, d\mu_n = \int g \, d\mu.$$

If $\{f_n: n = 1, ...\}$ *is a sequence of functions with the properties that* $\lim_n f_n = f$ *pointwise, each f_n is μ_n-measurable, and*

$$\forall n, \qquad |f_n| \leqslant g_n,$$

then $f \in L^1_\mu(X)$ and

$$\lim_{n \to \infty} \int f_n \, d\mu_n = \int f \, d\mu.$$

Generally we'll refer to Theorem 3.5, Theorem 3.7, Theorem 3.8, and Theorem 3.9 as LDC in the sequel. The hypothesis that μ in (3.13) is a measure raises the problem to find conditions so that (3.13) defines a measure. We shall study this question when we discuss weak-convergence of measures in Chapter 6 and we shall do it in the context of trying to find necessary and sufficient conditions for the conclusion of LDC to hold. For now we refer to Problem 3.8b and the following "outline" which we'll expand on later.

Proposition 3.2 *Given a measure space $(X, \mathscr{A}, \mu)$ and an element $f \in L^1_\mu(X)$. Then*

$$\forall \varepsilon > 0 \; \exists \delta > 0 \text{ such that } \forall A \in \mathscr{A}, \text{ for which } \mu A < \delta,$$

$$\int_A |f| \, d\mu < \varepsilon.$$

Proof. The result is obvious if $\|f\|_\infty < \infty$.
Define

$$|f_n|(x) = \begin{cases} |f|(x), & \text{if } |f(x)| \leq n \\ n, & \text{if } |f(x)| > n. \end{cases}$$

Then $\|f_n\|_\infty \leq n$ and $|f_n| \to |f|$ pointwise. From LDC

$$\exists N \text{ such that } \forall n \geq N, \quad \int (|f| - |f_n|) \, d\mu < \varepsilon/2.$$

Letting $\delta < \dfrac{\varepsilon}{(2N)}$ we have

$$\left| \int_A f \, d\mu \right| \leq \int_A |f| \, d\mu = \int_A (|f| - |f_N|) \, d\mu + \int_A |f_N| \, d\mu$$

$$\leq \int_X (|f| - |f_N|) \, d\mu + N\mu A < \varepsilon$$

if $\mu A < \delta$.

q.e.d.

A μ-measurable function f on a measure space $(X, \mathscr{A}, \mu)$ is **absolutely continuous** (with respect to μ) if

$$\forall \varepsilon > 0 \; \exists \delta > 0 \text{ such that } \forall A \in \mathscr{A}, \text{ for which } \mu A < \delta,$$

$$\left| \int_A f \, d\mu \right| < \varepsilon.$$

(Thus each element $f \in L_\mu^1(X)$ is absolutely continuous with respect to μ. In Chapter 5 we'll define a "measure v absolutely continuous with respect to μ" and show that such measures are actually characterized by $L_\mu^1(X)$.)

A collection $\{f_\alpha\} \subseteq L_\mu^1(X)$ is uniformly absolutely continuous if

$$\forall \varepsilon > 0 \; \exists \delta > 0 \text{ such that } \forall A \in \mathscr{A}, \text{ for which } \mu A < \delta, \text{ and } \forall \alpha$$

$$\left| \int_A f_\alpha \, d\mu \right| < \varepsilon.$$

If $\mathscr{F} \subseteq \mathscr{P}(X)$ and v is a scalar valued function on $\mathscr{F}$, we say that v is Vitali continuous if for each decreasing sequence $\{A_n : n = 1, \ldots\} \subseteq \mathscr{F}$ for which $\bigcap A_n = \emptyset$ we can conclude that

$$\lim v A_n = 0;$$

a sequence of Vitali continuous functions v_m on $\mathscr{F}$ is Vitali equicontinuous if for each decreasing sequence $\{A_n : n = 1, \ldots\} \subseteq \mathscr{F}$, for which $\bigcap A_n = \emptyset$, we have

$$\forall \varepsilon > 0 \; \exists N \text{ such that } \forall n > N \text{ and } \forall m$$
$$|v_m A_n| < \varepsilon.$$

If a measure space $(X, \mathscr{A}, \mu)$ is given (and $\mathscr{F} = \mathscr{A}$) we say that a sequence $\{v_m : m = 1, \ldots\}$ of scalar valued functions on $\mathscr{A}$ is uniformly absolutely continuous if

$$\forall \varepsilon > 0 \; \exists \delta > 0 \text{ such that } \forall A \in \mathscr{A}, \text{ for which } \mu A < \delta, \text{ and } \forall m$$
$$|v_m A| < \varepsilon.$$

This definition obviously generalizes the above definition of uniform absolute continuity.

Clearly if $\int |f - f_n| \, d\mu \to 0$ then

$$\forall \varepsilon > 0, \qquad \lim_{n \to \infty} \mu\{x : |f(x) - f_n(x)| \geq \varepsilon\} = 0. \tag{3.14}$$

We are basically interested in finding a converse to this observation in order to obtain the best possible LDC. We make the following definition: a sequence $\{f_n : n = 1, \ldots\}$ of μ-measurable functions on a measure space $(X, \mathscr{A}, \mu)$ converges in measure to a μ-measurable function f if (3.14) holds. Vitali made initial and deep progress in characterizing LDC with essentially the following results [121]:

Theorem 3.10 *Let $(X, \mathscr{A}, \mu)$ be a finite measure space and choose a sequence $\{f_n : n = 1, \ldots\} \subseteq L_\mu^1(X)$.*

$$\lim \int |f - f_n| \, d\mu = 0, \tag{3.15}$$

for some $f \in L^1_\mu(X) \Leftrightarrow$

a) $\{f_n : n = 1, ...\}$ *converges in measure to a μ-measurable function f, and*

b) $\{f_n : n = 1, ...\}$ *is uniformly absolutely continuous.*

Theorem 3.11 *Let $(X, \mathscr{A}, \mu)$ be a measure space and choose a sequence $\{f_n : n = 1, ...\} \subseteq L^1_\mu(X)$. (3.15) holds for some $f \in L^1_\mu(X) \Leftrightarrow$*

a) $\{f_n : n = 1, ...\}$ *converges in measure to a μ-measurable function f, and*

b) $\{v_n : \forall A \in \mathscr{A}, v_n A = \int_A |f_n| \, d\mu\}$ *is Vitali equicontinuous.*

Lebesgue proved that if $f_n \to f$, μ-a.e., on a measure space $(X, \mathscr{A}, \mu)$, where f is μ-measurable, and if $|f_n| \leqslant g$, μ-a.e., for some $g \in L^1_\mu(X)$, then $f_n \to f$ in measure (cf. Problem 3.19e and Proposition 6.1). Further, with these hypotheses it is straightforward to deduce b) in Theorem 3.11. Thus LDC is a corollary of Theorem 3.11.

We now close section 3.3 with some examples.

Example 3.8 In Problem 3.3b we show that $f(x) = 1/x^{1/2} \in L^1_m[0, 1]$. Using this result we now observe: if $f \geqslant 0$ is a Lebesgue measurable function defined on **R** and

$$\forall a < b, \forall r \in \mathbf{R}, \qquad m\{(a, b) \cap \{x : f(x) \geqslant r\}\} > 0$$

then it is not necessarily true that

$$\int_{\mathbf{R}} f(x) \, dx = \infty.$$

In fact, define

$$f_k(x) = \begin{cases} (x - r_k)^{-1/2}, & \text{if } x \in (r_k, r_k + 1) \\ 0, & \text{if otherwise} \end{cases}$$

and $\qquad f = \sum f_k / 2^k,$

where $\{r_k : k = 1, ...\} = \mathbf{Q}$. We then compute that

$$\int_{\mathbf{R}} f(x) \, dx = \tfrac{2}{3}.$$

Clearly, for any $a < b$ and $r > 0$ there is $q \in \mathbf{Q} \cap (a, b)$ such that $(q, c) \subseteq (q, q + 1)$ and $f \geqslant r$ on (q, c).

Theorem 3.12 *Let f be a non-constant bounded Lebesgue measurable function defined on **R** such that*

$$\forall x, \qquad f(x + 1) = f(x).$$

Set $f_n(x) = f(nx)$. There is no subsequence of $\{f_n : n = 1, ...\}$ which converges m-a.e. on any interval $[a, b]$, where $b > a$.

Proof. Take any interval $[\alpha, \beta]$. From the periodicity of f,

$$\int\limits_\alpha^\beta f(nx)\,dx = \frac{1}{n}\int\limits_{n\alpha}^{n\beta} f(x)\,dx = \frac{1}{n}\left\{\sum_{j=[n\alpha]}^{[n\beta]}\int\limits_j^{j+1} f\,dx + \int\limits_{[n\beta]}^{n\beta} f\,dx - \int\limits_{[n\alpha]}^{n\alpha} f\,dx\right\}$$

$$= \frac{1}{n}(1 + [n\beta] - [n\alpha])\int\limits_0^1 f\,dx + \frac{1}{n}\left(\int\limits_{[n\beta]}^{n\beta} f\,dx - \int\limits_{[n\alpha]}^{n\alpha} f\,dx\right),$$

where $[x]$ is the largest integer $k \leqslant x$.

Clearly, the last term in the last expression is bounded by $2\|f\|_\infty/n$ and so tends to 0 as $n \to \infty$.

Now note that

$$(1/n)([n\beta] - [n\alpha]) = (1/n)(n\beta - n\alpha + ([n\beta] - n\beta) + (n\alpha - [n\alpha]))$$

$$= \beta - \alpha + \frac{[n\beta] - n\beta}{n} + \frac{n\alpha - [n\alpha]}{n}.$$

Consequently,

$$\lim_{n\to\infty}\int\limits_\alpha^\beta f_n(x)\,dx = (\beta - \alpha)\int\limits_0^1 f(x)\,dx. \tag{3.16}$$

Assume that $f(n_k x) \to g(x)$, m-a.e., on $[a, b]$ as $k \to \infty$.

Take $[\alpha, \beta] \subseteq [a, b]$ and let $K = \int_0^1 f(x)\,dx$.

From LDC we compute

$$\int\limits_\alpha^\beta g(x)\,dx = \lim_k\int\limits_\alpha^\beta f(n_k x)\,dx = (\beta - \alpha)K,$$

and so

$$\int\limits_\alpha^\beta (g(x) - K)\,dx = 0.$$

Since $[\alpha, \beta]$ is an arbitrary subinterval of $[a, b]$ we conclude that $g = K$, m-a.e. (in $[a, b]$).

We now use $|f - g| = |f - K|$ and $[a, b]$ instead of f and $[\alpha, \beta]$ in (3.16). Thus

$$(b - a)\int\limits_0^1 |f(x) - K|\,dx = \lim_k\int\limits_a^b |f_{n_k}(x) - K|\,dx = 0.$$

The last equality follows by LDC, and so $f = K$, m-a.e., on $[0, 1]$, a contradiction.

q.e.d.

Example 3.9 Apply Theorem 3.12 to

a) $f(x) = x - [x]$ on $[0, 1]$ and

b) $f(x) = \sin x$ on $[0, 2\pi]$.

Remark Theorem 3.12 is interesting in light of Fejér's theorem (e.g. Problem 3.10) which tells us, in particular, that if f is a bounded Lebesgue measurable function on **R** with period 1 then

$$\lim_n \int_0^1 f(nx)\,dx = \int_0^1 f(x)\,dx.$$

(cf. Chapter 6.3).

3.4 The Riemann and Lebesgue integrals

We begin by proving a fact that we have implicitly assumed for a while now.

Proposition 3.3 *Let $f: [a, b] \to$ **R** be bounded function. If $R \int_a^b f$ exists then $f \in L_m^1[a, b]$*

and

$$R \int_a^b f = \int_a^b f(x)\,dx. \tag{3.17}$$

Proof. Let g and h denote simple functions. Then

$$R \int_{-a}^b f \leqslant \sup_{g \leqslant f} \int g \leqslant \inf_{h \geqslant f} \int h \leqslant R \overline{\int}_a^b f,$$

since, for example, $R \int_{-a}^b f = \sup \{R \int g : g = \sum a_j \chi_{[x_{j-1}, x_j)} \leqslant f$ and $a = x_0 < x_1 < ... < x_n = b\}$ and $\{g : g = \sum a_j \chi_{[x_{j-1}, x_j)}$ and $a = x_0 < x_1 < ... < x_n = b\}$ is a subfamily of the class of simple functions.

(3.17) follows because $R \int_a^b f$ exists, by Theorem 3.1′, and from the definition of the Lebesgue integral.

q.e.d.

Proposition 3.4 *A function* $f: \mathbf{R} \to \mathbf{R}$ *is continuous m-a.e.* $\Leftrightarrow$

$$\forall V \subseteq \mathbf{R}, \text{ open,} \qquad f^{-1}(V) = U \cup A,$$

where U is open and $A \in \mathcal{M}$ *has Lebesgue measure* $mA = 0$.

Proof. ($\Rightarrow$) Let $X = f^{-1}(V)$ and set $X = X_c \cup X_d$ where $X_c = X \cap C(f)$ and $X_d = X \cap D(f)$.

For each $x \in X_c$ choose an open neighborhood U_x of x such that $f(U_x) \subseteq V$; we can do this since f is continuous on X_c. Clearly U_x need not be contained in X_c (for example, take the ruler function).

Since $f^{-1}(V) = X$ and $U_x \subseteq f^{-1}(V)$ we have

$$X = X_c \cup X_d \subseteq \left(\bigcup_{x \in X_c} U_x \right) \cup X_d \subseteq X.$$

Let $A = X_d$ and $U = \bigcup_{x \in X_c} U_x$. $mA = 0$ by hypothesis, and U is obviously open.

($\Leftarrow$) For each $x \in D(f)$ there is an open set V such that $f(x) \in V$, and for each open neighborhood U of x, $f(U) \nsubseteq V$. Let $f(x) \in N(s, r) = \{y \in \mathbf{R}: y \in (s - r, s + r), s, r \in \mathbf{Q}\} \subseteq V$. Consequently, for all such sets $U, f(U) \nsubseteq N(s, r)$.

By hypothesis, $f^{-1}(N(s, r)) = U_{s,r} \cup A_{s,r}$ where $U_{s,r}$ is open and $mA_{s,r} = 0$.

Now, because of the above observations, $x \in f^{-1}(N(s, r))$ and $x \notin U_{s,r}$.

Thus, $x \in A_{s,r}$ and so $D(f) \subseteq \bigcup \{A_{s,r}: s, r \in \mathbf{Q}\}$.

Hence f is continuous *m-a.e.*

q.e.d.

Example 3.10 We'll show that if f and g are continuous *m-a.e.*, then $f \circ g$ is not necessarily continuous *m-a.e.* Define

$$g(x) = \begin{cases} 1, & \text{if } x = 0, \\ 1/|q|, & \text{if } x = p/q, \text{ where } (p, q) = 1, \end{cases}$$

and

$$f(x) = \begin{cases} 1, & \text{if } x = 1/n, \ n = 1, \dots, \\ 0, & \text{otherwise.} \end{cases}$$

Then $f \circ g = \chi_{\mathbf{Q}}$ and this function is not continuous anywhere. See Problem 3.15 in this regard.

We now sketch a "first approximation" to one direction of Lebesgue's characterization of Riemann integrable functions.

The proof is easy and does not involve Lebesgue measure.

Proposition 3.5 *Let f be a Riemann integrable function defined on* $[a, b]$. *Then f is continuous on a dense subset of* $[a, b]$.

Proof. Assume that f is a real-valued function.

Let $\{P_n : n = 1, \ldots\}$ be a sequence of partitions of $[a, b]$ such that each P_n divides $[a, b]$ into n segments of equal length. Thus, $S_{P_n} - s_{P_n} \to 0$. Letting "$\varepsilon = \frac{1}{2}(b - a)$" we have

$$\exists m \geqslant 4 \text{ such that } \forall n \geqslant m, \sum_1^n (M_i - m_i)(x_i - x_{i-1}) < \tfrac{1}{2}(b - a).$$

By the definition of $\{P_k : k = 1, \ldots\}$, and since $n \geqslant 2$,

$$\sum_1^n (M_i - m_i) < \tfrac{1}{2}(b - a) \frac{n}{(b - a)} = \frac{n}{2} \leqslant n - 2.$$

Consequently for each $n \geqslant m$ there are at least 3 integers, i, where $1 \leqslant i \leqslant n$, such that $M_i - m_i < 1$.

With these estimates we generate inductively a nested sequence of intervals $[a_n, b_n] \subseteq [a_{n-1}, b_{n-1}]$ such that $a_n \neq a_{n-1}$, $b_n \neq b_{n-1}$, $b_n - a_n \leqslant (b - a)/4^n$, and

$$\omega(f, [a_n, b_n]) < 1/n.$$

Thus $\bigcap [a_n, b_n] = \{x_0\}$ and f is continuous at x_0. Using the same technique we obtain continuity on a dense set.

(We needed three integers above to determine continuity instead of one-sided continuity.)

q.e.d.

Theorem 3.13 *Let $f : [a, b] \to \mathbf{R}$ be a bounded function.* $R \int_a^b f$ *exists $\Leftrightarrow f$ is continuous m-a.e.*

Proof. ($\Rightarrow$) Let

$$D_k = \{x : \lim_{\delta \to 0} \sup_{y, z \in [x - \delta, x + \delta)} |f(y) - f(z)| > 1/k\},$$

so that $D(f) = \bigcup D_k$.

We assume that $mD(f) > 0$ and prove that $R \int_a^b f$ does not exist.

Since $mD(f) > 0$ there is a k for which $mD_k = d_k > 0$.

Let $P : x_0 < \ldots < x_n$ be a partition of $[a, b]$. Then $\bigcup (x_{j-1}, x_j)$ covers all but a finite number of points of D_k, and so

$$\sum_{j \in \dot{P}} (x_j - x_{j-1}) \geqslant d_k,$$

where $j \in \dot{P}$ indicates that $(x_{j-1}, x_j) \cap D_k \neq \varnothing$. From the definition of D_k, if $j \in \dot{P}$ then

$$M_j - m_j > 1/k.$$

Thus,

$$S_P - s_P = \sum (M_j - m_j)(x_j - x_{j-1}) \geqslant \sum_{j \in \dot{P}} (M_j - m_j)(x_j - x_{j-1})$$

$$> \frac{1}{k} \sum_{j \in \dot{P}} (x_j - x_{j-1}) \geqslant \frac{d_k}{k} > 0. \tag{3.18}$$

The number k is fixed and (3.18) is true for all partitions P. Consequently $R\int_a^b f$ does

not exist.

($\Leftarrow$) Let $M = \|f\|_\infty$ and consider the functions $R\int_a^x f$ and $R\overline{\int_a^x} f$, where $R\underline{\int_a^x} f$ is defined

as $R\overline{\int_a^b} f$ if $x \geqslant b$.

For each $a \leqslant x_1 < x_2 \leqslant b$ we have

$$\left| R\overline{\int_a^{x_1}} f - R\overline{\int_a^{x_2}} f \right| \leqslant M(x_2 - x_1). \tag{3.19}$$

It is easy to check directly (e.g. Problem 1.24 and Problem 3.17) that

$$\overline{F}'(x) = f(x), \quad m\text{-}a.e., \tag{3.20}$$

where $\overline{F}(x) = R\overline{\int_a^x} f$. Here, of course, we use the hypothesis that f is continuous

m-$a.e.$

Define the functions

$$f_n(x) = n\left[\overline{F}\left(x + \frac{1}{n}\right) - \overline{F}(x)\right], \qquad n = 1, \dots .$$

Each f_n is continuous on $[a, b]$ and, because of (3.19), $\{f_n : n = 1, \dots\}$ is a uniformly bounded sequence of functions. It is also clear, from (3.20), that $f_n \to f$, m-$a.e.$
Consequently we apply LDC and obtain

$$\int_a^b f = \lim_n \int_a^b f_n. \tag{3.21}$$

Next observe that

$$\int_a^b f_n = n \int_b^{b+(1/n)} \overline{F} - n \int_a^{a+(1/n)} \overline{F} = \overline{F}(b) - n \int_a^{a+(1/n)} \overline{F}.$$

Also,

$$\left| n \int_a^{a+(1/n)} \bar{F} \right| \leq \sup_{x \in [a,\, a+\,(1/n)]} \left| R \int_a^x f \right| n \int_a^{a+(1/n)} 1 \cdot dx \leq M/n.$$

Hence

$$\lim_n \int_a^b f_n = \bar{F}(b) = R \int_a^b f.$$

Combining this with (3.21) yields

$$\int_a^b f = R \int_a^b f;$$

and we compute similarly that $\underline{\int_a^b} f = R \underline{\int_a^b} f$.

q.e.d.

For the ruler function $r \colon [0, 1] \to \mathbf{R}$, $R \int_0^1 r$ exists since $mD(r) = m(\mathbf{Q} \cap [0, 1]) = 0$.

$\chi_{[0,\,1] \cap \mathbf{Q}}$ is not Riemann integrable as we saw in Example 3.1. For the generalized ruler function r_y (of which r and $\chi_{[0,\,1] \cap \mathbf{Q}}$ are special cases) we observed that if $\gamma_q \to 0$ then $mD(r_y) = m(\mathbf{Q} \cap [0, 1]) = 0$ and so $R \int_0^1 r_y = 0$ (e.g. Example 1.5). A complete solution is given by the following result due to I. J. Schoenberg.

Proposition 3.6 *The Riemann integral of r_y on $[0, 1]$ exists $\Leftrightarrow \gamma_q \to 0$; in this case* $R \int_0^1 r_y = 0.$

Proof. Obviously we need only check that if $R \int_0^1 r_y$ exists then $\gamma_q \to 0$. Assume not.

We prove that $R \int_0^1 r_y$ does not exist by showing that r_y is not continuous at any irrational and applying Theorem 3.13.

Given n, consider the $\varphi(n)$ integers $r_1 < \ldots < r_{\varphi(n)} \leq n$ for which $(r_j, n) = 1$, and define the partition

$$P_n \colon 0 < r_1/n < \ldots < r_{\varphi(n)}/n < 1$$

with norm

$$g_n = \sup \left\{ \frac{r_j}{n} - \frac{r_{j-1}}{n} : j = 1, \ldots, \varphi(n) + 1, r_0 = 0, \text{ and } r_{\varphi(n)+1} = n \right\}.$$

φ is the Euler function mentioned in Problem 1.21. We now make an act of faith and state that

$$\lim g_n = 0. \tag{3.22}$$

The proof of (3.22) depends on a theorem due to G. Polya (e.g. [84, I, problem 188]) which proves that partitions P_n are asymptotically equidistributed (recall Problem 1.23 and see Problem 3.22 with regard to equidistribution).

Pick any irrational number $y \in (0, 1)$, and for each n let $k = k(n)$ be the integer for which

$$\frac{r_{k-1}}{n} < y < \frac{r_k}{n}.$$

From (3.22), $\lim_n r_{k(n)}/n = y$. On the other hand $r_y(y) = 0$ and $r_y(r_{k(n)}/n) = \gamma_n$. Since we have assumed that $\{\gamma_n : n = 1, \ldots\}$ does not tend to 0 we conclude that r_y is not continuous m-a.e.

q.e.d.

The following is left as an exercise (Problem 3.16).

Proposition 3.7 *Assume that the function $f: [a, b] \to \mathbf{R}$ has the property that*

$$\forall x \in (a, b), \qquad \lim_{y \to x\pm} f(y) \text{ exists.}$$

Then $\quad R \int_a^b f \text{ exists.}$

Example 3.11 We have seen that $f(x) = x^{-1/2}$ and $f(x) = \chi_{[0, 1] \cap \mathbf{Q}}$ are not Riemann integrable on $[0, 1]$, whereas they are both Lebesgue integrable. In some sense both of these functions provide unsatisfactory examples: in the first case f is unbounded, and, although the Riemann integral necessarily supposes f to be bounded, $\int_0^1 x^{-1/2} \, dx$ exists in an "improper" Riemann sense; similarly $f = \chi_{[0, 1] \cap \mathbf{Q}} = 0$, m-a.e., so that even though $R \int_0^1 \chi_{[0, 1] \cap \mathbf{Q}}$ doesn't exist the function g, which is identically 0 and equal to f, m-a.e., is Riemann integrable. We now give examples of bounded Lebesgue measurable functions f defined on $[0, 1]$ such that

$$\forall g = f, m\text{-a.e.,} \quad R \int_0^1 g \quad \text{does not exist.} \tag{3.23}$$

a) Let $E \subseteq [0, 1]$ be a perfect symmetric set with Lebesgue measure $mE > 0$. Setting $f = \chi_E$, we obtain (3.23) since $mD(f) > 0$.

b) Volterra's example f from Chapter 1.3 has the property that f' is a bounded Lebesgue measurable function and $mD(f') > 0$.

Example 3.12 In view of the function $f(x) = x^{-1/2}$ defined on $[0, 1]$ we now define a function g, unbounded on $[0, 1]$, such that $g \in L^1_m[0, 1]$ and the improper Riemann integral of g does not exist. Note that for $f(x) = (\sin x)/x$ on $(0, \infty)$ the opposite phenomenon is true (Problem 3.24d). Let $\sum a_k$ be a convergent series of positive terms, and set

$$g(x) = \sum_1^\infty a_k / |x - b_k|^{1/2},$$

where $\{b_k : k = 1, \ldots\}$ is a dense subset of $[0, 1]$. Of course we could even define some sort of improper Riemann integral in this case in terms of partial sums.

3.5 Some fundamental applications

The following two results are trivial variants of LDC.

Theorem 3.14 *Let* $(X, \mathscr{A}, \mu)$ *be a measure space and choose a subset* $A \subseteq \mathbf{R}$*. Pick* $t_0 \in A$*. Define the function* $f(x, t)$ *on* $X \times A$ *assuming that*

i) $\forall t \in A, f(\cdot, t)$ *is a* μ*-measurable function,*

ii) $\lim_{t \to t_0} f(x, t) = f_0(x)$ *exists* μ*-a.e.,*

and

iii) $\exists g \in L^1_\mu(X)$ *such that* $\forall t \in A, |f(x, t)| \leqslant g(x), \mu$*-a.e.*

Then $f_0 \in L^1_\mu(X)$ *and*

$$\lim_{t \to t_0} \int_X f(x, t) \, d\mu(x) = \int_X f_0 \, d\mu.$$

Theorem 3.14 is true for the case that A is a metric space (e.g. Appendix I.1).

Theorem 3.15 *Let* $(X, \mathscr{A}, \mu)$ *be a measure space and assume that* $\{f_n : n = 1, \ldots\}$ *is a sequence of* μ*-measurable functions for which*

$$\sum \int_X |f_n| \, d\mu < \infty.$$

Then $\sum f_n \in L^1_\mu(X)$ *and*

$$\int_X (\sum f_n) \, d\mu = \sum (\int f_n \, d\mu).$$

Another corollary of LDC (e.g. Problem 3.18) is

Theorem 3.16 *Given the measure space* $(X, \mathcal{M}^n, m^n)$*, where* $X \subseteq \mathbf{R}^n$ *is an open set, and let* $A \subseteq \mathbf{R}$ *be an interval. Define the function* $f(x, t)$ *on* $X \times A$ *assuming that*

i) $\forall x \in X, f(x, \cdot)$ *is m-measurable on* A,

ii) $\forall x \in X, D_{x_j} f(x, t)$ *exists everywhere on* A,

iii) $\forall x \in X \quad \exists g_x \in L^1_m(A)$ *such that* $|D_{x_j} f(x, t)| \leqslant g_x(t)$, *m-a.e., on* A, *and*

iv) $\int_A f(x, t) \, dt$ *exists for all* $x \in X$.

Then

$$D_{x_j} \int_A f(x, t) \, dt = \int_A D_{x_j} f(x, t) \, dt.$$

(In the above statement, j is fixed and $D_{x_j} = \partial/\partial x_j$.)

Remark Theorem 3.14, 3.15, and 3.16 generalize classical criteria which one proves with a uniform convergence hypothesis.

The R i e m a n n – L e b e s g u e l e m m a, which we now prove, is a fundamental result in Fourier series; we indicate another proof of it in Problem 3.10d.

Theorem 3.17 *(R i e m a n n – L e b e s g u e) For each* $f \in L^1_m(\mathbf{T})$

$$\lim_{|n| \to \infty} \hat{f}(n) = 0.$$

Proof. If $(a, b) \subseteq [0, 2\pi)$,

$$\lim_{|n| \to \infty} \left| \int_a^b e^{-inx} \, dx \right| = \lim_{|n| \to \infty} \left| \frac{1}{n} (e^{-inb} - e^{-ina}) \right| = 0.$$

Consequently, the result is true for characteristic functions of intervals; and the expected application of our approximation theorems yields the result for arbitrary elements of $L^1_m(\mathbf{T})$.

q.e.d.

Example 3.13 There is a sequence $\{f_n : n = 1, \ldots\} \subseteq L^1_m(\mathbf{T})$ such that

$$\forall A \in \mathcal{M}, \quad \lim_{n \to \infty} \int_A f_n(x) \, dx = 0$$

and

$$\overline{\lim_{n \to \infty}} \int_0^{2\pi} |f_n(x)| \, dx > 0.$$

In fact, $\lim\limits_{n \to \infty} \int_A \sin nx \, dx = 0$ by Theorem 3.17, whereas

$$\int_0^{2\pi} |\sin nx| \, dx = 4.$$

From the point of view of functional analysis this will serve as an example of weak convergence which is not strong convergence. We shall discuss this phenomenon more fully in Chapter 6.

Example 3.14 We shall construct an open set $S \subseteq [0, 2\pi)$ such that

$$\overline{\lim_{|n| \to \infty}} \, |n\hat{\chi}_S(n)| = \infty;$$

note that $\lim\limits_{|n| \to \infty} \hat{\chi}_S(n) \to 0$ because of Theorem 3.17. In particular χ_S is not a function of bounded variation (e.g. Problem 4.2d), a notion we study systematically in the next chapter.

For each j and each $k = 1, \ldots, 2j$ set

$$S_{jk} = (2\pi(k - \tfrac{1}{2})/(2j)!, \, 2\pi k/(2j)!).$$

These $2j$ intervals are obviously disjoint. We define

$$S_j = \bigcup_{k=1}^{2j} S_{jk} \quad \text{and} \quad S = \bigcup_1^\infty S_j,$$

noting that the sequence $\{S_j : j = 1, \ldots\}$ forms a disjoint family. We shall prove that

$$\lim_{m \to \infty} (2m)! \hat{\chi}_S(-(2m)!) = \infty.$$

We first calculate

$$\int_{S_{jk}} \exp i(2m)! x \, dx \tag{3.24}$$

$$= \{\exp [2\pi i k(2m)!/(2j)!] - \exp [2\pi i(k - \tfrac{1}{2})(2m)!/(2j)!]\}/2\pi i(2m)!.$$

Observe that the numerator in (3.24) vanishes if $j < m$ and takes the value 2 if $j = m$. Therefore

$$\hat{\chi}_S(-(2m)!) = \frac{1}{2\pi} \sum_{j=m}^{\infty} \int_{S_j} \exp i(2m)! x \, dx$$

$$= \frac{1}{2\pi} \sum_{m+1}^{\infty} \int_{S_j} \exp i(2m)! x \, dx + \frac{m}{i\pi(2m)!}.$$

Further $\bigcup\limits_{m+1}^{\infty} S_j \subseteq (0, 2\pi/(2m + 1)!)$ and so

$$\left| \sum_{m+1}^{\infty} \int_{S_j} \exp i(2m)! x \, dx \right| < 2\pi/(2m + 1)!$$

Consequently,

$$|(2m)! \hat{\chi}_S(-(2m)!)| \geq \frac{m}{\pi} - \frac{2\pi}{2m + 1} \to \infty \,.$$

Problems for Chapter 3

Some of the more elementary problems in this set are Problems 3.1, 3.2, 3.3, 3.5, 3.8, 3.9, 3.13, 3.14, 3.15, 3.17, 3.19, 3.20, 3.25.

3.1 Let $f_n(x) = (\pi n)^{-1/2} e^{-x^2/n}$ on $\mathbf{R}$ and consider $f(x) = \lim_n f_n(x)$. Compute f, and determine whether (and why) $\int f_n \to \int f$. (*Hint.* In order to prove that

$$\int_{-\infty}^{\infty} e^{-x^2} \, dx = \sqrt{\pi},$$

Laplace considered the product $\left(\int e^{-x^2} \, dx\right)\left(\int e^{-y^2} \, dy\right)$ and then used polar co-ordinates.)

3.2 Let $f \in L_m^1(\mathbf{R})$, $f(0) = 0$, and assume $f'(0)$ exists. Prove that

$$\int_{-\infty}^{\infty} \frac{f(x)}{x} \, dx$$

exists.

3.3 a) Let $f \geq 0$ on $[a, b]$ be Lebesgue measurable. Prove that

$$\int f = \lim_{\varepsilon \to 0+} \int f \chi_{[a+\varepsilon, b]},$$

where $\int f$ may be infinite.

b) Prove that $\int_0^1 x^{\alpha-1} \, dx = 1/\alpha$ if $\alpha > 0$.

c) With respect to a) and Problem 1.23a, find a function $f: (0, 1] \to \mathbf{R}$ such that

$$\forall \varepsilon > 0, \qquad f \in L^1_m(\varepsilon, 1)$$

and $\qquad \lim_{n \to \infty} (1/n) \sum_{k \leqslant n} f(k/n)$ exists,

but $\lim_{\varepsilon \to 0+} \int f \chi_{[\varepsilon, 1]}$ does not exist.

3.4 Let $(X, \mathscr{A})$ be a measurable space, and let μ and ν be measures on $(X, \mathscr{A})$. Assume that $\mu = \nu$ on some subset $\mathscr{D} \subseteq \mathscr{A}$. A reasonable problem is to consider non-trivial conditions in order that $\mu = \nu$ on $\mathscr{A}$. More specifically, prove the following: *assume that X and $\emptyset$ are elements of $\mathscr{D}$, that $\mathscr{D}$ is closed under finite unions and intersections, and that $\mathscr{A}$ is the smallest σ-algebra containing $\mathscr{D}$; if*

$$\exists\{D_n : n = 1, \ldots\} \subseteq \mathscr{D} \text{ such that } \bigcup D_n = X \text{ and } \forall n, \mu D_n < \infty$$

then $\mu = \nu$ on $\mathscr{A}$ (cf. Problem 3.27). Consequently we see that a measure defined on the Borel subsets of the line is uniquely determined there by its value on the $\frac{1}{2}$-open intervals $[a, b)$ (including $(-\infty, b)$ and $[a, \infty)$). This uniqueness issue leads to a problem which has still not been satisfactorily solved (e.g. Chapter 6): suppose that ν is not necessarily a measure but only finitely additive on $\mathscr{D}$ whereas the other hypotheses for the above problem are satisfied; when can we conclude that $\mu = \nu$ on $\mathscr{A}$?

3.5 Given $\{A_n : n = 1, \ldots\}$ for which $\bigcup A_n = (0, 1)$ and $\sum m A_n < \infty$; prove that

$$m\{x : x \text{ is in at most finitely many } A_n\} = 1.$$

(*Hint.* See the proof of Corollary 2.10.2.) Compare with Problem 2.8b.

3.6 a) For real-valued functions $f \in L^\infty_m(0, \infty)$ define

$$f \odot f(x) = \int_0^x f(t) f(x - t) \, dt.$$

It is not unreasonable to expect that $f \odot f \geqslant 0$ on some small interval $(0, \varepsilon)$. Show that this is not necessarily the case. (*Hint.* Take $f(t) = \sin(1/t)$.)

b) The c o n v o l u t i o n of the functions f and g, where $f, g \in L^1_m(\mathbf{R})$, is

$$f * g(x) = \int_{-\infty}^\infty f(t) g(x - t) \, dt.$$

As such $L^1_m(\mathbf{R})$ becomes a Banach algebra, a most elegant gateway to harmonic analysis. For this problem prove Steinhaus' result (mentioned in Problem 2.9): *if $X \in \mathscr{M}(\mathbf{R})$ and $mX > 0$, then $X - X = \{x - y : x, y \in X\}$ is a neighborhood of 0.* (*Hint.* Let $f(x) = \chi_X(x)$ and note that $f * f$ is continuous.) We can, of course, prove the result "set theoretically".

3.7 a) Let $X \subseteq \mathbf{R}$ have the property that finite linear combinations of elements from X with rational coefficients generate an interval, I. Prove that X contains an Hamel basis. (*Hint.* Let $0 \in I$; and so

$$\forall x \in \mathbf{R} \, \exists r \in \mathbf{Q} \text{ such that } (1/r)x \in I.$$

Hence every real number is a finite rational linear combination from X. Let $\mathscr{F}$ be the family of all subsets $Y \subseteq X$ with the same property. Apply Zorn's lemma.) Thus, *the Cantor set contains an Hamel basis* (e.g. Problem 1.12).

b) Take $A \in \mathscr{M}(\mathbf{R})$ for which $0 < mA < \infty$; prove that A contains an Hamel basis. (*Hint:* Use part a) and the method of proof from Problem 3.6b.) This is simpler to prove than Problem 2.10e since we do not have to deal with the likes of (P2.2); of course, it is also a weaker result.

3.8 a) Prove Theorem 3.9. (*Hint.* Generalize Theorem 3.6 appropriately.)

b) Show that there can be strict inequality in Theorem 3.6. (*Hint.* Let $f_n = \chi_{[n, n+1)}$.)

c) Prove that Theorem 3.7 can fail if we replace "$f_n \leqslant f$, μ-a.e." by the hypothesis that

$$\forall x \text{ and } \forall n, \quad f_n(x) \geqslant f_{n+1}(x).$$

(*Hint.* Let $f_n = \chi_{[n, \infty)}$.)

3.9 Find a function $f: [0, 1] \to \mathbf{R}$ such that f' exists on $[0, 1]$, $f''(0)$ exists, and $R \int_0^b f'$ does not exist for any $b > 0$. (*Hint.* Let g be the Volterra function defined in Chapter 1.3 and set $f(x) = x^2 g(x)$.)

3.10 a) Prove: For each $f \in L_m^1(\mathbf{R})$ and $\varepsilon > 0$ there is a continuous function g such that

$$m\{x: g(x) \neq 0\} < \infty$$

and

$$\int |f(x) - g(x)| \, dx < \varepsilon.$$

(*Hint.* Use LDC and the observation on this matter given after Example 2.15.)

b) Prove:

$$\forall f \in L_m^1(\mathbf{R}), \quad \lim_{h \to 0} \int_{\mathbf{R}} |f(x + h) - f(x)| \, dx = 0. \tag{P3.1}$$

(*Hint.* Use a) and a fact about uniformly continuous functions.) We shall later characterize functions of bounded variation (resp., constants) on $[a, b]$ as those elements in $L_m^1[a, b]$ which satisfy:

$$\int_a^b |f(x + h) - f(x)| \, dx = O(|h|)(\text{resp., } o(|h|)), \quad h \to 0.$$

The "O" and "o" notation is standard in classical analysis. $F(x) = O(G(x))$ (resp., $o(G(x))$), as $x \to r$, means that

$$\exists M > 0 \text{ and } \exists y \text{ such that } \forall x \in [r - y, r + y], |F(x)| \leqslant MG(x)$$

$$(\text{resp., } \lim_{x \to r} F(x)/G(x) = 0);$$

G is positive. For $r = \pm \infty$ there is an obvious adjustment in the definition.

c) A trigonometric polynomial on $\mathbf{T}$ has the form

$$t_N(x) = \sum_{|n| \leqslant N} a_n e^{inx}.$$

Using a) prove that

$$\forall \varepsilon > 0 \quad \text{and} \quad \forall f \in L_m^1(\mathbf{T}) \, \exists t_N \text{ such that } \int_{\mathbf{T}} |f - t_N| < \varepsilon.$$

(*Hint.* Find a function $g \in C^2(\mathbf{R})$ in terms of indefinite integrals such that g has period 2π and $\|f - g\|_1 < \varepsilon/2$; then find N for which $\|g(x) - \sum_{|n| \leqslant N} \hat{g}(n) e^{inx}\|_\infty < \varepsilon$.)

d) We proved the Riemann–Lebesgue lemma in section 3.5. We have the following proof which uses (P3.1):

$$|\hat{f}(n)| = |\tfrac{1}{2}(f - \tau_{-\pi/n}f)(n)| \leqslant \frac{1}{4\pi} \int_0^{2\pi} \left| f(x) - f\left(x + \frac{\pi}{n}\right) \right| dx,$$

where $\tau_y f(x) = f(x - y)$.

e) Prove Fejér's lemma: If $f \in L_m^1(\mathbf{T})$ and $g \in L_m^\infty(\mathbf{T})$ then

$$\lim_{|n| \to \infty} \frac{1}{2\pi} \int_0^{2\pi} f(x)g(nx) \, dx = \left(\frac{1}{2\pi}\right)^2 \int_0^{2\pi} f(x) \, dx \int_0^{2\pi} g(x) \, dx.$$

This result reduces to the Riemann–Lebesgue lemma for the case of $g(x) = e^{ix}$.

3.11 Let $f_n: [0, 1] \to \mathbf{R}$ be m-measurable; assume

$$\int_0^1 f_n = 1$$

and

$$\int_0^1 f_n^2 = o(1), \, n \to \infty.$$

Prove that there is a subsequence $\{f_{n_k}: k = 1, ...\}$ such that f_{n_k} converges to 1 pointwise m-a.e. Give an example to show that, generally, we cannot conclude that $f_n \to 1$ pointwise m-a.e.

3.12 Let $H \subseteq \mathbf{R}$ be an Hamel basis. Prove that

$$\forall a \in \mathbf{R}, \quad a \neq 1, \quad \exists h \in H \quad \text{such that} \quad ah \notin H.$$

(*Hint.* Take such an a and assume $aH \subseteq H$; for each $x = \sum r_\alpha h_\alpha$ define $f(x) = \sum r_\alpha$ and note that $f(ax) = f(x)$. Compute $f(x)$, where $x = h/(a-1)$ and $h \in H$.)

3.13 Let $f \in L^1_m(0, \infty)$ and define I_r to be a subset of $(0, \infty)$ parameterized by $r > 0$.

a) Prove that $\lim\limits_{r \to \infty} \int_{I_r} f(x) \cos rx \, dx = 0$ if I_r is an interval.

b) Show that a) is not generally true if I_r is taken to be a finite union of intervals. These results are interesting in light of the Riemann–Lebesgue lemma.

3.14 Compute

$$\lim_{n \to \infty} \int_0^n \left(1 - \frac{x}{n}\right)^n e^{x/2} \, dx.$$

(*Hint.* Transform the integral to $\int_{-n}^0$ and note, comparing like terms of the binomial expansions of $(1 + t/n)^n$ and $(1 + t/(n+1))^{n+1}$, that $(1 + t/n)^n e^{-t/2}$ increases to $e^{t/2}$; then use LDC.)

3.15 Prove that if $f: \mathbf{R} \to \mathbf{R}$ is a continuous function and $g: \mathbf{R} \to \mathbf{R}$ is continuous *m-a.e.* then $f \circ g$ is continuous *m-a.e.*, cf., Problem 4.41.

3.16 Prove Proposition 3.7. (*Hint.* Using the criterion of Theorem 3.13 there are several ways to prove the result; one method is to employ Proposition 3.4.)

3.17 Prove that if f is bounded on $[a, b]$ and continuous *m-a.e.*, then $\bar{F}' = f$, *m-a.e.*, where $\bar{F}(x) = R \int_a^x f$.

3.18 Verify Theorems 3.14, 3.15 and 3.16.

3.19 a) Given a finite measure space $(X, \mathscr{A}, \mu)$. Let Y be the space of real-valued μ-measurable functions defined on X; and let $\tilde{Y}$ be the space of equivalence classes of such functions where $f, g: X \to \mathbf{R}$ are defined to be equivalent if $f = g$, μ-*a.e.* Define

$$\forall f, g \in Y, \quad \rho(f, g) = \int_X \frac{|f - g|}{1 + |f - g|} \, d\mu.$$

Prove that ρ is a metric on $\tilde{Y} \times \tilde{Y}$. (*Hint.* If a and b are real, $|a + b|/(1 + |a + b|) \leqslant |a|/(1 + |a|) + |b|/(1 + |b|)$.)

b) Given the setting of part a). Prove that $\rho(f_n, f) \to 0 \Leftrightarrow f_n \to f$ in measure. (*Hint.* Let $A_n(\varepsilon) = \{x: |f_n(x) - f(x)| \geqslant \varepsilon\}$ and prove that

$$\frac{\varepsilon}{1 + \varepsilon} \mu A_n(\varepsilon) \leqslant \rho(f_n, f) \leqslant \mu A_n(\varepsilon) + \varepsilon \mu X.)$$

c) Find an example of a sequence $\{f_n: n = 1, ...\}$ of m-measurable functions (on the measure space $([0, 1], \mathcal{M}, m)$) such that $f_n \to 0$ in measure whereas $\{f_n: n = 1, ...\}$ does not converge pointwise at any point.

d) Let $(X, \mathcal{A}, \mu)$ be a measure space and let $\{f, f_n: n = 1, ...\}$ be a sequence of measurable functions. Prove F. Riesz's result:

let $f_n \to f$ in measure on $(X, \mathcal{A}, \mu)$; there is a subsequence $\{f_{n_k}: k = 1, ...\}$ which converges to f pointwise μ-a.e.

e) The following converse to Riesz's result is due to Lebesgue: *given a finite measure space $(X, \mathcal{A}, \mu)$ and a sequence $\{f, f_n: n = 1, ...\}$ of measurable functions on X; if $f_n \to f$, μ-a.e., then $f_n \to f$ in measure.* Prove it.

(*Hint.* Use Egoroff's theorem.) Show by example that the hypothesis, $\mu X < \infty$, is necessary generally.

3.20 a) Show that if $f \in L^1_c(\mathbf{Z})$, where c is counting measure, then

$$\lim_{|n| \to \infty} f(n) = 0.$$

b) Prove that there are functions $f \in L^1_m(\mathbf{R})$ such that

$$\overline{\lim_{|x| \to \infty}} |f(x)| = \infty.$$

3.21 The Fejér kernel $\{F_N: N = 1, ...\}$ is defined by

$$\forall x \in \mathbf{R} \quad \text{and} \quad \forall N \geqslant 0, \qquad F_N(x) = \sum_{|n| \leqslant N} \left(1 - \frac{|n|}{N + 1}\right) e^{inx}.$$

a) Prove that if $x/2\pi \in \mathbf{Z}$ then $F_N(x) = N + 1$, and that if x is any other real number then

$$F_N(x) = \frac{1}{N + 1} \left[\sin \frac{(N + 1)x}{2} \Big/ \sin \frac{x}{2}\right]^2.$$

b) Clearly $F_N \geqslant 0$. Prove that

$$\frac{1}{2\pi} \int_0^{2\pi} F_N = 1,$$

and that

$$\forall \delta \in (0, \pi), \qquad \lim_{N \to \infty} \frac{1}{2\pi} \int_{\delta}^{2\pi - \delta} |F_N| = 0.$$

c) An approximate identity in $L_m^1(\mathbf{T})$ is a sequence $\{K_N : N = 1, ...\} \subseteq L_m^1(\mathbf{T})$ which satisfies

$$\sup_N \frac{1}{2\pi} \int_0^{2\pi} |K_N| < \infty,$$

$$\lim_{N \to \infty} \frac{1}{2\pi} \int_0^{2\pi} K_N = 1,$$

and

$$\forall \delta \in (0, \pi), \qquad \lim_{N \to \infty} \frac{1}{2\pi} \int_{\delta}^{2\pi - \delta} |K_N| = 0.$$

Thus $\{F_N : N = 1, ...\}$ is an approximate identity. We defined convolution in $L_m^1(\mathbf{R})$ in Problem 3.6b; the same definition makes sense on $\mathbf{T}$ as long as we define our functions periodically. Hence, if $f, g \in L_m^1(\mathbf{T})$ we have

$$f * g(x) = \frac{1}{2\pi} \int_0^{2\pi} f(x - y) g(y) \, dy.$$

Prove that if $f \in C(\mathbf{R})$ has period 2π and $\{K_N : N = 1, ...\}$ is an approximate identity in $L_m^1(\mathbf{T})$ then

$$\lim_{N \to \infty} \|K_N * f - f\|_\infty = 0,$$

where "$\| \ \|_\infty$" designates the $L_m^\infty(\mathbf{T})$-norm. (*Hint.*

$$\|K_N * f - f\|_\infty \leqslant \frac{1}{2\pi} \int_0^{2\pi} |K_N(y)| \, \|\tau_y f - f\|_\infty \, dy = I.$$

Take $\varepsilon > 0$ and fix $0 < \delta < \pi$ such that $\|\tau_y f - f\|_\infty < \varepsilon$ if $y \in [0, \delta] \bigcup [2\pi - \delta, 2\pi]$. Compute

$$I \leqslant M\varepsilon + \|f\|_\infty \frac{1}{\pi} \int_{\delta}^{2\pi - \delta} |K_N|,$$

where M is independent of N.) In particular, we have proved that

$$f(0) = \lim_{N \to \infty} \frac{1}{2\pi} \frac{1}{N+1} \int_0^{2\pi} f(x) \left[\sin \frac{(N+1)x}{2} \Big/ \sin \frac{x}{2} \right]^2 dx,$$

where f is a bounded m-measurable function on $\mathbf{R}$ which is continuous at 0 and has period 2π.

d) Given a function $f: [a, b] \times [a, b] \to \mathbf{R}$ which is continuous in each variable separately. Prove that

$$\forall (x, y) \in [a, b] \times [a, b], \qquad \lim_N f_N(x, y) = f(x, y),$$

where $\{f_N : N = 1, ...\}$ is a sequence of continuous functions on $[a, b] \times [a, b]$. (*Hint:* Assume without loss of generality that $[a, b] = [0, 2\pi]$, that $\|f\|_\infty < \infty$, and that for each y, $f(0, y) = f(2\pi, y)$. Set

$$f_N(x, y) = \sum_{|n| \leqslant N} \left(1 - \frac{|n|}{N+1} \right) c_n(y) e^{inx},$$

where

$$c_n(y) = \frac{1}{2\pi} \int_0^{2\pi} f(x, y) e^{-inx} dx.$$

From part c),

$$\forall y \in \mathbf{T}, \qquad \lim_N \sup_{x \in \mathbf{T}} |f_N(x, y) - f(x, y)| = 0.$$

In particular, we have the desired pointwise convergence. To prove that f_N is continuous we need only check that each c_n is continuous, and this is clear from LDC.)

3.22 We have mentioned the concept of equidistribution in Problem 1.23 and Proposition 3.6. We shall now define it and prove one of its basic properties. A sequence $\{x_n : n = 1, ...\} \subseteq \mathbf{R}$ is equidistributed mod a, $a > 0$, if, when $y_n \in [0, a)$ and $(y_n - x_n)/a \in \mathbf{Z}$, we can conclude that

$$\forall I \subseteq [0, a), \qquad \lim_{N \to \infty} \frac{N_I}{N} = \frac{1}{a} mI,$$

where $I \subseteq [0, a)$ is an arbitrary interval and $N_I = \text{card } \{y_1, ..., y_N\} \cap I$. Prove that if $x/2\pi$ is irrational then

a) For each Riemann integrable function g defined on $[0, 2\pi)$,

$$\lim_{N \to \infty} \frac{1}{N} \sum_1^N g(nx (\text{mod } 2\pi)) = \hat{g}(0);$$

b) $\{nx : n = 1, ...\}$ is equidistributed mod 2π.

(*Hint.* Part b) is clear from part a) by taking $g = \chi_I$. To prove part a) first compute that

$$\left| \sum_{n=1}^{N} e^{imnx} \right| \leqslant 1 / \left| \sin \frac{mx}{2} \right|;$$

and use the hypothesis that $x/2\pi$ is irrational to prove that if $m \neq 0$ then

$$\lim_{N \to \infty} \frac{1}{N} \sum_{n=1}^{N} e^{imnx} = 0.$$

The rest of the demonstration follows from standard approximation results.)

3.23 Given $a > 0$ and a sequence $\{x_n : n = 1, ...\} \subset \mathbf{R}$; define $y_n \in \mathbf{R}$ as in Problem 3.22 by the condition that $y_n \in [0, a)$ and $(y_n - x_n)/a \in \mathbf{Z}$. The antithesis of equidistribution for a sequence $\{x_n : n = 1, ...\}$ occurs when $y_n \to 0$. Assume that the sequence $\{x_n : n = 1, ...\} \subseteq \mathbf{R}$ does not tend to 0 and let $A = \{a > 0 : y_n \to 0\}$. Prove that $mA = 0$. (*Hint.* Without loss of generality let $\lim |x_n| = \infty$. If $a \in A$ then $\lim_{n \to \infty} \exp(2\pi i x_n/a) = 1$. Consequently, $\lim_{n \to \infty} \exp(2\pi i x_n r) = 1$ for $r \in A^{-1} = \{a^{-1} : a \in A\}$. Thus

$$\forall b > 0, \quad \lim_n \int_{A_b^{-1}} \exp(2\pi i x_n r) \, dr = mA_b^{-1},$$

where $A_b^{-1} = A^{-1} \cap (0, b)$, by LDC. On the other hand, from the Riemann–Lebesgue lemma

$$\lim_n \int_{A_b^{-1}} \exp(2\pi i x_n r) \, dr = 0.$$

Therefore $mA^{-1} = mA = 0$.) Note that A can be uncountable.

3.24 Given the open interval (a, b), where b can be ∞, and a function $f : (a, b) \to \mathbf{R}$. Assume that $R \int_a^r f$ exists for each $r \in (a, b)$. The Cauchy–Riemann integral of f on (a, b) is

$$CR \int_a^b f = \lim_{r \to b-} R \int_a^r f$$

when the right-hand side exists. Naturally we can define the "Cauchy–Lebesgue" integral if we replace $R \int_a^r f$ by $\int_a^r f$.

a) Give an example of an unbounded function on (a, ∞) whose Cauchy–Riemann integral exists (e.g. Problem 3.20b).

b) Let f be a non-negative function on the interval (a, ∞) and assume that $R \int_a^r f$ exists for all $r > a$. Prove that $CR \int_a^\infty f$ exists if and only if $f \in L_m^1(a, \infty)$.

c) Prove that if $f \in L_m^1(a, \infty)$ and $R \int_a^r f$ exists for all $r > a$ then $\int_a^\infty f = CR \int_a^\infty f$.

d) Let $f(x) = (\sin x)/x$. Show that $CR \int_0^\infty f$ exists but that $f \notin L_m^1(0, \infty)$. (*Hint.*

$$\int_0^\infty \left| \frac{\sin x}{x} \right| dx = \sum_0^\infty \int_{k\pi}^{(k+1)\pi} \left| \frac{\sin x}{x} \right| dx$$

$$\geq \sum_0^\infty \frac{1}{(k+1)\pi} \int_{k\pi}^{(k+1)\pi} |\sin x| \, dx.)$$

Thus f can be integrated whereas $|f|$ cannot be integrated; this phenomenon is referred to as *conditional convergence*.

e) Show that the Cauchy–Riemann integral does not integrate every bounded Lebesgue measurable function on $[a, b]$, $b < \infty$.

f) Find a Cauchy–Riemann integrable function f on $[a, b]$, $b < \infty$, such that $f \notin L_m^1[a, b]$. (*Hint.* Example 3.6b.)

g) Let $f(x) = \dfrac{1}{x} \sin \dfrac{\pi}{x}$ on $(0, 1)$. Is f' Cauchy–Riemann integrable on $(0, 1)$? Does $f' \in L_m^1(0, 1)$?

h) In light of Problem 3.20 we would like to know conditions on a function f so that we could conclude that

$$\lim_{x \to \infty} f(x) = 0 \tag{P3.2}$$

when $CR \int_a^\infty f$ exists. In fact, the following is true: *let f be continuous on $[a, \infty)$ and assume that $CR \int_a^\infty f$ exists; then (P 3.2) is valid if and only if f is uniformly continuous.*

i) Let f be a twice continuously differentiable real-valued function defined on (a, ∞), and assume that CR $\int\limits_a^\infty f^2$ and CR $\int\limits_a^\infty (f'')^2$ exist. Prove that (P 3.2) is valid. (*Hint.* Integrate ff'' by parts and use part h).)

3.25 a) In light of Example 3.13 does there exist a sequence $\{f_n: n = 1,...\} \subseteq L_m^1[0, 1]$ such that $f_n \to 0$ pointwise and

$$\forall A \in \mathcal{M}, \quad \lim_n \int f_n(x)\chi_A(x) \, dx = 0,$$

but $\overline{\lim} \int |f_n(x)| \, dx > 0$?

b) Construct a sequence $\{f_n: n = 1,...\} \subseteq L_m^1(\mathbf{R})$ such that
i) f_n is continuous and $\lim\limits_{|x| \to \infty} f_n(x) = 0$,

ii) $\sup\limits_n \int |f_n(x)| \, dx < \infty$,

iii) $f_n \to f$ uniformly on $\mathbf{R}$,
but such that $\{f_n: n = 1,...\}$ contains no subsequence $\{f_{n_k}: k = 1,...\}$ for which

$$\int |f_{n_k}(x) - f(x)| \, dx \to 0.$$

(*Hint.* Start with $g_n = (1/n)\chi_{[n, 2n]}$.)

3.26 Given a sequence of functions f_n on $[a, b]$. The general problem for this exercise is to investigate pointwise convergence of subsequences on certain subsets of $[a, b]$.

a) Let $S_{n_k} = \{x: \sin n_k x \text{ converges}\}$. Show that even though $mS_{n_k} = 0$, it is possible to find subsequences $\{n_k: k = 1,...\} \subseteq \mathbf{Z}^+$ for which card $S_{n_k} = $ card $\mathbf{R}$.

b) Does there exist $\delta \in (0, 1]$ and $X \subseteq [0, 2\pi)$ such that $mX > 0$ and

$$\text{card } \{n: \sup_{x \in X} \sin nx \geq \delta\} < \infty?$$

c) Solve part b) for any non-constant periodic function $f \in L_m^\infty(\mathbf{R})$, where $\sin nx$ is replaced by $f(nx)$.

d) Let $A_{n_k} = S_{n_k} \cap [0, 2\pi)$. Give examples to show whether A_{n_k} can be closed and infinite, countable, or finite.
Find $A_{n!}$. See Remark 2 below.

e) Prove the Cantor–Lebesgue theorem: if $\sum\limits_{|n| \leq N} c_n e^{inx} \to 0$ pointwise on a set $X \subseteq [0, 2\pi)$, as $N \to \infty$, and $mX > 0$ then $\lim\limits_{|n| \to \infty} c_n = 0$. (*Hint.* Without loss of generality consider the series $\sum\limits_0^\infty r_n \cos(nx + a_n)$, and assume that there is a $\delta > 0$ and a subsequence $\{n_k: k = 1, \ldots\} \subseteq \mathbf{Z}^+$ such that for each k, $r_{n_k} > \delta > 0$. Thus $\lim_k \cos(n_k x + a_{n_k}) = 0$. Now use LDC and the fact that

$$\int_X \cos^2(kx + a)\, dx = \tfrac{1}{2}mX + \tfrac{1}{2}\int_X \cos 2(kx + a)\, dx. \tag{P 3.3}$$

Obtain the contradiction by employing the Riemann–Lebesgue lemma.) See the remarks in sections A3.1.1 and A3.1.2.

Remark 1 Using (P3.3) and LDC we have another proof (besides Theorem 3.12) that $mA_{n_k} = 0$.

Remark 2 S. Mazurkiewicz [78] proved that if $\{f_n: n = 1, \ldots\}$ is a uniformly bounded sequence of continuous functions on $[a, b]$ then there is a closed uncountable set $F \subseteq [a, b]$ (without isolated points) and integers $n_1 < n_2 < \ldots$ such that $\{f_{n_k}: k = 1, \ldots\}$ converges uniformly on F. Using the continuum hypothesis, Sierpiński [108] proved: *there is a sequence of (non-measurable) functions on $[a, b]$ which is uniformly bounded but such that no subsequence converges on an uncountable set.* Note, with regard to Problem 3.26d, that if $\{f_n: n = 1, \ldots\}$ is any sequence of functions and $D \subseteq [a, b]$ is countable then there is a subsequence $\{f_{n_k}: k = 1, \ldots\}$ which converges on D.

Remark 3 Some interesting positive results on the general problem stated at the beginning of this exercise have been given by K. Schrader [100; 101]. For example: let $f_k: [a, b] \to \mathbf{R}$ be continuous; assume that for some $N \geqslant 0$ the sequence $\{f_k: k = 1, \ldots\}$ has the property that

if $f_k = f_j$ for more than N values of $x \in [a, b]$ then f_k is identical to f_j on $[a, b]$;

then there is a subsequence $\{f_{n_k}: k = 1, \ldots\}$ which converges pointwise (possibly with infinite value) on $[a, b]$.

3.27 Given a set X and let $\mathscr{A} \subseteq \mathscr{P}(X)$ be an algebra. Assume that the set function $\mu: \mathscr{A} \to \mathbf{R}^+$, for which $\mu\emptyset = 0$, satisfies:

$$\forall\{A_j\} \subseteq \mathscr{A}, \text{ a disjoint sequence such that } \bigcup A_j \in \mathscr{A}, \quad \mu\left(\bigcup_1^\infty A_j\right) = \sum_1^\infty \mu A_j.$$

Define

$$\forall E \in \mathscr{P}(X), \qquad \mu^* E = \inf \sum_1^\infty \mu A_j,$$

where the infimum is taken over all collections $\{A_j: j = 1, \ldots\} \subseteq \mathscr{A}$ which cover E. $E \in \mathscr{P}(X)$ is μ^*-measurable if

$$\forall F \in \mathscr{P}(X), \qquad \mu^* F \geqslant \mu^*(F \cap E) + \mu^*(F \cap E^\sim).$$

(cf., the development in Chapter 2.2). Prove Caratheodory's theorem: the μ^*-measurable sets form a σ-algebra $\mathscr{C}$ containing $\mathscr{A}$ such that $(X, \mathscr{C}, \mu^*)$ is a measure space and $\mu^* = \mu$ on $\mathscr{A}$; if μ is σ-finite then μ^* is the unique extension of μ as a measure to the smallest σ-algebra containing $\mathscr{A}$ (this latter part is precisely Problem 3.4).

Appendix A3.1 Sets of Uniqueness and Measure Zero

A3.1.1 B. Riemann

Bernhard Riemann's life (September 17, 1826–July 20, 1866) has been documented by his friend Dedekind. Our interest for the sequel focuses on his Habilitationsschrift [88]; here he begins with an important historical note on Fourier series, defines the Riemann integral to provide a broader setting for an analytically precise theory of Fourier series, and develops the Riemann localization principle which is a key technique in the study of U-sets. $E \subseteq [0, 2\pi) = \mathbf{T}$ is a U-set if

$$\lim_{N \to \infty} \sum_{|n| < N} c_n \, e^{inx} = 0 \quad \text{off} \quad E \Rightarrow c_n = 0 \quad \text{for all } n.$$

The problem to determine U-sets is important since one would like to know if the representation of a function by a trigonometric series is unique or not. The first explicit results in this direction were given by Cantor (e.g. section A 3.1.2) although the following fundamental theorem (for uniqueness questions), first proved by Cantor, was apparently known by Riemann [69, p. 110; 88]:

(C-L) $\lim_{N \to \infty} \sum_{|n| \leqslant N} c_n \, e^{inx} = 0 \quad \text{on} \quad [a, b] \Rightarrow \lim_{|n| \to \infty} c_n = 0$

(e.g. Problem 3.26e).

It is interesting to observe the overlap between Dini and Riemann. Riemann convalesced and toured in Italy during the winter of 1862; and then came to Pisa during 1863. He became quite friendly with Betti and Beltrami (Betti was director of the Scuola Normale Superiore in Pisa from 1865–1892); there is a Betti–Riemann correspondence which has yet to be studied. Dini graduated from the Scuola Normale in 1864 when he was 19 years old; he then spent a year in Paris with Bertrand, and returned to the Scuola Normale where he spent the next 52 years. Dini became one of the 19th-century giants in real variable and Fourier analysis, and includes Volterra and Vitali among his students. Riemann returned to Germany for the winter of 1864–65, but then came back to Pisa. He died and was buried at Biganzolo in the northern part of Verbania (the Italian resort town on the western banks of Lago Maggiore just 15 miles south of the Swiss border).

A3.1.2 G. Cantor

Georg Cantor (March 3, 1845–January 6, 1918) wrote several important papers on U-sets during the early 1870's (*Crelle's Journal*, volumes 72 and 73, and *Math. Ann.*, volume 5). In the first, he proved $(C - L)$ (see section A3.1.1) and using this fact proved (in the second) that $\emptyset$ is a U-set. The subsequent papers gave simplifications of proof and extensions of the basic result, showing finally that certain countable infinite sets are U-sets. The study of special types of infinite sets in this work influenced his later research activity, and it was in 1874 that he gave his famous and controversial proof of only countably many algebraic numbers (cf. Problem 1.20). The remainder of his life was devoted to the study of the infinite, not only in a mathematical milieu, but often delving into various philosophical notions of infinity due to the Greeks, the scholastic philosophers, and his contemporaries. An interesting letter in this latter regard was sent by Cantor, in London at the time, to Bertrand Russell (at Trinity College, Cambridge); he writes: "... and I am quite an adversary of Old Kant, who, in my eyes has done much harm and mischief to philosophy, even to mankind; as you easily see by the most perverted development of metaphysics in Germany in all that followed him, as in Fichte, Schilling, Hegel, Herbart, Schopenhauer, Hartman, Nietzche, etc. etc. on to this very day. I never could understand why ... reasonable ... peoples ... could follow yonder sophistical philistine, who was so bad a mathematician."

Dedekind was also involved in work related to Cantor's, and many of his set theoretic contributions are found in their 27-year correspondence (edited by Jean Cavaillés and Emmy Noether and published by Hermann of Paris in 1937). F. A. Medvedev (1966) has studied these particular results. The marvelous Dedekind, by the way, taught at the Gymnasium in Brunswick for fifty years beginning in 1862 (cf. Chapter A6.1).

Cantor's U-set papers were preceded by H. Heine's uniqueness theorem (Crelle's Journal, volume 71) in 1870 which assumed that the given trigonometric series were uniformly convergent off arbitrary neighborhoods of a finite number of points. Heine was at the University of Halle with Cantor and attributes this approach to Cantor.

Cantor, of course, tried to prove that all countable sets are U-sets; and this was finally achieved by F. Bernstein (1908) and W. H. Young (1909). Actually Bernstein proved somewhat more when applied to general groups, showing that E is a U-set if it does not contain any non-$\emptyset$ closed sets without isolated points.

A3.1.3 D. Men'shoff

Dmitrii Men'shoff (April 18, 1892) proved a key result on U-sets in 1916 by finding a non-U-set E with Lebesgue measure $mE = 0$. He did this just after graduating from Moscow University, where he wrote his thesis under N. Lusin. His example has stimulated a great deal of study about sets of measure zero; and research about specific sets of measure zero now forms a significant part of modern Fourier analysis and potential theory. Actually, on the basis of Men'shoff's example, Lusin and Bari defined the notion "U-set" as such. Earlier, de la Vallée-Poussin had proved that if a trigonometric series converges to $f \in L^1_m(\mathbf{T})$ off a countable set E then the series is the Fourier series of f; and it was generally felt that the same would be true if $mE = 0$. Consequently, Men'shoff's example had a certain amount of shock value to say the least.

Since we'll be discussing Carleson's solution to Lusin's problem in Chapter A5.1

it is interesting to note that Men'shoff solved the analogue for measurable functions in 1940–41. Lusin, in 1915, had noted that if f is measurable on **T** and finite m-a.e. then there is a trigonometric series which converges to f by both Riemann and Abel summation. The problem was to see if such a series exists which converges pointwise m-a.e. to f; Men'shoff showed precisely this. Thus with the Carleson–Hunt theorem and Kolmogoroff's example of $f \in L_m^1(\mathbf{T})$ with Fourier series diverging everywhere (1926), "all that remains" (in the broad sense) of Lusin's problem is an investigation of the analogous situation for f measurable but taking infinite values on a set of positive measure. Actually Men'shoff has an affirmative answer on this latter problem for the case of convergence in measure instead of convergence m-a.e.

A3.1.4 N. Bari and A. Rajchman

What with Men'shoff's example, Alexander Rajchman (who died at Dachau in 1940) "seems to have been the first to realize that for sets of measure zero that occur in the theory of trigonometric series it is not so much the metric as the arithmetic properties that matter" [98 (from Zygmund's biography of Salem)]. Rajchman [87 (1922)] proved the existence of closed uncountable U-sets. He was motivated by some work of Hardy and Littlewood (Acta Math. 37 (1914)), and later (1920) Steinhaus, on diophantine approximation to introduce "H-sets" and proved that such sets are U-sets. Rajchman, in a letter to Lusin, thought that any U-set is contained in a countable union of H-sets, and it was only in 1952 that Pyatetskii–Shapiro proved this conjecture false. The Cantor set is H and therefore U. It is easy to verify that if $mE > 0$ then E is not U.

Actually, Nina Bari had proved the existence of closed uncountable U-sets in 1921 and presented her results at Lusin's seminar (at University of Moscow); they were unpublished at the time of Rajchman's paper, although they were communicated to him in [87 (1923)]. This does not minimize the importance of Rajchman's results since he established a large class of uncountable U-sets and illustrated the need for diophantine properties in constructing such sets.

Nina Bari (November 19, 1901–July 15, 1961) established her first results on U-sets as an undergraduate, and throughout her life, although she engaged in several other research areas, was an outstanding expositor and contributor on the tricky business of uniqueness. One of her major results is that the countable union of closed U-sets is a U-set; although the problem is open for the finite union of arbitrary U-sets. Another, which was proven in 1936–37 and which has an interesting sequel (e.g. section A3.1.6) shows that if α is rational and E is the perfect symmetric set determined by $\xi_k = \alpha$, then E is a U-set if and only if $1/\alpha$ is an integer; her theorem depends heavily on diophantine considerations.

A3.1.5 Number theoretic and spectral synthesis remarks

A 3.1.5.1 Kronecker Sets. Kronecker proved: *if* $\{x_1, \ldots, x_n, \pi\} \subseteq \mathbf{R}$ *is linearly independent over the rationals,* $\{\hat{y}_1, \ldots, y_n\} \subseteq \mathbf{R}$, *and* $\varepsilon > 0$, *then there is an integer* m

such that for each j,

$$|e^{ix_j m} - e^{iy_j}| < \varepsilon.$$

Because of this we say that a closed set $E \subseteq \mathbf{T}$ is a Kronecker set if for each $\varepsilon > 0$ and each continuous function $f: E \to \mathbf{C}$, satisfying $|f| = 1$, there is an integer m for which

$$\sup_{x \in E} |f(x) - e^{imx}| < \varepsilon.$$

A 3.1.5.2 The Spectral Synthesis Problem. Let $A(\mathbf{T})$ be the absolutely convergent Fourier series $f(x) = \sum a_n e^{inx}$ on $\mathbf{T}$ normed by $\|f\| = \sum |a_n|$, and let its dual be $A'(\mathbf{T})$. For each closed set $E \subseteq \mathbf{T}$ let $A'(E)$ be those elements of $A'(\mathbf{T})$ with support contained in E and let $A'_s(E)$ be those $T \in A'(E)$ such that $\langle T, f \rangle = 0$ for all $f \in A(\mathbf{T})$ vanishing on E. E is a spectral synthesis set or S-set if $A'(E) = A'_s(E)$.

Norbert Wiener and Arne Beurling posed the problem to determine if a given closed subset of $\mathbf{T}$ is an S-set or not, and Wiener proved (in his Tauberian theorem) that the empty set is an S-set in 1930. This result was extended extensively by Ditkin, Shilov, Kaplansky, Segal, and Helson, and their general Wiener theorem tells us, in particular, that if E is closed and has a countable boundary then E is an S-set. Using a different idea, Carl Herz proved that the Cantor set is an S-set.

In the other direction, Laurent Schwartz (1948) showed that the sphere's surface is a non-S-set in $\mathbf{R}^3$; and Paul Malliavin (1959) proved that every non-discrete locally compact abelian group has non-S-sets.

Since I have defined S-sets in such a cold-blooded fashion you may wonder from where they came. The story is exciting and complicated, but an amusing start would be a visit to the "Harmonic Analyzers and Synthesizers" exhibit at the Smithsonian's Museum of History and Technology. Although there was much mathematical stimulation for the S-set problem, Wiener also convincingly claims significant physical motivation.

A 3.1.5.3 Pisot Numbers. An algebraic integer is an algebraic number (cf. Problem 1.20 and section A 3.1.2) where the corresponding polynomial is monic. A Pisot number is a real algebraic integer $\alpha > 1$ with the property that all of the other roots of its minimal polynomial have modulus less than 1. Pisot numbers come into the picture for the investigation of badly distributed sequences, as opposed to Weyl's equidistribution (Kronecker's theorem can be proved by using Weyl's results on equidistribution).

Independently of each other, Thue (1912 in Norske Vid. Selsk. Skr.) and Hardy (1919 in J. of Ind. Math. Soc.) observed that

$$\alpha \text{ Pisot} \Rightarrow \lim_{n \to \infty} \alpha^n \pmod{1} = 0,$$

and proved (what is still one of the key properties of Pisot numbers): if $\alpha > 1$ is an algebraic integer and $\lim_n \alpha^n(\bmod 1) = 0$ then α is Pisot. The work of both Thue and Hardy does not seem to have been properly advertised until the early 1960's. Pisot numbers have been most extensively studied by Pisot beginning with his thesis in 1938 and a good bibliography on the subject up to 1962 is in Crelle's Journal 209 (1962).

Note that $\alpha_1 = (\frac{1}{2})(1 + \sqrt{5})$ is Pisot since $|(\frac{1}{2})(1 - \sqrt{5})| < 1$. For quadratic Pisot numbers α_1 with conjugate α_2 we trivially verify the above Thue–Hardy observation as follows:

$$(\alpha_1^n + \alpha_2^n) = (\alpha_1^{n-1} + \alpha_2^{n-1})(\alpha_1 + \alpha_2) - \alpha_1\alpha_2(\alpha_1^{n-2} + \alpha_2^{n-2}),$$
$$\alpha_1 + \alpha_2 = 1, \qquad \alpha_1\alpha_2 = -1,$$

and an induction argument shows that

$$\lim_{n \to \infty} \alpha_1^n (\bmod 1) = 0.$$

A3.1.6 R. Salem and the French School

How do Pisot numbers arise in Fourier analysis? Salem proved the following brilliant result: *let E be a perfect symmetric set with $\xi_k = \alpha \in (0, \frac{1}{2})$; E is a U-set if and only if $1/\alpha$ is a Pisot number.* Salem announced this result in 1943 and an error in the sufficiency conditions was found by members of the theory of functions seminar at University of Moscow in 1945. In 1948 Salem published some special cases in which the sufficiency is true, and in 1954 Pyatetskii–Shapiro proved that if β is a Pisot number of degree n and $\beta > 2^n$ then $E = E(\alpha)$ is a U-set where $\alpha = 1/\beta$. ($E(\alpha)$ is the perfect symmetric set determined by $\xi_k = \alpha$.) Finally, in 1955, Salem and Zygmund, using the Pyatetskii–Shapiro method, proved the full generality of the originally stated result.

Raphaël Salem (November 7, 1898–June 20, 1963) was the key figure in the revival of the now flourishing Paris (Orsay) school of Fourier analysis. Salem returned to Paris after the war and his lectures in 1948 on unsolved problems in Fourier series were the catalyst for this present activity. Salem's career is warmly sketched by Zygmund [98] from his birth in Saloniki, his banking profession (manager of the Banque de Paris et des Pays-Bas by 1938!), to the days in Cambridge, and to Paris.

Two of the striking results that have evolved from the study of U-sets and the notions of section A 3.1.5 are:

a) (*Malliavin*, 1962) *If every closed subset of a closed set $E \subseteq \mathbf{T}$ is an S-set then E is a U-set.*

b) (*Varopoulos*, 1965) *Measures are the only pseudo-measures supported by Kronecker sets; and so if E is a Kronecker set then the hypothesis of* a) *is satisfied and E is a U-set.*

4 The relationship between differentiation and integration on R

4.1 Functions of bounded variation and associated measures

A function $f: \mathbf{R} \to \mathbf{R}$ is increasing (resp. decreasing) if

$$\forall x \leqslant y, \quad f(x) \leqslant f(y) \quad (\text{resp. } f(x) \geqslant f(y)).$$

A real or complex-valued function f defined on an interval $[a, b]$ is of bounded variation on $[a, b]$, in which case we write $f \in BV[a, b]$, if

$$V(f, [a, b]) = \sup \left\{ \sum_1^n |f(x_j) - f(x_{j-1})| : a \leqslant x_0 \leqslant x_1 \leqslant \cdots \right.$$

$$\left. \leqslant x_n \leqslant b \right\} < \infty.$$

If $f \in BV[a, b]$ then $V(f, [a, b])$ is the total variation of f on $[a, b]$. When $f: \mathbf{R} \to \mathbf{R}$ is an element of $BV[a, b]$ for each interval $[a, b] \subseteq \mathbf{R}$, we define

$$Vf(x) = \sup \{V(f, [c, x]): c \leqslant x\}.$$

Clearly, Vf is a non-negative, increasing, possibly infinite function on $\mathbf{R}$.

The following result is the Jordan decomposition theorem which we shall prove in general measure theoretic form in Theorem 5.4.

Theorem 4.1 a) *Let the function $f: \mathbf{R} \to \mathbf{R}$ be of the form*

$$f = g_1 - g_2,$$

where each g_j is an increasing function. Then for each interval $[a, b] \subseteq \mathbf{R}, f \in BV[a, b]$ and

$$V(f, [a, b]) \leqslant g_1(b) - g_1(a) + g_2(b) - g_2(a).$$

b) *If the function $f: \mathbf{R} \to \mathbf{R}$ is an element of $BV[a, b]$ for each interval $[a, b] \subseteq \mathbf{R}$ and if $Vf(c) < \infty$ for some c, then the following hold:*

$$P(x) = \tfrac{1}{2}(Vf(x) + f(x)) \quad \text{and} \quad N(x) = \tfrac{1}{2}(Vf(x) - f(x))$$
are increasing functions, \hfill (4.1)

$$f = P - N,$$ \hfill (4.2)

$$Vf = P + N.$$ \hfill (4.3)

Proof. a) Take an interval $[a, b]$ and a partition $a \leqslant x_0 \leqslant x_1 \leqslant \ldots \leqslant x_n \leqslant b$. Clearly

$$\sum_1^n |f(x_j) - f(x_{j-1})| = \sum_1^n |g_1(x_j) - g_1(x_{j-1}) - g_2(x_j) + g_2(x_{j-1})|$$

$$\leqslant \sum_1^n |g_1(x_j) - g_1(x_{j-1})| + \sum_1^n |g_2(x_j) - g_2(x_{j-1})|$$

$$= g_1(x_n) - g_1(x_0) + g_2(x_n) - g_2(x_0)$$

$$\leqslant g_1(b) - g_1(a) + g_2(b) - g_2(a).$$

(4.4)

b) (4.2) and (4.3) are obvious from (4.1).

Using the additivity of $V(f, [a, b])$ (e.g. Problem 4.1) observe that

$$\forall [a, b] \subseteq \mathbf{R}, \qquad Vf(b) - Vf(a) = V(f, [a, b]). \tag{4.5}$$

Consequently,

$$P(b) - P(a) = \tfrac{1}{2}[V(f, [a, b]) + (f(b) - f(a))] \geqslant 0,$$

where the inequality follows since $V(f, [a, b])$ is obviously greater than or equal to $f(b) - f(a)$. Thus P is an increasing function. A similar argument works for N.

q.e.d.

If f and g are functions $\mathbf{R} \to \mathbf{R}$ which are of bounded variation on each interval $[a, b]$ and which differ from each other by a constant, i.e.

$$\exists r \in \mathbf{R} \text{ such that } \forall x \in \mathbf{R}, \qquad f(x) = g(x) + r,$$

then $V(f, [a, b]) = V(g, [a, b])$ for each interval $[a, b]$. Thus, if $f \in BV[a, b]$ for each interval $[a, b]$ and $\lim_{x \to -\infty} f(x)$ exists then there is a constant r for which $\lim_{x \to -\infty} (f(x) + r)$ $= 0$. Consequently, we define the class BV_{loc} of functions $f: \mathbf{R} \to \mathbf{R}$ for which $f \in BV[a, b]$ for each interval $[a, b]$ and

$$\lim_{x \to -\infty} f(x) = 0. \tag{4.6}$$

The subclass $BV \subseteq BV_{\mathrm{loc}}$ will consist of all those functions $f \in BV_{\mathrm{loc}}$ for which

$$V(f) = V(f, \mathbf{R}) = \sup \{Vf(x): x \in \mathbf{R}\} < \infty.$$

Obviously,

$$\forall f \in BV_{\mathrm{loc}}, \qquad V(f) = \lim_{x \to \infty} Vf(x).$$

A complex-valued function $f: \mathbf{R} \to \mathbf{C}$ is defined to be an element of (the complex-valued class) BV_{loc} if both $\mathrm{Re}\, f$ and $\mathrm{Im}\, f$ are elements of BV_{loc}. Similarly we define a space BV for complex-valued functions. If $g \in BV[a, b]$ then the function

$$f(x) = \begin{cases} g(x) - g(a), & \text{if } x \in [a, b] \\ 0, & \text{if } x \leqslant a \\ g(b) - g(a), & \text{if } x \geqslant b \end{cases}$$

is an element of BV and $V(f, \mathbf{R}) = V(g, [a, b])$. As such, an element of $BV[a, b]$, properly normalized, can be considered as an element of BV.

Proposition 4.1 BV_{loc} is a vector space and $BV \subseteq BV_{\text{loc}}$ is a subspace.

Example 4.1 Let $f(x) = x^2 \sin(1/x^2)$ if $x \neq 0$ and let $f(0) = 0$. We showed in Example 3.6d that $f \notin BV[-1, 1]$. In this regard note Problem 4.2. Observe that f' exists on $[-1, 1]$ but that $f' \notin L^1_m[-1, 1]$ (e.g. Theorem 4.15).

We denote $BV[0, 2\pi]$ by $BV(\mathbf{T})$; and if $f \in BV(\mathbf{T})$, we denote $(1/(2\pi))V(f, [0, 2\pi])$ by $V(f, \mathbf{T})$. We give the following ingenious proof due to Taibleson [113] of the classical result that

Theorem 4.2 If $f \in BV(\mathbf{T})$ then

$$\forall n \in \mathbf{Z}, \qquad |n\hat{f}(n)| \leqslant V(f, \mathbf{T}).$$

Proof. Note that

$$\forall k \in \mathbf{Z} \text{ and } \forall n \in \mathbf{Z}\backslash\{0\}, \qquad \int\limits_{2\pi k/|n|}^{2\pi(k+1)/|n|} e^{-inx}\, dx = 0. \tag{4.7}$$

When $n \neq 0$, set $a_k = 2\pi k/|n|$ for $k = 0, \ldots, |n|$. Let g be the function

$$g(x) = \sum_{k=1}^{n} f(a_k)\chi_{(a_{k-1}, a_k)}(x).$$

Then by (4.7)

$$\forall n \neq 0, \qquad \hat{g}(n) = 0.$$

Thus, for $n \neq 0$,

$$|\hat{f}(n)| = \frac{1}{2\pi}\left|\int\limits_{0}^{2\pi} (f(x) - g(x))\, e^{-inx}\, dx\right| \leqslant \frac{1}{2\pi}\sum_{k=1}^{|n|} \int\limits_{a_{k-1}}^{a_k} |f(x) - f(a_k)|\, dx$$

$$\leqslant \frac{1}{|n|}\sum_{k=1}^{|n|} V(f, [a_{k-1}, a_k]) = \frac{1}{|n|} V(f, \mathbf{T}),$$

where the final step follows from the additivity of $V(f, [a, b])$.

q.e.d.

Remark The converse to Theorem 4.2 is not true. In fact (e.g. Problem 4.3), there is an element $f \in L_m^1(\mathbf{T})$ such that

$$|n\hat{f}(n)| = o(1), \quad |n| \to \infty,$$

(in particular, $|n\hat{f}(n)| = O(1), |n| \to \infty$) and $f \notin BV(\mathbf{T})$.

Proposition 4.2 *Let* $(\mathbf{R}, \mathcal{B}, \mu)$ *be a finite measure space and set*

$$f(x) = \mu(-\infty, x]. \tag{4.8}$$

Then f *is an increasing element of BV which is continuous from the right at each point* $x \in \mathbf{R}$.

Proof. Clearly, f is increasing and

$$\mu(a, b] = \mu(-\infty, b] - \mu(-\infty, a] = f(b) - f(a).$$

Now $\bigcap_1^\infty (-\infty, -n] = \emptyset$ so that since $f(-1) = \mu(-\infty, -1] < \infty$ we have

$$\lim_{x \to -\infty} f(x) = \lim_{n \to \infty} \mu(-\infty, -n] = 0.$$

Since $\mu\mathbf{R} < \infty$ we conclude that $f \in BV$.

Finally we must show that $f(b) = \lim_{x \to b+} f(x)$ for each $b \in \mathbf{R}$.

We write

$$(-\infty, b] = \bigcap_{n=1}^\infty \left(-\infty, b + \frac{1}{n}\right],$$

so that, by the properties of measures,

$$f(b) = \mu(-\infty, b] = \lim_{n \to \infty} \mu\left(-\infty, b + \frac{1}{n}\right] = \lim_{n \to \infty} f\left(b + \frac{1}{n}\right).$$

q.e.d.

In the following result a) is Proposition 4.2 and the statement of uniqueness is contained in Problem 3.4. We postpone the "hard" part, viz., the existence of μ in part b) (cf. Theorem 5.16 and the remark following it).

Theorem 4.3 *Given the measurable space* $(\mathbf{R}, \mathcal{B})$.

a) *If* μ *is a finite measure on* $(\mathbf{R}, \mathcal{B})$ *then* f *defined by* (4.8) *is an increasing function of bounded variation, continuous from the right.*

b) *If* f *satisfies the conclusions of part* a) *there is a unique finite measure* μ *on* $(\mathbf{R}, \mathcal{B})$ *such that*

$$\forall (a, b] \subseteq \mathbf{R}, \quad \mu(a, b] = f(b) - f(a). \tag{4.9}$$

Remark For perspective note the relation between Theorem 4.3 and the **Riesz representation theorem** (F. Riesz, 1909):

> μ is a continuous linear functional on $C[a, b]$, where $C[a, b]$ is taken with the $\|\ \|_\infty$ norm (see Appendix I.1 and I.2 for definitions), if and only if there is a function $g \in BV[a, b]$ such that   (4.10)

$$\forall f \in C[a, b], \qquad \mu(f) = \int_a^b f \, dg$$

(cf. Appendix III.1).

In (4.10), $\int_a^b f \, dg$ is the Riemann–Stieltjes integral. This theorem has been generalized (by **Bourbaki**) in the most perfect way by becoming the definition of measure; and has been a powerful influence on the formation of modern analysis.

We now give an application of (4.10) (and several other results). The statement (of Proposition 4.3) is true by a simpler proof for the cases that $|\alpha| < 1$ and α is a root of unity, and it is false for $|\alpha| > 1$; the proof need not be assimilated the first time around.

Proposition 4.3 *Let* $|\alpha| = 1$. *There is a non-zero function* $f \in BV[0, 1]$ *such that*

$$\forall x \in [0, 1], \qquad \alpha f(x) = f(x/2) + f((x + 1)/2).$$

Proof. For this result let $C[0, 1]$ be the space of **C**-valued continuous functions g on $[0, 1]$ for which $g(0) = g(1)$. From Sidon's theorem (e.g. Problem 4.4),

$$\sup_{|z|=1} \left| \sum_1^N a_n z^{2^n} \right| \geq \tfrac{1}{4} \sum_1^N |a_n|. \tag{4.11}$$

We'll use this fact to prove that the set

$$S = \{\alpha g(x) - g(2x) : g \in C[0, 1]\}$$

is not dense in $C[0, 1]$; in fact, we'll show that

$$\inf \{\|e - f\|_\infty : f \in S\} \geq \tfrac{1}{4}, \tag{4.12}$$

where $e(x) = \exp 2\pi i x$.
Set $\alpha g(x) - g(2x) = e(x) + \delta(x)$. Then

$$\alpha^2 g(x) - g(2^2 x) = \alpha^2 g(x) - \alpha g(2x) + \alpha g(2x) - g(2^2 x)$$
$$= \alpha(e(x) + \delta(x)) + e(2x) + \delta(2x).$$

Consequently, when we iterate N times,

$$\alpha^N g(x) - g(2^N x) = \sum_{n=1}^N \alpha^{N-n} e(2^{n-1} x) + \sum_{n=1}^N \alpha^{N-n} \delta(2^{n-1} x). \tag{4.13}$$

From (4.13) we compute

$$2\|g\|_\infty \geqslant |\alpha^N g(x) - g(2^N x)| \geqslant \left| \sum_{n=1}^{N} \alpha^{N-n} e(2^{n-1} x) \right| - N\|\delta\|_\infty,$$

so that from (4.11),

$$2\|g\|_\infty \geqslant \tfrac{1}{4} \sum_{n=1}^{N} |\alpha|^{N-n} - N\|\delta\|_\infty = N(\tfrac{1}{4} - \|\delta\|_\infty).$$

Thus,

$$\|\delta\|_\infty \geqslant \frac{1}{4} - \frac{2\|g\|_\infty}{N},$$

and since for any $g \in C[0, 1]$ this is valid for all N, we have (4.12).

We use (4.10), (4.12), and the Hahn–Banach theorem (see Appendix I.8) to conclude that a function $h \in BV[0, 1]$ exists such that

$$\forall g \in C[0, 1], \qquad \int_0^1 (\alpha g(x) - g(2x)) \, dh(x) = 0.$$

Computing

$$\int_0^1 g(2x) \, df(x) = \int_0^1 g(x) \, d\left\{ h\left(\frac{x}{2}\right) + h\left(\frac{x+1}{2}\right) \right\},$$

we obtain

$$\forall g \in C[0, 1], \qquad \int_0^1 g(x) \, d\left\{ \alpha h(x) - h\left(\frac{x}{2}\right) + h\left(\frac{x+1}{2}\right) \right\} = 0;$$

and thus $\alpha h(x) - h(x/2) - h((x+1)/2)$ is a constant c.
Set $f = h - d$, where $d = c/(\alpha - 2)$.

q.e.d.

Proposition 4.4 *Given* $f \in BV[a, b]$.

a) *For each* $c \in [a, b]$, $f(c \pm)$ *exist and* card $D(f) \leqslant \aleph_0$.
b) f *is Lebesgue measurable and bounded.*

Proof. b) is clear from a).
a) $f(c \pm)$ exist from Theorem 4.1.
For each n let $D_n = \{c \in (a, b): |f(c+) - f(c-)| > 1/n\}$.
Clearly, card $D_n < \infty$ since $f \in \overline{BV}[a, b]$. Thus, card $D(f) \leqslant \aleph_0$.

q.e.d.

Example 4.2 We construct a bounded increasing function f on $(0, 1)$ which is discontinuous precisely on the set $\mathbf{Q} \cap (0, 1) = \{r_n : n = 1, ...\}$. Define

$$f_n(x) = \begin{cases} 0, & \text{if} \quad x \in [0, r_n) \\ 1/2^n, & \text{if} \quad x \in [r_n, 1). \end{cases}$$

$f = \sum f_n$ is an increasing function since each f_n is increasing; and $0 \leqslant f \leqslant 1$ from definition. For any fixed r_k, the fact that $\sum_{n \neq k} f_n$ is increasing implies that

$$\lim_{x \to r_k-} \sum_{n \neq k} f_n(x) \leqslant \lim_{x \to r_k+} \sum_{n \neq k} f_n(x);$$

consequently from the definition of $f_k(r_k \pm)$ we have that $f(r_k-) < f(r_k+)$.

An interesting discussion of total variation and related concepts is found in [25].

The following discussion began in Problem 3.10b. For all $f \in L^1_m[a, b]$,

$$\|\tau_{-h} f - f\|_1 = o(1), \qquad |h| \to 0. \tag{4.14}$$

We can show further (e.g. Problem 4.15) that: if $f \in L^1_m[a, b]$ and

$$\|\tau_{-h} f - f\|_1 = o(|h|), \qquad |h| \to 0, \tag{4.15}$$

then f is a constant k, m-a.e. Consider the condition

$$\|\tau_{-h} f - f\|_1 = O(|h|), \qquad |h| \to 0. \tag{4.16}$$

Clearly $(4.15) \Rightarrow (4.16) \Rightarrow (4.14)$.

$(BV[a, b] \subseteq L^1_m[a, b]$ since each element of $BV[a, b]$ is bounded and Lebesgue measurable). We write $f \in BV[a, b]$, m-a.e., if there is a function $g \in BV[a, b]$ such that $g = f$, m-a.e. The following characterization of bounded variation is due to Hardy and Littlewood (1928); we shall use some of basic results from section 4.3–section 4.5 in the proof.

Theorem 4.4 Let $f \in L^\infty_m[a, b]$. $f \in BV[a, b]$, m-a.e. $\Leftrightarrow$ (4.16) *is valid.*

Proof. ($\Rightarrow$) Let $f = g_1 - g_2$, where each g_j is increasing and $f = 0$ on $[a, b]^\sim$. For each $h > 0$,

$$\int_a^b |f(x + h) - f(x)| \, dx \leqslant \int_a^b (g_1(x + h) - g_1(x)) \, dx$$

$$+ \int_a^b (g_2(x + h) - g_2(x)) \, dx$$

$$= \int_b^{b+h} g_1 - \int_a^{a+h} g_1 + \int_b^{b+h} g_2 - \int_a^{a+h} g_2$$

$$\leqslant 2h(g_1(b) + g_2(b)).$$

($\Leftarrow$) Assume (4.16) and let K be the constant associated with the "O" hypothesis. Set

$$\varphi_n(x) = n \int_x^{x+(1/n)} f(t) \, dt.$$

Then, using Fubini's theorem (Appendix II),

$$\int_a^b |\varphi_n(x + h) - \varphi_n(x)| \, dx = n \int_a^b \left| \int_0^{1/n} (f(x + h + t) - f(x + t)) \, dt \right| dx$$

$$\leqslant n \int_0^{1/n} \left(\int_a^b |f(x + h + t) - f(x + t)| \, dx \right) dt$$

$$\leqslant (K + \|f\|_\infty)|h|.$$

Thus,

$$\|\tau_{-h}\varphi_n - \varphi_n\|_1 = O(|h|), \qquad |h| \to 0, \quad \text{uniformly in } n. \tag{4.17}$$

Because of the differentiation theory we'll develop in section 4.3 and 4.4, especially Theorem 4.9, φ_n' exists m-a.e. and $\varphi_n' \in L_m^1[a, b]$. Also, from Fatou's lemma,

$$\int_a^b |\varphi_n'| \leqslant \varliminf_{h \to 0} \int_a^b \left| \frac{\varphi_n(x + h) - \varphi_n(x)}{h} \right| dx,$$

and from (4.17) the right-hand side is uniformly bounded (in n) as h tends to 0. Hence, there is a constant $M > 0$ such that for each n and for each finite collection $\{(x_j, x_j + h_j) : j = 1, \dots, m\}$ of non-overlapping intervals,

$$\sum_{j=1}^m |\varphi_n(x_j + h_j) - \varphi_n(x_j)| \leqslant M; \tag{4.18}$$

this follows since

$$\sum_{j=1}^m |\varphi_n(x_j + h_j) - \varphi_n(x_j)| = \sum_{j=1}^m \left| \int_{x_j}^{x_j+h_j} \varphi_n' \right| \leqslant \int_a^b |\varphi_n'|,$$

by Theorem 4.11.

Using Theorem 4.9 again, we see that $\varphi_n \to f$, m-a.e. Consequently, aside from a set S of Lebesgue measure $mS = 0$ in which the points x_j, $x_j + h_j$ cannot belong,

$$\sum_1^m |f(x_j + h_j) - f(x_j)| \leqslant M.$$

Now define functions $V_S f$, P_S, and N_S analogous to the functions Vf, P, and N, but excluding values from S in the partitions. Then we can extend P_S and N_S to increasing functions on $[a, b]$, and in this way we form an element of $BV[a, b]$ which is equal to f, m-a.e.

q.e.d.

4.2 Decomposition into discrete and continuous parts

Let $f: \mathbf{R} \to \mathbf{R}$ be an increasing function. We shall try to find the "continuous part" of f. If $D(f) = \{x_n : n = 1, ...\}$, define

$$s_n(x) = \begin{cases} 0, & \text{if } x < x_n \\ f(x_n) - f(x_n-), & \text{if } x = x_n \\ f(x_n+) - f(x_n-), & \text{if } x > x_n. \end{cases}$$

Clearly, s_n is a non-negative increasing function. (The "s" is for "saltus", which, in turn, is for old times' sake.)

Proposition 4.5 *Assume that* $f: \mathbf{R} \to \mathbf{R}$ *is an increasing function, and let* $D(f) = \{x_n : n = 1, ...\}$. *Define*

$$g_n = f - \sum_{k=1}^{n-1} s_k.$$

Then
a) g_n *is an increasing function.*
b) g_n *is continuous on the set* $C(f) \cup \{x_1, ..., x_{n-1}\}$.

Proof. a) If $x < y < x_1$ then $g_1(y) = f(y) \geqslant f(x) = g_1(x)$. If $x < x_1$ then $g_1(x_1) = f(x_1) - (f(x_1) - f(x_1-)) = f(x_1-) \geqslant f(x) = g_1(x)$.
For $x_1 < x, g_1(x) = (f(x) - f(x_1+)) + f(x_1-) \geqslant f(x_1-) = g_1(x_1)$.
Finally, when $x_1 < x < y$, $g_1(y) = f(y) - (f(x_1+) - f(x_1-)) \geqslant f(x) - (f(x_1+) - f(x_1-)) = g_1(x)$.
Consequently, g_1 is increasing; and for arbitrary g_n the result follows by induction.
b) It is sufficient to check that for $n > 1$

$$\lim_{x \to x_1 \pm} g_n(x) = g_n(x_1).$$

Note that $g_n(x_1) = f(x_1-) + \sum_{2}^{n-1} s_j(x_1)$. We'll check that

$$\lim_{x \to x_1 -} g_n(x) = g_n(x_1);$$

a similar argument works for $x \to x_1 +$.

Clearly,

$$\lim_{x \to x_1-} g_n(x) = f(x_1-) - \lim_{x \to x_1-} s_1(x) - \lim_{x \to x_1-} \sum_{2}^{n-1} s_j(x);$$

in this case $s_1(x) = 0$ since $x < x_1$, and

$$\lim_{x \to x_1-} \sum_{2}^{n-1} s_j(x) = \sum_{2}^{n-1} s_j(x_1)$$

by the definition of s_j.

q.e.d.

Remark Assume that $f: \mathbf{R} \to \mathbf{R}$ is an element of BV. With the previous notation observe that $\sum s_j(x)$ converges. In fact, if all but finitely many x_n are greater than x then the sum is finite and if infinitely many x_n are less than x we have convergence by the definition of s_j and the fact that f increases. For example, if $x_n < x$ for each n then

$$\sum s_j(x) = \sum (f(x_j+) - f(x_j-)),$$

and the second sum is finite since $Vf(x) < \infty$. Besides the pointwise convergence we also have that $\sum s_j$ converges uniformly on any closed bounded interval $[a, b]$. To verify this latter assertion note that if $s = \sum s_j$ and $\varepsilon > 0$ is given, then

$$\exists N \text{ such that } \forall n > N, \quad \left| \sum_n^\infty s_j(a) \right| < \varepsilon \quad \text{and} \quad \left| \sum_n^\infty s_j(b) \right| < \varepsilon.$$

Consequently, for all $x \in [a, b]$ and for all $n > N$

$$-\varepsilon < \sum_n^\infty s_j(a) \leqslant \sum_n^\infty s_j(x) = s(x) - \sum_1^{n-1} s_j(x) \leqslant \sum_n^\infty s_j(b) < \varepsilon.$$

If f is an increasing element of BV then $s = \sum s_j$ is the discrete (or discontinuous part of f. If $f \in BV$ and $f = P - N$, with notation as in Theorem 4.1, the discrete part of f is

$$s = s_P - s_N,$$

where s_P (resp. s_N) is the discrete part of P (resp. N).

From Proposition 4.5 and the above remark we conclude with
Theorem 4.5 *Let $f \in BV$ have discrete part s. Then*

$$g = f - s \in BV$$

is a continuous function on $\mathbf{R}$. *Further, if f is an increasing function then g is an increasing function, and if f is a continuous function then $s = 0$.*

The function g in Theorem 4.5 is the continuous part of f. In this regard and taking f and μ as in Proposition 4.2, note that

$$\forall x \in \mathbf{R}, \ \mu\{x\} = \lim_{n \to \infty} \mu\left(x - \frac{1}{n}, x\right] = f(x) - f(x-),$$

and so $f \in BV$ *is continuous at x if and only if $\mu\{x\} = 0$.* Also note that the function f in Example 4.2 is a discrete function which is discontinuous precisely on the rational numbers in $[0, 1]$.

Example 4.3 Let $E \subseteq [0, 1]$ be a perfect symmetric set determined by $\{\xi_k : k = 1\} \subseteq (0, \frac{1}{2})$ and let C_E be its associated Cantor function (e.g. Example 1.3). If $mE = 0$ then $C_E' = 0$, *m-a.e.* C_E has no discrete part and the measure μ_E associated with C_E (by Theorem 4.3b) is the Cantor–Lebesgue continuous measure for E. In light of the following example note that C_E takes constant values on a set of positive measure, and, in particular, it is not a strictly increasing function.

Example 4.4 (cf. Example 4.9) We now present Hellinger's example (1907). This is a continuous, strictly increasing function $H : [0, 1] \to \mathbf{R}$ such that $H' = 0$, *m-a.e.* H will, in fact, depend on a fixed $t \in (0, 1)$ and we sometimes write $H = H_t$. H will be of the form $\lim_{n \to \infty} H_n$ and we begin by setting $H_0(x) = x$ on $[0, 1]$. Assume that H_{n-1} is constructed. We shall take H_n to be continuous and linear in each interval $[k/2^n, (k + 1)/2^n]$, $k = 0, ..., 2^n - 1$. Besides that, we set

$$H_n\left(\frac{k}{2^{n-1}}\right) = H_{n-1}\left(\frac{k}{2^{n-1}}\right), \qquad k = 0, \cdots, 2^{n-1},$$

and for our fixed $t \in (0, 1)$ we define

$$H_n\left(\frac{2k + 1}{2^n}\right) = \frac{1 - t}{2} H_{n-1}\left(\frac{k}{2^{n-1}}\right) + \frac{1 + t}{2} H_{n-1}\left(\frac{k + 1}{2^{n-1}}\right),$$

$k = 0, ..., 2^{n-1} - 1$. Note that $(2k + 1)/2^n$ is the midpoint of $[k/2^{n-1}, (k + 1)/2^{n-1}]$. For example, take $t = \frac{1}{2}$ and observe that $H_1(\frac{1}{2}) = \frac{3}{4}$ (e.g. Fig. 4). From this construction we see that H_n is continuous and strictly increasing, and that

$$\forall x \in [0, 1], \qquad 0 \leqslant H_n(x) \leqslant H_{n+1}(x) \leqslant 1.$$

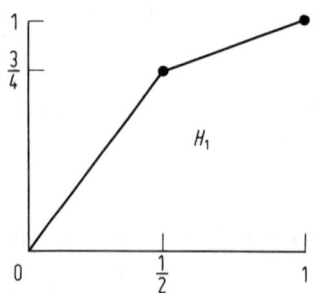

Fig. 4

Consequently, $H = \lim H_n$ exists pointwise and is an increasing function. Now let $x < y$ and choose an n and a corresponding $k \in \{0, \ldots, 2^n - 1\}$ such that

$$x < k/2^n < y.$$

Then $\qquad H(x) \leqslant H(k/2^n) = H_n(k/2^n) < H_n(y) \leqslant H(y),$

so that H is *strictly increasing*. We now prove that H is continuous at each $x \in [0, 1]$. Take

$$\{x\} = \bigcap [\alpha_n, \beta_n], \qquad [\alpha_n, \beta_n] = [k_n/2^n, (k_n + 1)/2^n],$$

where $[\alpha_{n+1}, \beta_{n+1}] \subseteq [\alpha_n, \beta_n]$. Clearly, $\alpha_{n+1} = \alpha_n$ and $\beta_{n+1} = \beta_n - (1/2^{n+1})$, or $\alpha_{n+1} = \alpha_n + (1/2^{n+1})$ and $\beta_{n+1} = \beta_n$.
In the first case,

$$
\begin{aligned}
H(\beta_{n+1}) - H(\alpha_{n+1}) &= H_{n+1}(\beta_{n+1}) - H_{n+1}(\alpha_{n+1}) \\
&= H_{n+1}(\beta_{n+1}) - H_n(\alpha_n) \\
&= \frac{1-t}{2} H_n(\alpha_n) + \frac{1+t}{2} H_n(\beta_n) - H_n(\alpha_n) \\
&= \frac{1+t}{2} H_n(\beta_n) - \frac{1+t}{2} H_n(\alpha_n);
\end{aligned}
\tag{4.19}
$$

for the second case the computation yields "$(1 - t)/2$" instead of "$(1 + t)/2$", and so in either situation we have

$$H(\beta_{n+1}) - H(\alpha_{n+1}) = \frac{1 \pm t}{2}(H_n(\beta_n) - H_n(\alpha_n)).$$

Continuing this process we obtain

$$H(\beta_{n+1}) - H(\alpha_{n+1}) = \prod_{j=1}^{n} \frac{1 + \varepsilon_j t}{2}, \qquad \varepsilon_j = \pm 1, \tag{4.20}$$

and hence

$$|H(\beta_{n+1}) - H(\alpha_{n+1})| < \left(\frac{1+t}{2}\right)^n, \tag{4.21}$$

since $1 + \varepsilon_j t \leqslant 1 + t$. (As a gift, (4.20) also yields another proof that H is strictly increasing). The *continuity of H at x follows* from (4.21) which in fact tells us that $H(x+) = H(x-)$. Now, $H'(x)$ exists *m-a.e.* as we shall prove in Theorem 4.8. For any such x choose $\{[\alpha_n, \beta_n]\}$ as before, and observe that

$$\frac{H(\beta_{n+1}) - H(\alpha_{n+1})}{\beta_{n+1} - \alpha_{n+1}} = 2^{n+1} \prod_{1}^{n} \frac{1 + \varepsilon_j t}{2} = 2 \prod_{1}^{n}(1 + \varepsilon_j t), \qquad \varepsilon_j = \pm 1.$$

The product $\prod_1^\infty (1 + \varepsilon_j t)$ diverges to 0 since we are assuming that $H'(x)$ exists, and so $H'(x) = 0$, *m-a.e.*

A similar function was introduced by Minkowski in 1912 in his study of quadratic irrationals. Further examples have been given by Jessen, Salem, Wiener, and Wintner (e.g. [98, pp. 282–294]).

The following result is straightforward to prove (cf. Problem 4.16).

Proposition 4.6 Given $f \in BV$. If f is continuous at x then Vf is continuous at x.

Because of Proposition 4.6, if $f \in BV$ is continuous then the corresponding functions P and N are continuous.

We also omit the details of

Proposition 4.7 Let $f: [a, b] \to$ **R** be a continuous function. Then

$$\lim_{|P|\to 0} \sum |f(x_j) - f(x_{j-1})| = V(f, [a, b]),$$

where $V(f, [a, b])$ may be infinite and $|P|$ is defined as $\max\{|x_j - x_{j-1}|: j = 1, ..., n\}$ for the partition $P: a = x_0 \leqslant x_1 \leqslant ... \leqslant x_n = b$.

In order to determine sets of continuity in Proposition 1.7 we dealt with the oscillations $\omega(f, I_j)$, $I_j = [x_j, x_{j+1}]$. If f is continuous on $[a, b]$ we can easily check, using Proposition 4.7 (and the notation there), that

$$\lim_{|P|\to 0} \sum_{j=0}^{n-1} \omega(f, I_j) = V(f, [a, b]).$$

We shall now give a criterion in order that a continuous function be a function of bounded variation. Let $f: [a, b] \to$ **R** be continuous and define a function

$$\forall y \in \mathbf{R}, \qquad R_y = \{x \in [a, b]: f(x) = y\}.$$

The Banach indicatrix $B: \mathbf{R} \to \mathbf{R}^+ \cup \{\infty\}$ is defined as

$$\forall y \in \mathbf{R}, \qquad B(y) = \text{card } R_y.$$

Theorem 4.6 (*Banach–Vitali*) Let $f: [a, b] \to$ **R** be a continuous function.
a) B is Lebesgue measurable and

$$\int_c^d B(y)\, dy = V(f, [a, b]),$$

where $c = \inf\{f(x): x \in [a, b]\}$ and $d = \sup\{f(x): x \in [a, b]\}$.
b) $f \in BV[a, b] \Leftrightarrow B \in L_m^1[c, d]$.
c) If $f \in BV[a, b]$ then $m\{y: \text{card } R_y \geqslant \aleph_0\} = 0$.

Proof. b) is clear from a), and c) is clear from b).

a) Given the partition $P_n: a = x_0 < a + (b - a)/2^n = x_1 < ... < a + j(b - a)/2^n = x_j < ... < x_{2^n} = b$, we have $|P_n| = (b - a)/2^n$. "$|P_n|$" is defined in the statement of Proposition 4.7. For each $j = 1, ..., 2^n$ define

$$\forall y \in \mathbf{R}, \quad S_j(y) = \begin{cases} 1, & \text{if} \quad \exists x \in (x_{j-1}, x_j] \text{ such that } f(x) = y, \\ 0, & \text{if} \quad \forall x \in (x_{j-1}, x_j], f(x) \neq y. \end{cases}$$

Observe that S_j is a bounded Lebesgue measurable function (there are at most two points of discontinuity), and that

$$\int_c^d S_j(y) \, dy = \omega(f, J_j), \quad \text{where} \quad J_j = (x_{j-1}, x_j].$$

Define $B_n = \sum_1^{2^n} S_j$. From the above remarks,

$$\lim_{n \to \infty} \int_c^d B_n(y) \, dy = V(f, [a, b]).$$

Now, $\{B_n : n = 1, ...\}$ is an increasing sequence so that by LDC

$$\int_c^d \tilde{B} \, dy = \lim_{n \to \infty} \int_c^d B_n \, dy,$$

where $\tilde{B}$ is the pointwise limit of the sequence $\{B_n : n = 1, ...\}$.

Clearly $\tilde{B} \leqslant B$ since $B_n \leqslant B$ and so we shall complete the proof of a) once we prove that $B \leqslant \tilde{B}$.

Let $m \leqslant B(y)$ and choose $z_1 < z_2 < ... < z_m$ where each $f(z_k) = y$. Take n large enough so that

$$\forall k = 2, \cdots, n, \quad \frac{b - a}{2^n} < z_k - z_{k-1}.$$

Consequently, no two z_k's will be in the same $[x_{j-1}, x_j]$ for the partition P_n. Thus $B_n(y) \geqslant m$ and so $\tilde{B}(y) \geqslant m$. Since this is true for each $m \leqslant B(y)$ we conclude that $B(y) \leqslant \tilde{B}(y)$.

q.e.d.

Banach is usually given full credit for the above result; the relevant Vitali paper is [123].

4.3 The Lebesgue differentiation theorem

In 1904, Lebesgue proved the differentiation theorem: *continuous functions of bounded variation have a derivative m-a.e.* In 1907, while trying to extend the fundamental theorem of calculus (FTC) to $\mathbf{R}^2$, Vitali proved what is now known as the Vitali covering theorem. This became the basis for proving general differentiation theorems. In 1932, F. Riesz gave a different proof of Lebesgue's result using the so-called "rising sun lemma"; the extension to all functions of bounded variation was easy. Riesz's proof is found in [4; 89]. Riesz's approach to prove the result without measure theory was anticipated by Faber (1910) and the Youngs (1911). L. W. Cohen [28] has recently made a simplification of Riesz's technique and used it to obtain the differentiation theorem for df/dg where f and g are increasing; this is interesting since $g(x) = x$ is the usual result and because it provides a means to compute a Radon–Nikodym derivative without using the integration theory centered about the Radon–Nikodym theorem (cf. Problem 5.33). Vitali's approach has been studied extensively and we refer to [22] for an important survey. There are also proofs of the differentiation theorem by D. Austin [5] and G. Letta [72]. None of the proofs is transparent and Letta's seems simpler than most. Vitali's proof is long but readable, and we shall give Banach's proof [6] of Vitali's theorem. Banach's technique is brilliant, and the expositions of it in [56; 92] are perfect; as such we have no other choice but to engage in a copying exercise.

A collection $\mathscr{V}$ of intervals is a Vitali covering of $X \subseteq \mathbf{R}$ if

$$\forall \varepsilon > 0 \text{ and } \forall x \in X \; \exists I \in \mathscr{V}, \text{ for which } mI < \varepsilon, \text{ such that } x \in I.$$

Vitali's result below does not preclude the possibility that if $X \subseteq \bigcup_{1}^{N} I_n$ then $m^* \left(\bigcup_{1}^{N} I_n \setminus X \right)$ is large.

Theorem 4.7 (*Vitali*) *Given $X \subseteq \mathbf{R}$ with $m^* X < \infty$, and let $\mathscr{V}$ be a Vitali covering of X. Then for each $\varepsilon > 0$ there is a disjoint family $\{I_n : n = 1, ..., N\} \subseteq \mathscr{V}$, such that*

$$m^* \left(X \setminus \bigcup_{1}^{N} I_n \right) < \varepsilon. \tag{4.22}$$

Proof. i) Without loss of generality we can take each interval $I \in \mathscr{V}$ to be closed since

$$m^* \left(X \setminus \bigcup_{1}^{N} I_n \right) = m^* \left(X \setminus \bigcup_{1}^{N} \bar{I}_n \right).$$

Also without loss of generality let U be an open set, with $mU < \infty$, such that

$$\forall I \in \mathscr{V}, \quad I \subseteq U.$$

We can do this since $m^*X < \infty$ and $\mathscr{V}$ is a Vitali covering, even though we may be throwing away some of our original elements from $\mathscr{V}$.

ii) We now choose a disjoint family $\{I_n : n = 1, ...\} \subseteq \mathscr{V}$ which "almost" covers X (in the following way).

Take any $I_1 \in \mathscr{V}$ and assume that a disjoint family $\{I_1, ..., I_n\} \subseteq \mathscr{V}$ has been chosen.

If $X \subseteq \bigcup_1^N I_j$ we stop the procedure. Assume otherwise.

Define $r_n = \sup \{mI : I \cap I_j = \emptyset, j = 1, ..., n, I \in \mathscr{V}\}$, noting that

$$\forall n, \qquad r_n \leqslant mU < \infty.$$

r_n is positive since $\mathscr{V}$ is a Vitali covering and $\bigcup_1^n I_j$ does not cover X.

Thus we can take $I_{n+1} \in \mathscr{V}$ for which

$$mI_{n+1} > \tfrac{1}{2}r_n$$

and $\qquad \forall j = 1, ..., n, \qquad I_{n+1} \cap I_j = \emptyset.$

Obviously, without loss of generality, we can choose $\{mI_j : j = 1, ...\}$ to be a decreasing sequence.

iii) Using ii) we see that

$$\exists N > 0 \text{ such that } \qquad \sum_{N+1}^{\infty} mI_j < \varepsilon/5;$$

this follows since $\{I_j : j = 1, ...\}$ is a disjoint family and $mU < \infty$. We'll prove that

$$m^*\left(X \backslash \bigcup_1^N I_j\right) < \varepsilon.$$

iv) If $y \in X \backslash \bigcup_1^N I_j$ we can choose $I_y \in \mathscr{V}$, such that $y \in I_y$ and

$$\forall j = 1, ..., N, \qquad I_y \cap I_j = \emptyset; \tag{4.23}$$

this follows from the definition of $\mathscr{V}$ (including the hypothesis that each of its elements is a closed interval).

Next observe that $I_y \cap I_n \neq \emptyset$ for some n. To see this, take any n (larger than N in light of (4.23)) and assume that $I_y \cap I_j = \emptyset$ for each $j \leqslant n$. Then, from the definition of r_n and the construction of $\{I_j : j = 1, ...\}$,

$$mI_y \leqslant r_n < 2mI_{n+1}. \tag{4.24}$$

Since $\sum mI_j < \infty$, we have $mI_j \to 0$; and thus $I_y \cap I_j \neq \emptyset$ for some j.

v) Let $n = n(y)$ be the smallest integer $(n > N)$ for which $I_y \cap I_n \neq \emptyset$, where $y \in X \backslash \bigcup_1^N I_j$, and let x_n be the midpoint of I_n. We shall prove that

$$|y - x_n| \leqslant \tfrac{5}{2}mI_n. \tag{4.25}$$

Clearly (from Fig. 5, for example)

$$|y - x_n| \leqslant mI_y + \tfrac{1}{2}mI_n. \tag{4.26}$$

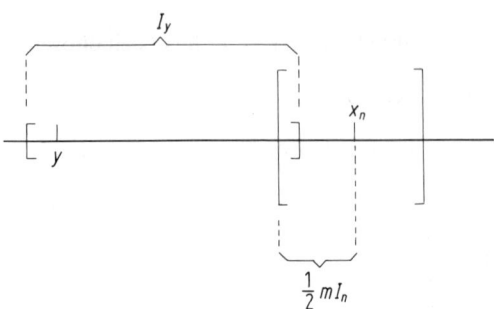

$$\tfrac{1}{2}mI_n \qquad\qquad \text{Fig. 5}$$

Because of (4.24) and the fact that $\{mI_j : j = 1, ...\}$ decreases, we obtain (4.25) from (4.26).

vi) For each n let J_n be the closed interval with center x_n and length five times that of I_n. From (4.25), $y \in J_n$. Since $n = n(y) > N$ and y is arbitrary in $X \setminus \bigcup_1^N I_j$, we have

$$X \setminus \bigcup_1^N I_j \subseteq \bigcup_{N+1}^{\infty} J_n$$

and

$$m\left(\bigcup_{N+1}^{\infty} J_n \right) \leqslant \sum_{N+1}^{\infty} mJ_n \leqslant 5 \sum_{N+1}^{\infty} mI_n < \varepsilon.$$

This yields (4.22).

q.e.d.

The Dini derivates of a function $f: \mathbf{R} \to \mathbf{R}$ are

$$D^+ f(x) = \overline{\lim_{h \to 0+}} \, \frac{f(x + h) - f(x)}{h},$$

$$D_+ f(x) = \underline{\lim_{h \to 0+}} \, \frac{f(x + h) - f(x)}{h},$$

$$D^- f(x) = \overline{\lim_{h \to 0-}} \, \frac{f(x + h) - f(x)}{h} = \overline{\lim_{h \to 0+}} \, \frac{f(x) - f(x - h)}{h},$$

$$D_- f(x) = \underline{\lim_{h \to 0-}} \, \frac{f(x + h) - f(x)}{h} = \underline{\lim_{h \to 0+}} \, \frac{f(x) - f(x - h)}{h}.$$

These numbers always exist and it is obvious that

i) $D^+ f(x) \geqslant D_+ f(x), \qquad D^- f(x) \geqslant D_- f(x),$

ii) $\exists f'(x) \Leftrightarrow D^+ f(x) = ... = D_- f(x) \neq \pm\infty.$

Now for the differentiation theorem.

Theorem 4.8 (*Lebesgue differentiation theorem*) Let $f \in BV[a, b]$. *Then*
a) f' *exists m-a.e.*
b) $f' \in L_m^1[a, b]$.

c) *If f is increasing then* $\int_a^b f' \leqslant f(b) - f(a)$.

Proof. Without loss of generality we assume that f is increasing.
i) To prove a) we must show that the set of points where any two derivates are unequal has Lebesgue measure 0. We'll do the computation for

$$A - \{x : D^+f(x) > D_-f(x)\},$$

proving that

$$r = m^* A_{a, b} = 0, \tag{4.27}$$

where $\quad A_{a, b} = \{x : D^+f(x) > a > b > D_-f(x)\}, \quad a, b \in \mathbf{Q}.$
It is sufficient to prove (4.27) since $A = \bigcup \{A_{a,b} : a, b \in \mathbf{Q}\}$.
In order to verify (4.27) we assume $r \neq 0$ and show that

$$\forall \varepsilon > 0, \quad b(r + \varepsilon) > a(r - 2\varepsilon); \tag{4.28}$$

this tells us that $b \geqslant a$, the desired contradiction.
ii) We construct a Vitali covering $\mathscr{V}$ of $A_{a, b}$.
Let U be an open set containing $A_{a, b}$ such that $mU < r + \varepsilon$.
For each $x \in A_{a, b}$ choose $h > 0$ for which

$$[x - h, x] \subseteq U, \quad f(x) - f(x - h) < bh; \tag{4.29}$$

this can be done from the definition of $A_{a, b}$ and because U is open.
$\mathscr{V}$ is the collection of all possible intervals $[x - h, x]$, where $x \in A_{a, b}$ and h satisfies (4.29).
$\mathscr{V}$ is obviously a Vitali covering of $A_{a, b}$ and so from Theorem 4.7 there is a disjoint family $\{I_j = [x_j - h_j, x_j] \in \mathscr{V} : j = 1, ..., N\}$ satisfying

$$m^* \left(A_{a, b} \backslash \bigcup_1^N I_j \right) < \varepsilon.$$

iii) Let

$$B = A_{a, b} \cap \left(\bigcup_1^N I_j \right),$$

and observe that

$$\left(A_{a,b}\backslash\bigcup_1^N I_j\right) \cup B = A_{a,b};$$

thus, from ii),

$$m^*A_{a,b} < \varepsilon + m^*B. \tag{4.30}$$

We now proceed to define a Vitali covering $\mathscr{U}$ of B in terms of $\{I_1,\ldots, I_N\}$.
For each $y \in B$ there is $j \in \{1,\ldots, N\}$ such that $y \in I_j$; choose $k \in \mathbf{R}$ such that $[y, y + k] \subseteq I_j$ and

$$f(y + k) - f(y) > ak. \tag{4.31}$$

(4.31) is possible by the definition of $A_{a,b}$.
$\mathscr{U}$ is defined to be the collection of intervals $[y, y + k]$, for $y \in B$ and $k \in \mathbf{R}$, for which $[y, y + k]$ is contained in some $I_j, j = 1,\ldots, N$, and (4.31) is satisfied.
$\mathscr{U}$ is obviously a Vitali covering of B and so from Theorem 4.7 there is a disjoint family $\{J_j = (y_j, y_j + k_j): j = 1,\ldots, M\}$ satisfying

$$m^*\left(B\backslash\bigcup_1^M J_j\right) < \varepsilon.$$

iv) Let

$$D = B \cap \left(\bigcup_1^M J_j\right)$$

and observe that

$$B = \left(B\backslash\bigcup_1^M J_j\right) \cup D;$$

thus $m^*B < \varepsilon + m^*D,$

which when combined with (4.30) and (4.31), yields

$$a(r - 2\varepsilon) < am^*D \leqslant am^*\left(\bigcup_1^M J_j\right) = a\sum_1^M k_j < \sum_1^M (f(y_j + k_j) - f(y_j)). \tag{4.32}$$

v) From the definition of $\mathscr{U}$, each $J_j, i = 1,\ldots, M$, is contained in some $I_i, i = 1,\ldots, N$. Hence, for each fixed $n = 1,\ldots, N$,

$$\sum_{J_j \subseteq I_n} (f(y_j + k_j) - f(y_j)) \leqslant f(x_n) - f(x_n - h_n),$$

since f is increasing and $\{J_j: j = 1,\ldots, M\}$ is a disjoint family.

This observation combined with (4.32) and (4.29) implies (4.28):

$$a(r - 2\varepsilon) \leqslant b \sum_1^N h_j \leqslant bmU < b(r + \varepsilon).$$

a) is complete.

b) and c). Define

$$g_n(x) = n\left[f\left(x + \frac{1}{n}\right) - f(x)\right],$$

where we define $f(x) = f(b)$ if $x \geqslant b$.

From part a), $g_n \to f'$ pointwise m-a.e.

g_n, f', and $|f'|$ are measurable since f is measurable.

By Fatou's lemma and the fact that f is increasing,

$$\int_a^b |f'| \leqslant \underline{\lim} \int_a^b |g_n| = \underline{\lim} \int_a^b g_n = \underline{\lim}\left[n \int_b^{b+(1/n)} f - n \int_a^{a+(1/n)} f\right]$$

$$\leqslant \underline{\lim}\,[f(b) - f(a)].$$

q.e.d.

Proposition 4.8 *Let $P: \mathbf{Z}^+ \setminus \{0\} \to \mathbf{R}^+$ be a function and let A be a set of irrational numbers x such that*

$$|x - (p/q)| < 1/(qP(q)), \qquad (p, q) = 1,$$

for infinitely many integers $q > 1$. If

$$\sum_{q=1}^{\infty} 1/P(q) < \infty$$

then $mA = 0$.

Proof. Take $A \subseteq [0, 1]$ and define

$$f(x) = \begin{cases} 0, & \text{if } x \text{ is irrational in } [0, 1] \\ 1/(qP(q)), & \text{if } x = p/q \in [0, 1], \text{ where } (p, q) = 1. \end{cases}$$

We first prove that $f \in BV[0, 1]$.

Observe that for each $q > 1$ there are fewer than q rational numbers $p/q \in [0, 1]$, where $(p, q) = 1$.

Take any partition $0 = x_0 < x_1 < \ldots < x_n = 1$, and observe that if each $x_j = p_j/q$ then $\sum |f(x_j)| < 1/P(q)$. This is the "worst" possible situation in the sense that no matter what the x_j are,

$$\sum_1^n |f(x_j) - f(x_{j-1})| \leqslant 2 \sum_0^n |f(x_j)| \leqslant 2 \sum_{q=1}^m 1/P(q) \leqslant 2 \sum_1^{\infty} 1/P(q) < \infty.$$

Thus $f \in BV[0, 1]$ and by the Lebesgue differentiation theorem, f' exists m-a.e. Consequently, f' exists for almost all irrationals. Take such an irrational x. Then $f'(x) = 0$ since in the difference quotient we can approximate x by irrationals. Observe that in this case

$$\frac{f(p/q) - f(x)}{(p/q) - x} = \frac{1}{qP(q)((p/q) - x)},$$

and so for $\varepsilon = 1$ there is a (largest possible) $\delta(1)$ for which

$$\frac{1}{qP(q)} < \left| \frac{p}{q} - x \right|, \tag{4.33}$$

when $|(p/q) - x| < \delta(1)$.

Clearly there are only finitely many q with the property that

$$\left| \frac{p}{q} - x \right| \geq \delta(1)$$

because, for such a q, (4.33) implies

$$\frac{1}{qP(q)} \geq \left| \frac{p}{q} - x \right| \geq \delta(1) > 0,$$

and the left-hand side tends to 0.

Therefore $x \notin A$, and so an irrational number $x \in [0, 1]$ is in A only if $f'(x)$ does not exist; and this latter possibility can occur only on a set of Lebesgue measure 0. Thus, $mA = 0$.

q.e.d.

4.4 FTC-I

If $f \in L_m^1[a, b]$, we set

$$\forall x \in [a, b], \qquad F(x) = \int_a^x f$$

for the next two results.

Proposition 4.9 *If $f \in L_m^1[a, b]$ then F is a continuous function of bounded variation.*

Proof. The fact that F is continuous follows immediately from Proposition 3.2. For the bounded variation observe that

$$\sum_1^n |F(x_j) - F(x_{j-1})| \leq \sum_1^n \int_{x_{j-1}}^{x_j} |f| = \int_a^b |f| < \infty.$$

q.e.d.

Once we define absolute continuity of point functions we shall see immediately (using Proposition 3.2) that F is absolutely continuous; this notion is stronger than that of bounded variation and continuity.

Proposition 4.10 *Given* $f \in L^1_m[a, b]$ *and assume that* F *is identically* 0. *Then* $f = 0$, *m-a.e.*

Proof. Without loss of generality assume that $f > 0$ on a set A of positive Lebesgue measure $mA > 0$.

By the properties of Lebesgue measure, there is a closed set $K \subseteq A$ such that $mK > 0$. Let $U = [a, b] \setminus K$.

Then
$$0 = \int_a^b f = F(b) = \int_K f + \int_U f$$

and so
$$\int_U f = - \int_K f < 0.$$

$U = \bigcup (a_n, b_n)$, a disjoint union of open intervals, so that by LDC,

$$0 \neq \int_U f = \Sigma \int_{a_n}^{b_n} f,$$

and consequently

$$\int_{a_n}^{b_n} f \neq 0$$

for some n.

Thus, either $\int_a^{a_n} f$ or $\int_a^{b_n} f$ is non-zero, and this contradicts the hypothesis on F.

q.e.d.

Theorem 4.9 (FTC—I) *Let* $f \in L^1_m[a, b]$ *and take* $r \in \mathbf{R}$.
Define the function $F: [a, b] \to \mathbf{R}$ *as*

$$F(x) = r + \int_a^x f$$

(*so that* $F(a) = r$). *Then*

$$F' = f, \text{ m-a.e.}$$

Proof. We consider the case that $\|f\|_\infty < \infty$. The unbounded case is left as an exercise (Problem 4.18).

$F \in BV[a, b]$ from Proposition 4.9 and so F' exists *m-a.e.* because of Theorem 4.8.

Define $F(x) = F(b)$ for $x \geq b$ and set

$$f_n(x) = (F(x + (1/n)) - F(x))/(1/n) = n \int_x^{x+(1/n)} f.$$

Consequently, $f_n \to F'$, m-a.e. Since $\|f\|_\infty < \infty$ we can employ LDC to obtain

$$\int_a^c F' = \lim_{n \to \infty} \int_a^c f_n = \lim_{h \to 0} \frac{1}{h} \int_a^c (F(x + h) - F(x)) \, dx$$

$$= \lim_{h \to 0} \frac{1}{h} \left[\int_c^{c+h} F - \int_a^{a+h} F \right] = F(c) - F(a) = \int_a^c f$$

for each $c \in [a, b]$. The penultimate equality follows by the continuity of F and the observation that

$$|F(c) - \frac{1}{h} \int_c^{c+h} F| = |\frac{1}{h} \int_c^{c+h} (F(c) - F)| \leq \sup_{x \in [c, c+h]} |F(c) - F(x)|.$$

Thus,

$$\forall c \in [a, b], \quad \int_a^c (F' - f) = 0,$$

and we can apply Proposition 4.10 to conclude that $F' = f$, m-a.e.

q.e.d.

Example 4.5 Let $f(x) = \sin(1/x)$ for $x > 0$ and let $f(0) = 0$; define the function

$$F(x) = \begin{cases} 0, & \text{if } x = 0 \\ x^2 \cos(1/x) - 2 \int_0^x t \cos(1/t) \, dt, & \text{if } x > 0. \end{cases}$$

Then $F' = f$ in $[0, \infty)$.

Proposition 4.11, below, is a useful modification of the following advanced calculus result: assume that the functions

$$f_n, g: (a, b) \to \mathbf{R}, \quad n = 1, \ldots,$$

have the properties that the sequence $\{f_n(x_0): n = 1, \ldots\}$ converges for some $x_0 \in (a, b)$ and $f_n' \to g$ uniformly on (a, b); then there is a function $f: (a, b) \to \mathbf{R}$ such that
i) $f_n \to f$ uniformly on (a, b),
and
ii) $\forall x \in (a, b), \exists f'(x) = g(x)$.

Proposition 4.11 *Given the functions* $f_n, f, g: (a, b) \to \mathbf{R}$, $n = 1, ...$, *where* g *and each* f_n' *are continuous on* (a, b). *Assume that for all* $x \in (a, b)$

$$\lim_{n \to \infty} f_n(x) = f(x) \quad \text{and} \quad \lim_{n \to \infty} f_n'(x) = g(x).$$

Then $f' = g$, *m-a.e., on* (a, b), *and if* f' *is continuous then* $f' = g$ *on* (a, b).
The proof of Proposition 4.11 is left as an exercise (Problem 4.23).

Let's look at the differentiation of indefinite integrals as an averaging procedure. Thus, from Theorem 4.9, if $f \in L_m^1[a, b]$ and

$$F(x) = \int_a^x f,$$

we have $F' = f$, *m-a.e.* and so

$$\lim_{h \to 0} \frac{1}{h} \int_x^{x+h} (f(t) - f(x)) \, dt = 0, \quad m\text{-a.e.} \tag{4.34}$$

Now (4.34) could be valid due to the cancellation caused by change of sign in the integration; or the stronger result

$$\lim_{h \to 0} \frac{1}{h} \int_x^{x+h} |f(t) - f(x)| \, dt = 0, \quad m\text{-a.e.} \tag{4.35}$$

could be true. In fact, (4.35) is true and is a corollary of the following even more general theorem.

Theorem 4.10 *Let* $f \in L_m^1[a, b]$ *be real valued. Then there is a subset* $L \subseteq [a, b]$, *for which* $mL = b - a$, *such that*

$$\forall r \in \mathbf{R} \quad \text{and} \quad \forall x \in L, \quad \lim_{h \to 0} \frac{1}{h} \int_x^{x+h} |f(t) - r| \, dt = |f(x) - r|.$$

Proof. Let $\{r_n : n = 1, ...\} \subseteq \mathbf{R}$ be dense, and define the elements $g_n \in L_m^1[a, b]$ as

$$g_n(x) = |f(x) - r_n|.$$

Because of Theorem 4.9, for each n there is a subset $L_n \subseteq [a, b]$, for which $mL_n = b - a$, such that

$$\forall x \in L_n, \quad \lim_{h \to 0} \frac{1}{h} \int_x^{x+h} g_n = g_n(x). \tag{4.36}$$

Setting $L = \bigcap L_n$, we have $mL^{\sim} = m(\bigcup L_n^{\sim}) \leqslant \sum mL_n^{\sim} = 0$, and so $mL = b - a$.

Given $\varepsilon > 0$ and $r \in \mathbf{R}$. Choose $n = n(\varepsilon, r)$ such that $|r - r_n| < \varepsilon/3$. Then

$$\forall x \in [a, b], \qquad ||f(x) - r| - |f(x) - r_n|| < \varepsilon/3;$$

and so

$$\left| \frac{1}{h} \int_x^{x+h} |f(t) - r| \, dt - |f(x) - r| \right| < \frac{2\varepsilon}{3} + \left| \frac{1}{h} \int_x^{x+h} (g_n(t) - g_n(x)) \, dt \right|.$$

If $x \in L$ then x is in some L_n; consequently, from (4.36) there is $h_\varepsilon > 0$ for which

$$\left| \frac{1}{h} \int_x^{x+h} (g_n(t) - g_n(x)) \, dt \right| < \varepsilon/3$$

for all $|h| < h_\varepsilon$.

q.e.d

We rewrite Theorem 4.10 as

Corollary 4.10.1 *Let $f \in L_m^1[a, b]$. Then there is a subset $L \subseteq [a, b]$, for which $mL = b - a$, such that for all $x \in L$,*

$$\int_0^h |f(x + t) - f(x)| \, dt = o(h), \quad h \to 0. \tag{4.37}$$

The largest set of points $x \in [a, b]$ for which (4.37) holds, for a given $f \in L_m^1[a, b]$, is the Lebesgue set $L(f)$ of f. It is clear from (4.37) that

$$C(f) \subseteq L(f)$$

and that

$$mL(f) = b - a.$$

Example 4.6 The Fejér kernel $\{F_n : n = 1, ...\}$ was defined in Problem 3.21 and we saw that $\|F_n * f - f\|_\infty \to 0$ for 2π-periodic functions $f \in C(\mathbf{R})$. (Observe that $F_N * f$ is the arithmetic mean of the Nth partial sums of the Fourier series for f.) In particular,

$$\lim_{N \to \infty} |F_N * f(x) - f(x)| = 0 \tag{4.38}$$

for each $x \in [0, 2\pi)$. The question arises whether (4.38) has any meaningful generalization for arbitrary elements $f \in L_m^1(\mathbf{T})$. The answer is contained in Lebesgue's result: *if $f \in L_m^1(\mathbf{T})$ and $x \in L(f)$ then (4.38) is valid* (cf. Appendix A5.1.1).

4.5 Absolute continuity and FTC-II

The problem now is to find a converse to Theorem 4.9; that is, given a continuous function $F: [a, b] \to \mathbf{R}$ of bounded variation, is it true that

$$F(x) - F(a) = \int_a^x F'(t)\, dt$$

(recall that $F' \in L_m^1[a, b]$ if $F \in BV[a, b]$).

Example 4.7 Note that $C_C \in BV[0, 1]$ is continuous and $C_C' = 0$, m-a.e. Hence

$$\int_0^1 C_C' = 0,$$

whereas $C_C(1) - C_C(0) = 1$! (This fact tells us that C_C is not absolutely continuous (see below for definition)).

Thus the above question is answered negatively and so, if a converse is to hold we need F to satisfy a requirement that is stronger than both bounded variation and continuity. Such a notion exists and was given explicitly by Vitali [118] (although A. Harnack had used a similar notion in a different context during the late 1890's). Vitali's concept is absolute continuity, and with it he characterized the relation between differentiation and integration in the context of Lebesgue's theory; this characterization is the most general form of FTC (i.e. Theorem 4.11). Lebesgue's role in this matter is important. His major work was available to Vitali (cf. Appendix A5.1.2 for an historical remark on this point), and although Vitali published the first proof of Theorem 4.11 (in [118]), Lebesgue produced a quite efficient proof shortly thereafter in 1907.

A function $F: [a, b] \to \mathbf{R}$ is absolutely continuous on $[a, b]$ if

$$\forall \varepsilon > 0 \; \exists \delta > 0 \text{ such that } \forall \{(x_j, y_j) \subseteq [a, b]: j = 1, \ldots, n\},$$

a disjoint family,

$$\sum_1^n (y_j - x_j) < \delta \Rightarrow \sum_1^n |F(y_j) - F(x_j)| < \varepsilon.$$

Proposition 4.12 a) *Let* $f \in L_m^1[a, b]$ *and set*

$$F(x) = r + \int_a^x f, \qquad x \in [a, b].$$

Then $r = F(a)$ *and* F *is absolutely continuous on* $[a, b]$.

b) *If* $G: [a, b] \to \mathbf{R}$ *is absolutely continuous on* $[a, b]$ *then* G *is a continuous function of bounded variation; and, in particular,* G' *exists* m-a.e. *and is an element of* $L_m^1[a, b]$.

Proof. a) This is immediate from Proposition 3.2, as was Proposition 4.9.

b) Given $\varepsilon = 1$ and the corresponding δ from the definition of absolute continuity. Then

$$V(G, [a, b]) \leqslant 1 + (b - a)/\delta.$$

q.e.d.

Remark Thus FTC-I says that each $f \in L^1_m[a, b]$ is the derivative m-a.e. of an absolutely continuous function. We know that the converse of this is true: if $F: [a, b] \to \mathbf{R}$ is absolutely continuous then $F' \in L^1_m[a, b]$. Lusin [76] proved: *each Lebesgue measurable function $f: [a, b] \to \mathbf{R}$ is the derivative m-a.e. of a continuous function.* What sort of converse would you have in mind for this?

With regard to C_E and Hellinger's example H, note that

Proposition 4.13 *Let F be absolutely continuous on $[a, b]$ and assume that $F' = 0$, m-a.e. Then F is a constant.*

Proof. Take $c \in (a, b]$. To prove $F(c) = F(a)$. By hypothesis there is a subset $A \subseteq [a, c]$ such that $mA = c - a$ and $F' = 0$ on A.
Given $\varepsilon > 0$. Since $F' = 0$ on A we have that

$$\forall x \in A \ \exists [x, y] \subseteq (a, c) \text{ such that } |F(y) - F(x)| < \frac{\varepsilon(y - x)}{2(c - a)}.$$

From the Vitali covering theorem we can find

$$\{[x_j, y_j] : j = 1, ..., n \quad \text{and} \quad x_j \in A\},$$

a disjoint family, for which

$$|F(y_j) - F(x_j)| < \varepsilon(y_j - x_j)/2(c - a) \tag{4.39}$$

and

$$m\left(A \setminus \bigcup_{j=1}^{n} [x_j, y_j]\right) < \delta, \tag{4.40}$$

where $\delta = \delta(\varepsilon/2)$ is determined from the absolute continuity of F.
Without loss of generality (a labeling problem) take $x_j < x_{j+1}$. Using (4.40), we have

$$\sum_{j=0}^{n} |x_{j+1} - y_j| < \delta, \qquad x_{n+1} = c, \qquad y_0 = a,$$

so that from the absolute continuity

$$\sum_{1}^{n} |F(x_{j+1}) - F(y_j)| < \varepsilon/2;$$

and because of (4.39),

$$\sum_1^n |F(y_j) - F(x_j)| < \frac{\varepsilon}{2(c - a)} \sum_1^n |y_j - x_j|.$$

Combining the last two inequalities gives

$$|F(c) - F(a)| < \varepsilon.$$

q.e.d.

Theorem 4.11 (FTC-II) *A function $F: [a, b] \to \mathbf{R}$ is absolutely continuous on $[a, b]$ $\Leftrightarrow$ there is an element $f \in L_m^1[a, b]$ such that*

$$\forall x \in [a, b], \qquad F(x) - F(a) = \int_a^x f.$$

Proof. ($\Leftarrow$) This is Proposition 4.12a.
($\Rightarrow$) Define the function $H = F - G$ where

$$G(x) = \int_a^x F'$$

(note that $F' \in L_m^1[a, b]$ from Theorem 4.8).
By hypothesis and Proposition 4.12, H is absolutely continuous; and

$$H' = F' - G' = 0, \quad m\text{-}a.e.$$

Consequently, $F = G + r$, m-a.e. by Proposition 4.13; and so

$$F(x) - r = \int_a^x F'.$$

Thus, $r = F(a)$ and we set $f = F'$.

q.e.d.

Example 4.8 We prove that if f is a continuous function in $BV(\mathbf{T})$ and

$$\overline{\lim_{|n| \to \infty}} |n\hat{f}(n)| > 0,$$

then f is not absolutely continuous (thus, *if f is absolutely continuous on $\mathbf{T}$ then* $n\hat{f}(n) = o(1)$, $|n| \to \infty$). In fact, if f is absolutely continuous then $f' \in L_m^1(\mathbf{T})$ and

$$\hat{f'}(n) = in\hat{f}(n); \tag{4.41}$$

An application of the Riemann–Lebesgue lemma to f' concludes the proof. FTC is used to prove (4.41). Recall Problem 4.3 and Theorem 4.2 for perspective.

Remark Generalizations of FTC to $\mathbf{R}^n$ and to quite arbitrary measure spaces constitute one of the most important facets of the theory and leads through the Radon–Nikodym theorem to some of the most significant properties of measures.

4.6 Absolutely continuous functions

We have seen that every absolutely continuous function is a continuous function of bounded variation, but that the converse is not true. Theorem 4.12, below, gives necessary and sufficient conditions in order that a continuous element of $BV[a, b]$ is absolutely continuous. In this regard, recall that the Banach–Vitali theorem (Theorem 4.6) characterized those continuous functions which have bounded variation.

With regard to the criterion of Theorem 4.12, compare Problem 2.23.

The following lemma is Problem 4.27 and a proof can be found in [97, p. 226] or in Varberg's interesting exposition on absolutely continuous functions [117].

Lemma *Given a function $F: [a, b] \rightarrow \mathbf{R}$ and let $A \subseteq [a, b]$ be any subset on which F' exists. If*

$$\exists K \quad such \ that \quad \forall x \in A, \quad |F'(x)| \leqslant K$$

then

$$m^*F(A) \leqslant Km^*A. \tag{4.42}$$

Theorem 4.12 *(Banach–Zarecki, 1925) Let $F \in BV[a, b]$ be a continuous function. F is absolutely continuous on $[a, b] \Leftrightarrow$*

$$mA = 0 \Rightarrow mF(A) = 0.$$

Proof. ($\Rightarrow$) This is the easy direction. Without loss of generality take a subset $A \subseteq (a, b)$ with Lebesgue measure $mA = 0$. Let $\varepsilon > 0$. We'll prove that $m^*F(A) < \varepsilon$. From the hypothesis of absolute continuity there is $\delta > 0$ such that no matter what disjoint family $\{(a_k, b_k): k = 1, ...\}$ we take,

$$\sum (b_k - a_k) < \delta \Rightarrow \sum (M_k - m_k) < \varepsilon,$$

where $M_k = \sup \{F(x): x \in (a_k, b_k)\}$ and $m_k = \inf \{F(x): x \in (a_k, b_k)\}$ (e.g. Problem 4.31).

Choose an open set U for which $A \subseteq U \subseteq (a, b)$ and $mU < \delta$.

Set $U = \bigcup (a_k, b_k)$, a disjoint union, and note that since

$$F(A) \subseteq F(U) = \bigcup F((a_k, b_k)),$$

we have

$$m^*F(A) \leqslant \sum m^*[m_k, M_k] < \varepsilon.$$

($\Leftarrow$) a) We first use the Lemma to prove that if F' exists on $A \in \mathcal{M}$ then

$$m^* F(A) \leqslant \int_A |F'|. \tag{4.43}$$

Assume that there is an integer p such that

$$\forall x \in A, \qquad |F'(x)| < p,$$

and define

$$A_{k,n} = \left\{ x \in A : \frac{k-1}{2^n} \leqslant |F'(x)| < \frac{k}{2^n} \right\},$$

where $k = 1, \ldots, p2^n$ and $n = 1, \ldots$.
Using the Lemma we compute

$$m^* F(A) = m^* (\bigcup_k F(A_{k,n})) \leqslant \sum_k m^* F(A_{k,n}) \leqslant \sum_k \frac{k}{2^n} m A_{k,n}$$

$$= \sum_k \frac{k-1}{2^n} m A_{k,n} + \frac{1}{2^n} \sum_k m A_{k,n}.$$

From the definition of the Lebesgue integral and since $m^* F(A)$ is independent of n, we have (4.43).
The case for an unbounded function F' uses the bounded case and sets of the form

$$A_k = \{ x \in A : k - 1 \leqslant |F'(x)| < k \}$$

in the obvious manner.
b) Given a disjoint family $\{(a_k, b_k) : k = 1, \ldots, n\}$ in $[a, b]$ and let

$$A_k = \{ t \in [a_k, b_k] : \exists F'(t) \}.$$

Since F' exists m-a.e. we have that $m([a_k, b_k] \setminus A_k) = 0$; by hypothesis, then, $mF[a_k, b_k] = mF(A_k)$.
From part a) we compute

$$\sum_{k=1}^n |F(b_k) - F(a_k)| \leqslant \sum_{k=1}^n mF[a_k, b_k]$$

$$= \sum_{k=1}^n mF(A_k) \leqslant \sum_{k=1}^n \int_{A_k} |F'| = \sum_{k=1}^n \int_{a_k}^{b_k} |F'|. \tag{4.44}$$

Since $F' \in L_m^1[a, b]$, if $\sum (b_k - a_k)$ is small then the last sum in (4.44) is small.

q.e.d.

We now prove the integration by parts formula.

Theorem 4.13 *Given* $f, g \in L_m^1[a, b]$ *with corresponding functions*

$$F(x) = r + \int_a^x f \quad \text{and} \quad G(x) = s + \int_a^x f.$$

Then

$$\int_a^b fG + \int_a^b gF = F(b)G(b) - F(a)G(a).$$

Proof. Note that

$$|F(x)G(x) - F(y)G(y)| = |F(x)(G(x) - G(y)) + G(y)(F(x) - F(y))|$$
$$\leqslant \|F\|_\infty |G(x) - G(y)| + \|G\|_\infty |F(x) - F(y)|.$$

Thus FG is absolutely continuous on $[a, b]$; and so FG is differentiable m-a.e. and $(FG)' = F'G + G'F$, m-a.e.
From FTC-I, $F' = f$ and $G' = g$, m-a.e. Therefore

$$\int_a^b (FG)' = \int_a^b fG + \int_a^b gF.$$

On the other hand from the absolute continuity of FG and FTC-II,

$$\int_a^b (FG)' = F(b)G(b) - F(a)G(a).$$

q.e.d.

If $F \in BV[a, b]$ and $F' = 0$, m-a.e. then F is singular. It is obvious that the discrete part of $F \in BV[a, b]$ is singular. What is more interesting is that there are functions $F \in BV[a, b]$ with singular continuous parts (e.g. C_C). With regard to the following result, compare Theorem 5.17b). The sufficient conditions in Theorem 4.14 are found in [102].

Theorem 4.14 *Given a function* $F: [a, b] \to$ **R** *and assume that* F' *exists on* $A \subseteq [a, b]$. $F' = 0$, m-a.e. on $A \Leftrightarrow mF(A) = 0$.

Proof. $(\Rightarrow)$ Let $A_k = \{t \in A : |F'(t)| \leqslant k\}$. Then from (4.42) and hypothesis,

$$m^*F(A) \leqslant \sum_{k=0}^\infty m^*F(A_k) \leqslant \sum_{k=0}^\infty km^*A_k = 0.$$

$(\Leftarrow)$ Let $B = \{t \in A : |F'(t)| > 0\}$ and define

$$B_n = \{t \in B : |F(s) - F(t)| \geqslant |s - t|/n \quad \text{for} \quad |s - t| < 1/n\}.$$

$B = \bigcup B_n$ and we must prove that $mB_n = 0$ for each n. Fix an n and let I be an interval with Lebesgue measure $mI < 1/n$; set $D = I \cap B_n \subseteq A$. Thus it is sufficient to prove that

$$\forall \varepsilon > 0, \qquad m^*D \leqslant n\varepsilon. \tag{4.45}$$

We use our hypothesis now to choose a sequence $\{I_j : j = 1, \ldots\}$ of intervals such that $F(D) \subseteq \bigcup I_j$ and $\sum mI_j < \varepsilon$.
Set $D_j = F^{-1}(I_j) \cap D$. From the definition of B_n and the fact that $m^*D \leqslant \sum m^*D_j$, we compute (4.45).

q.e.d.

Relative to the decompositions earlier in this chapter we have .

Proposition 4.14 *If* $F \in BV[a, b]$ *then* $F - F_a + F_s$ *where* F_a *is absolutely continuous and* F_s *is singular.*

Proof. Set $F_a(x) = \int_a^x F'$. We know that F_a is absolutely continuous on $[a, b]$ and that $F_a' = F'$, m-a.e. Consequently, the proof is completed when we define $F_s = F - F_a$.

q.e.d.

It is this result which we generalize to the Radon–Nikodym theorem.
Recall that Hellinger's example (Example 4.4) is a strictly increasing continuous singular function $H = H_t$, where $t \in (0, 1)$ is fixed. A simpler construction is

Example 4.9 (cf. Problem 4.24) Given an interval $[a, b] \subseteq [0, 1]$ and a corresponding positive constant k_{ab}. Then

$$k_{ab}C_C\left(\frac{x - a}{b - a}\right), \qquad x \in [a, b] \tag{4.46}$$

is continuous on $[a, b]$ and increases from 0 to k_{ab} there. Enumerate all the intervals $I_n = [a_n, b_n]$, where $a_n, b_n \in \mathbf{Q} \cap [0, 1]$ and $a_n < b_n$; and define f_n on I_n, analogous to (4.46), such that f_n is continuous, increasing from 0 to $(\frac{1}{2})^n$, and $f_n' = 0$, m-a.e. on I_n. We set $f = \sum f_n$ and see that f is continuous and strictly increasing. From Fubini's differentiation theorem (given in Problem 4.16), $f' = 0$, m-a.e.

Remark Part of the technical difficulty arising in the construction of H_t, $t \in (0, 1)$, is compensated for by the following fact (which will be developed in Chapter 5). Not only is H_t, $t \in (0, 1)$, singular "with respect to Lebesgue measure", but, if μ_t corresponds to H_t (where the correspondence is in the sense of Theorem 4.3), we have a continuum of continuous singular measures μ_t such that if $t \neq s$ then μ_t and μ_s are "mutually singular", i.e., μ_t and μ_s are concentrated on disjoint sets. $((X, \mathscr{A}, \mu)$ is concentrated on $A \subseteq X$ if

$$\forall B \in \mathscr{A}, \qquad \mu B = \mu(A \cap B)).$$

As we have seen (for $F = C_C$, say), we cannot conclude that a function F is absolutely continuous if

$$F \text{ is continuous, } \exists F', \text{ } m\text{-a.e., and } F' \in L_m^1. \tag{4.47}$$

As far as positive results on this question go, we can conclude that *a function F:* $[a, b] \to \mathbf{R}$ *is absolutely continuous if it satisfies* (4.47) *and either*

$$\text{card } \{t : \nexists F'(t)\} \leqslant \aleph_0 \tag{4.48}$$

or

$$mA = 0 \quad \text{implies} \quad mF(A) = 0. \tag{4.49}$$

The latter theorem (i.e. "(4.49) and (4.47) imply absolute continuity") generalizes the Banach–Zarecki theorem—since we assumed that $F \in BV[a, b]$ there—and has the same proof since the only use of bounded variation in Theorem 4.12 was to obtain $F' \in L_m^1[a, b]$. For the proof of the first theorem (i.e. "(4.48) and (4.47) imply absolute continuity") it is sufficient to prove that (4.48) implies (4.49). Let $mA = 0$ and $A = B \cup D$ where card $D \leqslant \aleph_0$ and F' exists on B.
We must prove that $mF(B) = 0$. Set

$$B_{n,k} = \{t \in B : \forall J, \text{ for which } mJ < 1/k \text{ and } t \in J, \text{ we have } mF(J) < nmJ\},$$

where J is an open interval. Since F' exists on B,

$$B = \bigcup B_{n,k},$$

and so we only have to check that

$$\forall n, k \quad \text{and} \quad \forall \varepsilon > 0, \quad mF(B_{n,k}) < \varepsilon. \tag{4.50}$$

To verify (4.50) we cover $B_{n,k} \subseteq A$ by non-overlapping intervals each with length less than $1/k$ and with total length less than ε/n. This does it.
As a special case

Theorem 4.15 *Let $F : [a, b] \to \mathbf{R}$ be an everywhere differentiable function. If $F' \in L_m^1[a,b]$ then F is absolutely continuous on $[a, b]$.*

The requirement, $F' \in L_m^1[a, b]$, is necessary, as the example

$$F(x) = \begin{cases} 0, & \text{if } x = 0 \\ x^2 \sin(1/x^2), & \text{if } x \in [0, 1] \end{cases}$$

shows.

An interesting application of some of our previous results and an amusing source of counter-example attempts is

Theorem 4.16 *Given $f, f' \in L_m^1(\mathbf{R})$ and assume that*

$$\forall x \in \mathbf{R}, \qquad f'(x) \text{ exists.}$$

Then $\displaystyle\int_{\mathbf{R}} f' = 0.$

Proof. From Theorem 4.15 f is absolutely continuous on each finite interval. Take $\{F_n : n = 1, \ldots\} \subseteq C^\infty(\mathbf{R})$ such that $\sup_n \|F_n'\|_\infty \leqslant K$, $0 \leqslant F_n \leqslant 1$, $F_n = 1$ on $[-n, n]$, and $F_n = 0$ off of $[-(n+1), n+1]$. Note that each F_n is absolutely continuous on finite intervals. Now $F_n f' \to f'$ pointwise and $|F_n f'| \leqslant |f'| \in L_m^1(\mathbf{R})$. Consequently, by LDC, $\int f' = \lim_n \int F_n f'$.

Since each Γ_n and f is absolutely continuous we use Theorem 4.13 to compute

$$\int_{-(n+1)}^{n+1} F_n f' = - \int_{-(n+1)}^{n+1} F_n' f.$$

Observe that $F_n' f \to 0$ pointwise and $|F_n' f| \leqslant K|f| \in L_m^1(\mathbf{R})$ so that by LDC, $\lim \int F_n' f = 0$. Therefore

$$\int f' = \lim_{n \to \infty} \int F_n f' = 0.$$

q.e.d.

Problems for Chapter 4

Some of the more elementary problems in this set are Problems 4.1, 4.2, 4.11, 4.12, 4.13, 4.14, 4.16, 4.17, 4.18, 4.26, 4.32.

4.1 Let $f : \mathbf{R} \to \mathbf{R}$ be an element of BV.
a) Prove that if $a \leqslant b \leqslant c$ then

$$V(f, [a, c]) + V(f, [c, b]) = V(f, [a, b])$$

(cf. Theorem 5.15a).
b) Let $f = g_1 - g_2$, where each g_j is increasing and $g_2(-\infty) = 0$. Prove that

$$\forall a \leqslant b, \qquad 0 \leqslant P(b) - P(a) \leqslant g_1(b) - g_1(a),$$
$$0 \leqslant N(b) - N(a) \leqslant g_2(b) - g_2(a)$$

(cf. Theorem 5.5b).
Part b) is a uniqueness condition for the representation of $f \in BV$ as the difference of increasing functions.

4.2 a) Let $f\colon [a, b] \to$ **R** be everywhere differentiable and assume that $\|f'\|_\infty < \infty$. Give a simple proof that f is absolutely continuous.

b) Given $\alpha, \beta > 0$ and define the function $f\colon [0, 1] \to$ **R** as

$$f(x) = \begin{cases} 0, & \text{if} \quad x = 0, \\ x^\alpha \sin(1/x^\beta), & \text{if} \quad x \in (0, 1]. \end{cases}$$

Prove that $f \in BV[0, 1]$ if and only if $\alpha > \beta$.

c) It is trivial to find functions which are both continuous and of bounded variation; in this exercise, for example, let $(\alpha, \beta) = (2, 1)$. Find f, as in b), which is discontinuous at a point and which is not an element of $BV[0, 1]$ (resp. which is continuous and which is not an element of $BV[0, 1]$). (*Hint.* Take $(\alpha, \beta) = (0, 1)$ (resp. $(\alpha, \beta) = (1, 1)$).)

d) Prove that the function χ_S in Example 3.14 is not a function of bounded variation.

e) Show that if $f(x) = a_n x^n + a_{n-1} x^{n-1} + \ldots + a_0$ then $f \in BV[a, b]$. Compute $V(f, [a, b])$ given the roots of f'. (*Hint.* Use part a).)

4.3 With regard to the remark after Theorem 4.2 prove that

$$f \sim \sum_2^\infty \frac{1}{n[\log n]} \exp i x n[\log n] \qquad \text{and} \qquad g(x) = x \sin(1/x)$$

are elements of $L_m^1(\mathbf{T}) \setminus BV(\mathbf{T})$, and that $n\hat{f}(n)$, $n\hat{g}(n) = o(1)$, $|n| \to \infty$. "$[\log n]$" denotes the largest integer $k \leqslant \log n$. Note that g is continuous (when defined on $[-\pi, \pi)$ and extended periodically). Also, compare this with (1.11).

We now give an example of a continuous function $f \in BV[0, 2\pi)$ such that $n\hat{f}(n) \neq o(1)$, $|n| \to \infty$. Construct the Cantor function C_C (so that $C_C(0) = 0$ and $C_C(2\pi) = 1$). Setting $f(x) = C_C(x) - x/2\pi$ on $[0, 2\pi]$, f can be defined as a continuous periodic function on **R** with $f(0) = f(2\pi) = 0$ and period 2π. Prove that

$$\forall n, \qquad |3^n \hat{f}(3^n)| = |\hat{f}(1)| > 0.$$

(*Hint.* The calculation for the equality is routine, and to see that $\hat{f}(1) \neq 0$ it is sufficient to compute the real part.) For a more extensive treatment of finding and analyzing Fourier coefficients of singular measures we refer to [10; 60]. Further remarks and examples are given in Chapter 5.

4.4 Prove Sidon's theorem: let f be a bounded Lebesgue measurable function with Fourier series expansion

$$f \sim \sum_{k=1}^\infty a_k \cos n_k x,$$

where $\{n_k\colon k = 1, \ldots\}$ is an Hadamard set; prove that $\sum_1^\infty |a_k| \leqslant 2\|f\|_\infty$ (e.g. [128]).

Thus we have a criterion in order that a bounded Lebesgue measurable function have an absolutely convergent Fourier series.

4.5 a) Construct a continuous function $f\colon [a, b] \to \mathbf{R}$ such that for each interval $[c, d] \subseteq [a, b]$, $f \notin BV[c, d]$.

This is immediate from Theorem 4.8 and the existence of *ecnd* functions, but you should be able to give much simpler examples.

b) Given a function $f\colon [a, b] \to \mathbf{R}$. Define

$$\forall x \quad \text{and} \quad \forall \delta > 0, \qquad V(x, \delta) = 1/(1 + V(f, [x - \delta, x + \delta])),$$
$$v(x) = \lim_{\delta \to 0} V(x, \delta),$$

and $\qquad S_{BV} = \{x \in [a, b]: \exists \delta > 0 \text{ such that } V(f, [x - \delta, x + \delta]) < \infty\}.$

Show that S_{BV} is open. Prove

i) if f is continuous at $x \in S_{BV}$ then $v(x) = 1$,

ii) if $v(x) = 1$ then f is continuous at x.

Observe that if f is continuous at x then x need not be in S_{BV}.

4.6 With regard to Example 3.14 and Problem 4.3, find an open set $S \subseteq [0, 2\pi)$, with $m([0, 2\pi) \setminus S) > 0$, such that $\chi_S \notin BV[0, 2\pi)$ whereas $\sup_n |n\hat{\chi}_S(n)| < \infty$.

4.7 Van der Waerden's *ecnd* function W was mentioned in Chapter 1.3.2. Setting

$$W_n(x) = \sum_0^{n-1} 10^{-j} f(10^j x), \qquad x \in (0, 1),$$

where $\qquad f(x) = |x - k|, \qquad x \in [k - \tfrac{1}{2}, k + \tfrac{1}{2}], \qquad k \in \mathbf{Z},$

W can be defined as $W = \lim W_n$. From Theorem 4.8, $V(W, [0, 1]) = \infty$. On the other hand there is the following control on the variation:

$$\lim_{n \to \infty} \frac{1}{\sqrt{n}} V(W_n, [0, 1]) = \sqrt{2/\pi}. \tag{P 4.1}$$

Prove (P4.1). (*Hint.* Calculate that

$$\lim_n \frac{1}{\sqrt{n}} V(W_n, [0, 1]) = \lim_n \sum_{m=0}^n \left| \frac{m - n/2}{\sqrt{n/2}} \right| \binom{n}{m} \frac{1}{2^n}$$

$$= \frac{1}{\sqrt{2\pi}} \int_{-\infty}^{\infty} |x| \, e^{-x^2/2} \, \mathrm{d}x = \sqrt{2/\pi}.)$$

4.8 a) Prove that L'Hospital's rule is not necessarily true for complex-valued functions. (*Hint.* Let $f(x) = x$, $g(x) = x \, e^{-i/x}$, and take the limit as $x \to 0$.)

b) Given functions $f, g\colon (0, a) \to \mathbf{C}$ such that

i) f', g' are continuous on $(0, a)$,

ii) $\lim_{x \to 0} f(x) = \lim_{x \to 0} g(x) = 0$,

iii) $\forall x \in (0, a)$, $g'(x) \neq 0$,

iv) $\lim_{x \to 0} f'(x)/g'(x) = c$;

define $\delta(x) = -c + f'(x)/g'(x)$, and assume that $|g(x)|$ is monotone and Re δ, Im $\delta \in BV(0, a)$. Prove that $\lim_{x \to \infty} f(x)/g(x) = c$. (*Hint.* Define

$$R(x) = \left(\int_0^x \delta(y) g'(y) \, dy \right) / g(x)$$

and compute that

$$\lim_{x \to 0} R(x) = 0.$$

Use this fact and an integration of

$$g'(x) \delta(x) = f'(x) - cg'(x)$$

to complete the proof.)

4.9 Prove Proposition 4.7, and show by example that the continuity hypothesis is necessary.

4.10 Let $f: [a, b] \to$ **R** and $g: [c, d] \to$ **R** be continuous functions of bounded variation. Assume that $g([c, d]) \subseteq [a, b]$. Then $f \circ g$ is continuous. Find conditions to ensure that $f \circ g \in BV[c, d]$ (e.g. [85]).

4.11 Compute P, N, and Vf for $f(x) = e^{-x^2}$.

4.12 Given $\{a_n: n = 1, ...\}$, $\{b_n: n = 1, ...\} \subseteq$ **R** and define the function $f_{ab} = \Sigma f_n$ where

$$f_n(x) = \begin{cases} b_n & \text{if } x \geqslant a_n \\ 0, & \text{if } x < a_n. \end{cases}$$

a) Fix a sequence $\{b_n: n = 1, ...\} \subseteq$ **R**; and find necessary and sufficient conditions such that for each sequence $\{a_n: n = 1, ...\} \subseteq$ **R** which is bounded below,

$$f_{ab} \in BV_{\text{loc}}.$$

b) Do there exist sequences $\{a_n: n = 1, ...\}$, $\{a'_n: n = 1, ...\}$, $\{b_n: n = 1, ...\}$ for which $f_{ab} \in BV_{\text{loc}}$ whereas $f_{a'b} \notin BV_{\text{loc}}$?

4.13 Define the function $f: \mathbf{R} \to \mathbf{R}$ as

$$f(x) = \begin{cases} 0, & \text{if } x \text{ is irrational or } x = 0 \\ 1/(p^2 q^2), & \text{if } x = p/q \text{ and } (p, q) = 1. \end{cases}$$

a) Prove that $f \in BV$ (and so f' exists m-a.e.)

b) Prove that f' does not exist on a dense subset of $\mathbf{R}$.

4.14 Construct a continuous real-valued function $f \in BV[0, 1]$ such that

$$\forall q \in \mathbf{Q} \cap [0, 1], \quad f'(q) \text{ does not exist.}$$

(*Hint.* Let $\mathbf{Q} \cap [0, 1] = \{r_n: n = 1, \ldots\}$ and define

$$f_n(x) = \begin{cases} 0, & \text{if } x \in [0, r_n] \\ \dfrac{x - r_n}{2^n (1 - r_n)}, & \text{if } x \in (r_n, 1].\end{cases})$$

4.15 Assume that $f \in L_m^1[0, 1]$ satisfies the condition that

$$\|\tau_{-h} f - f\|_1 = o(|h|), \qquad |h| \to 0;$$

prove that $f = k$, a constant, m-a.e. (*Hint.* Use Corollary 4.10.1.)

4.16 a) (Fubini's differentiation theorem) Given a sequence $\{f_n: n = 1, \ldots\}$ of monotone increasing functions $[a, b] \to \mathbf{R}$, and assume that

$$f(x) = \sum f_n(x)$$

exists for each $x \in [a, b]$. Prove that $f' = \sum f_n'$ exists m-a.e.

b) Let $f \in BV$. Prove that $(Vf)' = |f'|$, m-a.e. (*Hint.* Use part a.)

Remark Thus if $f \in BV$, then not only does f' exist m-a.e. but also $(Vf)'$ exists m-a.e. The set of points where f' and $(Vf)'$ exist need not be the same (e.g. $f(x) = x$).

4.17 If $f(x) = e^{-n^2 x^2}$, then $V(f, \mathbf{R}) = 2$ (see Problem 4.11). Let $f_n(x) = e^{-n^2 x^2}/n^2$ and set $g = \sum f_n$. Prove that $g \in BV_{\text{loc}}$, whereas $\sum f_n'$ does not converge uniformly on $[-1, 1]$. (Consequently, g' exists m-a.e. but this fact cannot be proved by the standard advanced calculus criterion). Compute g'.

4.18 Prove FTC-I for the case of unbounded f.

4.19 Let $f, f_n \in BV$.

a) Assume that

$$\forall x, \; \lim_{n \to \infty} V(f - f_n)(x) = 0; \tag{P 4.2}$$

prove that there is a subsequence $\{f_{n_k}: k = 1, ...\} \subseteq \{f_n: n = 1, ...\}$ such that

$$\lim_k f'_{n_k} = f', \quad m\text{-}a.e.$$

b) Does a) hold when (P4.2) is weakened to almost everywhere pointwise convergence?

c) Show that the conclusion of a) cannot be replaced in general by

$$\lim f'_n = f', \quad m\text{-}a.e.$$

4.20 a) In light of Example 4.4 find a simple example of a strictly increasing function f such that $f' = 0$, m-a.e. (*Hint.* Use Example 4.2 and Problem 4.16). Note that we are not asking that f be continuous.

b) Does there exist a function $f: [a, b] \to$ **R** such that $\|f\|_\infty < \infty$, $f \notin BV[a, b]$, and $f' = 0$, m-a.e.?

4.21 a) Let $A \subseteq$ **R**. Prove that $mA = 0$ if and only if there is a countably infinite sequence $\{I_n: I_n = (a_n, b_n), n = 1, ...\}$, which satisfies $\sum mI_n < \infty$ and $A \subseteq \bigcup I_n$, such that for each $x \in A$ there is a countably infinite subsequence $\{I_{n_k}: k = 1, ...\} \subseteq \{I_n: n = 1, ...\}$ for which

$$\forall k, \quad x \in I_{n_k}.$$

b) Given a set $A \subseteq [a, b]$, for which $mA = 0$. Does there exist a function $f \in BV[a, b]$ such that f' exists precisely on $A^\sim$? (*Hint.* Use a).)

c) Given the Cantor function C_E for a perfect symmetric set E. Prove that C'_E does not exist for boundary points of contiguous intervals (cf. [13, pp. 128–134]).

4.22 With regard to Problem 4.9 (Proposition 4.7) we would like to find conditions on the partitions P such that if $f \in BV[a, b]$ then

$$\lim_{|P| \to 0} \sum_P |f(x_j) - f(x_{j-1})| = V(f, [a, b]).$$

Let f have jump discontinuities at $\{y_j: j = 1, ...\} \subseteq [a, b]$ and let $P(N, \delta)$ be a partition including $y_1, ..., y_N$ and with norm $|P(N, \delta)| < \delta$. Prove that

$$\forall \varepsilon > 0 \quad \exists N \text{ and } \exists \delta \quad \text{such that} \quad \forall P(N, \delta)$$

$$\sum_{P(N, \delta)} |f(x_j) - f(x_{j-1})| > V(f, [a, b]) - \varepsilon.$$

4.23 Prove Proposition 4.11.

4.24 We give an example of a strictly increasing absolutely continuous function $f: [0, 1] \to$ **R** with a continuous derivative such that

$$m\{f(t): f'(t) = 0\} > 0$$
$$m\{t: f'(t) = 0\} > 0. \tag{P 4.3}$$

Thus such sets are uncountable. Note that Hellinger's example H (Example 4.4), which is not absolutely continuous, gives the result that $m\{H(t): H'(t) = 0\} = 1$. Define $g = 0$ on E, a perfect symmetric set with $mE > 0$, and

$$g(x) = (x - a)(b - x), \quad \text{for} \quad x \in (a, b),$$

where (a, b) is a contiguous interval of E; prove that

$$f(x) = \int_0^x g(t)\, dt$$

satisfies (P4.3). (*Hint*. Prove that f is strictly increasing.)

4.25 Consider the function f of Problem 1.22 which is increasing, continuous on the irrationals, and discontinuous on the rationals. Define

$$F(x) = \int_0^x f.$$

Prove that $F' = f$ at each irrational (and F' does not exist at any rational). Compare this with Problem 4.13 and Problem 4.14.

4.26 A function $f: [a, b] \to \mathbf{R}$ is **locally recurrent** if

$$\forall x \in [a, b] \quad \text{and} \quad \forall \varepsilon > 0 \quad \exists y \quad \text{such that} \quad f(y) = f(x) \quad \text{and}$$
$$0 < |y - x| < \varepsilon.$$

Prove that if f is continuous, non-constant, and locally recurrent on $[a, b]$ then $f \notin BV[a, b]$. (*Hint*. Use Theorem 4.6.)

4.27 Prove the lemma for Theorem 4.12.

4.28 **Change of variable** Define the functions $f: [c, d] \to \mathbf{R}$, $g: [a, b] \to \mathbf{R}$, and $F(x) = \int_c^x f$, and assume that $g[a, b] \subseteq [c, d]$.

a) Suppose that $f \in L_m^1[c, d]$ and that g and $F \circ g$ are differentiable *m-a.e.* on their respective domains. Prove that

$$(F \circ g)' = (f \circ g)g', \quad m\text{-a.e.}$$

b) Assume that $f \in L_m^1[c, d]$ and that g' exists *m-a.e.* Prove that $F \circ g$ is absolutely continuous on $[a, b]$ if and only if

i) $\qquad (f \circ g)g' \in L_m^1[a, b]$

and ii) $\qquad \forall [\alpha, \beta] \subseteq [a, b], \quad \int_{g(\alpha)}^{g(\beta)} f = \int_\alpha^\beta (f \circ g)g'.$ $\hfill$ (P 4.4)

(*Hint.* The sufficient conditions for absolute continuity are clear, and we use part a) for the necessary conditions.)

c) Prove (P4.4) assuming any one of the following sets of conditions:

i) $f \in L^1_m[c, d]$ and g is an increasing absolutely continuous function;

ii) $\|f\|_\infty < \infty$ and g is absolutely continuous;

iii) g is absolutely continuous and $f, (f \circ g)g' \in L^1_m$.

(*Hint.* For i) and ii) show that $F \circ g$ is absolutely continuous; and for iii), approximate by bounded functions, apply ii), and use LDC.)

Remark The change of variable formula can be proved in a more general setting (e.g. [56, pp. 342–343]).

4.29 a) Given any set $A \subseteq [0, 1]$ for which $mA = 0$. Does there exist an increasing absolutely continuous function $f: [0, 1] \to \mathbf{R}$ such that f' exists on $A^\sim$ and

$$\forall x \in A, \; \lim_{h \to 0} \frac{f(x + h) - f(x)}{h} = \infty ?$$

(*Hint.* Cf. Problem 4.21.)

b) Find a continuous function f on $[0, 1]$ which is absolutely continuous on each interval $[\varepsilon, 1]$, $\varepsilon > 0$, but which is not absolutely continuous on $[0, 1]$. (*Hint.* Consider $x \sin(1/x)$.)

c) Along with the hypotheses of b) assume that $f \in BV[0, 1]$. Prove that f is absolutely continuous on $[0, 1]$. (*Hint.* Use the Banach–Zarecki theorem.)

d) Construct an increasing continuous function f on $[0, 1]$ such that for each interval $[\alpha, \beta] \subseteq [0, 1]$, f is not absolutely continuous on $[\alpha, \beta]$. (*Hint*—with a capital H!)

e) Find a continuous and strictly increasing function f on $[0, 1]$ and a set $A \subseteq [0, 1]$ such that $mA = 0$ and $mf(A) = 1$.

f) Find an absolutely continuous function f on $[0, 1]$ such that f is not monotonic on any interval in $[0, 1]$. (*Hint.* Find a set $E \subseteq [0, 1]$ such that, for each interval $I, m(E \cap I) > 0$ and $m(E^\sim \cap I) > 0$ (as in Problem 2.22); set $f(x) = \int\limits_0^x (\chi_E - \chi_{E^\sim}).)$

4.30 Given a function $f: \mathbf{R} \to \mathbf{R}$. Assume that

$$\forall x \in \mathbf{R} \quad \exists N_x \subseteq \mathbf{R}, \; \text{for which} \quad mN_x = 0, \; \text{such that}$$
$$\forall y \notin N_x, \; f(y) = f(y + x).$$

Prove that there is a constant c such that $f = c$, *m-a.e.*

4.31 Prove that the following statements are equivalent:

a) F is absolutely continuous on $[a, b]$.

b) For each $\varepsilon > 0$ there is $\delta > 0$ such that for all disjoint families $\{(a_j, b_j) \subseteq [a, b]$: $j = 1, ..., n\}$,

$$\sum_1^n (b_j - a_j) < \delta \Rightarrow \left| \sum_1^n (F(b_j) - F(a_j)) \right| < \varepsilon.$$

c) For each $\varepsilon > 0$ there is $\delta > 0$ such that for all disjoint families $\{(a_j, b_j) \subseteq [a, b]$: $j = 1, ..., n\}$,

$$\sum_1^n (b_j - a_j) < \delta \Rightarrow \sum_1^n \omega(F, [a_j, b_j]) < \varepsilon.$$

4.32 Let $f \in L^1_m[a, b]$ and set $F(x) = \int_a^x f$; prove that $V(F, [a, b]) = \|f\|_1$. (*Hint.*

The inequality $V(F, [a, b]) \leqslant \int_a^b |f|$ is clear as in the proof of Proposition 4.9. For the

opposite inequality define sgn $f = \bar{f}/|f|$ if $f(x) \neq 0$ and sgn $f = 0$ if $f(x) = 0$; choose simple functions s_j, such that $s_j \to$ sgn f, m-a.e. and $|s_j| \leqslant 1$ (such an s should have the form $\sum a_j \chi_{I_j}$, $I_j = (c_j, d_j)$). Check that

$$\left| \int_a^b f s_j \right| \leqslant V(F, [a, b])$$

and apply LDC.)

Remark This result is important because of the bijection between $BV[a, b]$ and the bounded measures on $[a, b]$ (i.e. the continuous linear functionals on $C[a, b]$); the bijection is given by taking the first distributional derivative (cf. Appendix III.1). We'll see that the total variation of $F \in BV[a, b]$ is the canonical Banach space norm of the corresponding measure. The bijection identifies the absolutely continuous functions on $[a, b]$ (as a subset of $BV[a, b]$) with $L^1_m[a, b]$ (as a subset of the bounded measures). The above Banach space norm on the measures reduces to $\| \ \|_1$ for $L^1_m[a, b]$.

4.33 With regard to Problem 4.32 and the remark there, prove the following: if $F \in BV[a, b]$ then

$$V(F, [a, b]) \geqslant \int_a^b |F'|. \tag{P 4.5}$$

(*Hint.* First prove that if G is increasing and $A = \{x: \exists G'(x)\}$ then

$$\int_a^b G' = m^* G(A).$$

Next calculate $\int\limits_a^b |F'|$ in terms of $(VF)'$, observing that VF is increasing.) Related to (P4.5) we can also prove that

$$\forall B \in \mathcal{M}, \qquad m^*(VF)(B) \geqslant \int\limits_B |F'|, \tag{P 4.6}$$

when $F \in BV[a, b]$. Naturally there is equality in both (P4.5) and (P4.6) when F is absolutely continuous.

4.34 a) Let $f: [a, b] \to \mathbf{R}$ be an m-measurable function and take $\varepsilon, \delta > 0$. Prove that there is an absolutely continuous function G and a set $A \in \mathcal{M}$ such that $mA < \delta$ and

$$\sup_{x \in [a, b] \setminus A} |f(x) - G(x)| < \varepsilon.$$

b) If $f \in L_m^\infty[a, b]$ is an increasing, non-negative function and $g \in L_m^1[a, b]$ then prove that there is $\xi \in (a, b)$ such that

$$\int\limits_a^b fg = f(b-) \int\limits_\xi^b g.$$

We also refer to [56, p. 420] on this matter.

4.35 The functional equation for a function $f: \mathbf{R} \to \mathbf{R}$ is

$$\forall x, y \in \mathbf{R}, \qquad f(x + y) = f(x) + f(y).$$

a) If f is continuous on $\mathbf{R}$ and satisfies the functional equation prove that $f(x) = f(1)x$ on $\mathbf{R}$. (*Hint.* Check first that $f(nx) = nf(x)$ if $n \in \mathbf{Z}$ and then that $f(rx) = rf(x)$ if $r \in \mathbf{Q}$; the conclusion follows since f is continuous.)

b) Prove that if f is continuous at some point of $\mathbf{R}$ and the functional equation is satisfied then $f(x) = f(1)x$ on $\mathbf{R}$.

c) Let $S \subseteq \mathbf{R}$ have the property that $S - S$ contains an interval (cf. Steinhaus' theorem in Problem 3.6b). If f is bounded on S and satisfies the functional equation prove that $f(x) = f(1)x$ on $\mathbf{R}$.

d) Let f be a measurable function on $\mathbf{R}$ which satisfies the functional equation. Prove that $f(x) = f(1)x$ on $\mathbf{R}$. (*Hint.* Let $g = e^{if}$ and use Theorem 4.9 and the hypothesis to compute that

$$\forall x \in \mathbf{R}, \qquad g'(x) = g'(0)g(x).$$

Solving the differential equation we find that

$$f(x) = Ax + 2\pi n(x),$$

where $n: \mathbf{R} \to \mathbf{Z}$ and $n(x + y) = n(x) + n(y)$. Use the fact that the graph of n is not dense in $\mathbf{C}$ to obtain that $n(x) = Bx$, and, from this, conclude that $B = 0$.)

e) The question now arises as to whether the functional equation has any discontinuous solutions. Prove that, in fact, it does. (*Hint.* Let H be an Hamel basis in $\mathbf{R}$. Then, for $x, y \in \mathbf{R}$,

$$x = \sum r_i(x)h_i(x), \qquad y = \sum r_i(y)h_i(y),$$
$$x + y = \sum r_i(x + y)h_i(x + y),$$

where "h" $\in H$ and "r" $\in \mathbf{Q}$. By the uniqueness of representation, $r_i(x) + r_i(y) = r_i(x + y)$. If r_i were a continuous function then $r_i(x) = Cx$ for each $x \in \mathbf{R}$ and this contradicts the fact that $r_i(x) \in \mathbf{Q}$.)

4.36 Let $f, f' \subset L_m^1(\mathbf{R})$.

a) Prove that if f is absolutely continuous on each interval $[a, b] \subseteq \mathbf{R}$ then $\lim_{|x| \to \infty} f(x) = 0$. (*Hint.* Use FTC over finite intervals and obtain a contradiction to the hypothesis that $f \in L_m^1(\mathbf{R})$.)

b) Give an example to show that the hypothesis of absolute continuity is necessary in part a). (*Hint.* "Put the Cantor set C in $I_n = [n, n + (1/2^n)]$". Then define the function f_n to have value 0 outside of I_n and to be a modified Cantor function ranging from 0 to 1 to 0 in I_n; set $f = \sum f_n$.) If we don't require f to remain continuous then $f = \chi_{\mathbf{Z}}$ provides a trivial counterexample.

4.37 Observe that if $f \in L_m^1(\mathbf{T})$ and $\sum |n\hat{f}(n)| < \infty$ then f is absolutely continuous; in fact

$$|\sum (f(a_{j+1}) - f(a_j))| \leqslant 2(\sum (a_{j+1} - a_j)) \sum_n |n\hat{f}(n)|.$$

Prove that the absolute continuity follows by the weaker hypotheses that $f \in L_m^1(\mathbf{T})$ and $\sum |n\hat{f}(n)|^2 < \infty$. Actually, what you prove is that f is equal *m-a.e.* to an absolutely continuous function F and that $F' \in L^2(\mathbf{T})$.

4.38 Suppose that $f \in L_m^1(\mathbf{R})$ and $g \in L_m^\infty(\mathbf{R})$. Find general conditions on f and g so that

$$\lim_{x \to \infty} f * g'(x) = 0 \qquad \text{when} \qquad \lim_{x \to \infty} f * g(x) = 0$$

(where the convolution $f * g$ was defined in Problem 3.6b).

4.39 Let f and g be continuous elements $\mathbf{R} \to \mathbf{R}$ in $L_m^1(\mathbf{R})$. Assume that f is continuously differentiable and vanishes outside of some finite interval. Prove that $(f * g)'$ exists and that

$$(f * g)' = f' * g. \tag{P 4.7}$$

To what extent can the conditions on f and g be relaxed so that (P4.7) is still valid.

4.40 Let $Y \subseteq \mathbf{R}$ have the property that $m(\mathbf{R} \setminus Y) = 0$ and let $f: \mathbf{R} \times \mathbf{R} \to \mathbf{R}$ be a function such that for each $y \in Y$ $D_x f(x, y)$ exists m-a.e. Give reasonable hypotheses and a rigorous proof to Leibnitz's rule for differentiating an integral:

$$D_x \int_{g(x)}^{h(x)} f(x, y) \, dy = [f(x, h(x))h'(x) - f(x, g(x))g'(x)]$$

$$+ \int_{g(x)}^{h(x)} D_x f(x, y) \, dy.$$

If $g(x) = a$ and $h(x) = b$ then we are in the setting of Theorem 3.16. If $f(x, y)$ is of the form $f(y)$, and if $g(x) = a$ and $h(x) = x$, then we are in the framework of Theorem 4.9.

4.41 Prove that the composition of two absolutely continuous functions need not be absolutely continuous or even of bounded variation, cf. Problem 4.10 and Problem 4.28.

4.42 Let $f: (0, \infty) \to \mathbf{R}$ be a positive strictly increasing function which is absolutely continuous on each finite interval $[a, b] \subseteq (0, \infty)$. Prove that if $f(x) = O(x^2)$, as $x \to \infty$, then $\int_0^\infty dx/f'(x)$ diverges. (*Hint.* For each $s > r > 0$ we have

$$(s - r)^2 = \left(\int_r^s dx \right)^2 \leqslant (f(s) - f(r)) \int_r^s dx/f'(x)$$

and so there is a constant A such that for each $r > 0$

$$A \leqslant \int_r^\infty dx/f'(x).)$$

5 Spaces of measures and the Radon–Nikodym theorem

5.1 Signed and complex measures, and the basic decomposition theorems

Let $(X, \mathscr{A})$ be a measurable space. A function

$$\mu: \mathscr{A} \to \mathbf{R}^* \quad (\text{resp.,} \quad \mathbf{C})$$

is a signed measure (resp. complex measure) if $\mu(\emptyset) = 0$ and

$$\mu\left(\bigcup_1^\infty A_n\right) = \sum_1^\infty \mu A_n$$

for every disjoint sequence $\{A_n: n = 1, ...\}$. If μ is a signed (resp. complex) measure we shall refer to the symbol $(X, \mathscr{A}, \mu)$ as a signed (resp. complex) measure or a signed (resp. complex) measure space. Note that every measure on $(X, \mathscr{A})$ is a signed measure.

Proposition 5.1 Let $(X, \mathscr{A}, \mu)$ be a signed measure. Then μ cannot take both $+\infty$ and $-\infty$ as values.

Remark Note that if $(X, \mathscr{A}, \mu)$ is a measure space and $f \in L^1_\mu(X)$ then

$$\forall A \in \mathscr{A}, \qquad \rho A = \int_A f \, d\mu$$

defines a complex measure ρ, and this canonical representation is one of the initial reasons we study complex and signed measures. The Radon–Nikodym theorem, which we designate by R–N, is the means to obtain a converse of this observation as well as being a major step in determining an FTC for abstract measure spaces. In fact we shall characterize "absolutely continuous measures" in terms of the above formula.

Let $(X, \mathscr{A}, \mu)$ be a signed measure. $A \in \mathscr{A}$ is non-negative (resp. non-positive) if
$$\forall E \subseteq A, \text{ for which } E \in \mathscr{A}, \text{ we have } \mu E \geqslant 0 \text{ (resp. } \mu E \leqslant 0).$$
For each $A \in \mathscr{A}$ we define

$$\mu^+ A = \sup \{\mu E: E \subseteq A, E \in \mathscr{A}\}$$

and $\qquad \mu^- A = \sup \{-\mu E: E \subseteq A, E \in \mathscr{A}\}.$

μ^+ (resp. μ^-) is the positive (resp. negative) variation of μ.

Theorem 5.1 Let $(X, \mathscr{A}, \mu)$ be a signed measure. $(X, \mathscr{A}, \mu^+)$ and $(X, \mathscr{A}, \mu^-)$ are measure spaces.

Proof. Clearly, $\mu^{\pm}(\varnothing) = 0$ since $\mu(\varnothing) = 0$.

For each $A \in \mathscr{A}$, $\varnothing \subseteq A$, and so $\mu^{\pm} A \geqslant 0$.

Given a disjoint sequence $\{A_n: n = 1, \ldots\} \subseteq \mathscr{A}$. We'll prove that $\mu^+(\bigcup A_n) = \sum \mu^+ A_n$; the same argument works for μ^-.

We need to know that if $A \subseteq B$, where $A, B \in \mathscr{A}$, then

$$\mu^+ A \leqslant \mu^+ B.$$

Take $E \subseteq A$, where $E \in \mathscr{A}$; then $\mu E \leqslant \mu^+ B$ by the definition of $\mu^+ B$ and since $A \subseteq B$. Because this holds for all such E we conclude that $\mu^+ A \leqslant \mu^+ B$.

If $\mu^+ A_n = \infty$ for some n then $\mu^+ A_n \leqslant \mu^+(\bigcup A_j)$ implies that $\mu^+(\bigcup A_j) = \infty$; consequently, $\mu^+ \geqslant 0$ and $\mu^+ A_n = \infty$ yield the countable additivity of μ^+.

Therefore, assume that $\mu^+ A_n < \infty$ for each n.

Let $\varepsilon > 0$ and choose $E_n \subseteq A_n$, where $E_n \in \mathscr{A}$, such that $\mu E_n \geqslant \mu^+ A_n - (\varepsilon/2^n)$.

$\{E_n: n = 1, \ldots\}$ is a disjoint sequence since $\{A_n: n = 1, \ldots\}$ is disjoint. Thus, by the countable additivity of μ,

$$\mu(\bigcup E_n) = \sum \mu E_n \geqslant \sum \mu^+ A_n - \varepsilon;$$

and by the definition of μ^+,

$$\mu^+(\bigcup A_n) \geqslant \sum \mu^+ A_n.$$

For the opposite inequality take any $\lambda < \mu^+(\bigcup A_n)$; it is sufficient to prove that $\lambda < \sum \mu^+ A_n$.

Pick $E \subseteq \bigcup A_n$ for which $\mu E > \lambda$. Since μ is countably additive,

$$\mu E = \sum \mu(E \cap A_n).$$

Because $\mu(E \cap A_n) \leqslant \mu^+ A_n$ we conclude that

$$\lambda < \mu E \leqslant \sum \mu^+ A_n.$$

q.e.d.

For any signed measure $(X, \mathscr{A}, \mu)$, the total variation $|\mu|$ is defined as:

$$\forall A \in \mathscr{A}, \quad |\mu|(A) = \mu^+ A + \mu^- A.$$

Theorem 5.2 Let $(X, \mathscr{A}, \mu)$ be a signed measure.

a) $(X, \mathscr{A}, |\mu|)$ is a measure space.

b) The non-negative (resp. non-positive) sets are closed under countable unions.

c) If $A \in \mathscr{A}$ is non-negative (resp. non-positive) then $\mu A = \mu^+ A$ (resp. $\mu A = -\mu^- A$).

d) If $\mu A = \mu^+ A < \infty$ (resp. $-\mu A = \mu^- A < \infty$) then A is non-negative (resp. non-positive).

Proof. a) is clear from Theorem 5.1 and b) is immediate.

c) Let A be non-negative so that $\mu B \geqslant 0$ for each $B \in \mathscr{A}$ contained in A. For such B we have $\mu B \leqslant \mu A$ (since $A \setminus B \subseteq A$); consequently, by the definition of μ^+, $\mu^+ A \leqslant \mu A$. Because A is such a B, $\mu^+ A = \mu A$.

d) If A is not non-negative there is $B \subseteq A$, for which $B \in \mathscr{A}$, such that $\mu B < 0$. Because μ is additive and $A = (A \setminus B) \cup B$, $\mu A = \mu B + \mu(A \setminus B)$. Thus, if $\mu A < \infty$ (our hypothesis) we have $\mu A < \mu(A \setminus B)$ since $\mu B < 0$.

On the other hand $A \setminus B \subseteq A$ implies $\mu^+ A \geqslant \mu(A \setminus B)$, and so $\mu^+ A > \mu A$ which is contrary to our hypothesis.

q.e.d.

Proposition 5.2 *Let $(X, \mathscr{A}, \mu)$ be a signed measure space. For each $A \in \mathscr{A}$ there is $E \subseteq A$, for which $E \in \mathscr{A}$, such that*

$$\mu E = \mu^+ E \geqslant \mu A.$$

Proof. For the first case let $\mu^+ A < \infty$ (and so $\mu A < \infty$). We shall inductively define a sequence $\{A_n : n = 1, \ldots\} \subseteq \mathscr{A}$. Let $A_0 = A$ and note that

$$\mu^+ A_0 \geqslant \mu A_0 \geqslant \mu A.$$

If $\mu^+ A_0 = \mu A_0$ take $E = A_0$ and we are done. If $\mu^+ A_0 \neq \mu A_0$ then there is $A_1 \subseteq A_0$ such that

$$\mu^+ A_0 \geqslant \mu A_1 > \mu A_0; \tag{5.1}$$

further, take A_1 such that $\mu A_1 > \mu^+ A_0 - 1$ (we can do this by the definition of μ^+ and since $\mu^+ A_0 < \infty$). Since $\mu^+ A_1 \geqslant \mu A_1$,

$$\mu^+ A_1 \geqslant \mu A_1 > \mu A_0 \tag{5.2}$$

by (5.1). If $\mu^+ A_1 = \mu A_1$ take $E = A_1$ and we are done by (5.2).

If $\mu^+ A_1 > \mu A_1$, there is $A_2 \subseteq A_1$ such that

$$\mu^+ A_1 \geqslant \mu A_2 > \mu A_1, \qquad \mu A_2 > \mu^+ A_1 - 1,$$

and again, since $\mu^+ A_2 \geqslant \mu A_2$,

$$\mu^+ A_2 \geqslant \mu A_2 > \mu A_1 > \mu A_0.$$

Proceeding in this way we are either done at the n-th step or else we determine a decreasing sequence $\{A_n : n = 1, \ldots\} \subseteq \mathscr{A}$ such that

$$\mu^+ A_n \geqslant \mu A_n > \mu A \tag{5.3}$$

and $\quad\quad \mu A_n > \mu^+ A_{n-1} - \dfrac{1}{n-1}. \tag{5.4}$

Set $E = \bigcap A_n$.

Since μ^+ is a measure and $\mu^+ A_0 < \infty$ we have

$$\mu^+ E = \lim \mu^+ A_n \geqslant \mu A.$$

Thus it remains to show that $\mu E \geqslant \mu^+ E$.

If E is non-negative we are done by Theorem 5.2c. Assume not and take $F \subseteq E$ satisfying $\mu F < 0$. We'll obtain a contradiction.

Choose n for which $\mu F < -1/n$. Note that

$$\mu(A_{n+1} \setminus F) = \mu A_{n+1} - \mu F > \mu A_{n+1} + (1/n). \tag{5.5}$$

To prove this observe that $A_{n+1} = (A_{n+1} \setminus F) \cup F$, $\mu A_{n+1} = \mu F + \mu(A_{n+1} \setminus F)$, and $\mu A_{n+1} \in \mathbf{R}$ by (5.3) and the fact that $\infty > \mu^+ A \geqslant \mu^+ A_{n+1} \geqslant \mu A_n$. Consequently, $\mu F \in \mathbf{R}$ and we can perform the algebra to obtain (5.5).

Next we observe that from the monotonicity of $\{A_n : n = 1, \ldots\}$,

$$\mu^+ A_n \geqslant \mu^+ A_{n+1} \geqslant \mu^+(A_{n+1} \setminus F) \geqslant \mu(A_{n+1} \setminus F).$$

Thus, by (5.5), $\mu^+ A_n > \mu A_{n+1} + (1/n)$, and this contradicts (5.4); consequently, the result is true if $\mu^+ A < \infty$.

For the second case let $\mu^+ A = \infty$. As in the first case we form $\{A_n : n = 1, \ldots\} \subseteq \mathscr{A}$, a decreasing sequence, such that $A_n \subseteq A$ and

$$\mu^+ A_n > \mu A_{n+1} \geqslant \mu A_n \geqslant \mu A \tag{5.6}$$
$$\mu A_{n+1} \geqslant n + 1;$$

further, $\mu^+ A_n = \infty$, for otherwise we can argue as in the first part.

Again, set $E = \bigcap A_n$.

For each n,

$$A_n = E \cup \bigcup_{j=n}^{\infty} (A_j \setminus A_{j+1}),$$

so that from (5.6),

$$\infty > \mu A_n = \mu E + \sum_{n}^{\infty} \mu(A_j \setminus A_{j+1}) \geqslant n. \tag{5.7}$$

Thus μE is a fixed finite number and since $\mu A_n \to \infty$ we have that $\lim_{n} \sum_{n}^{\infty} \mu(A_j \setminus A_{j+1}) = \infty$.

On the other hand $\mu(A_j \setminus A_{j+1}) \leqslant 0$ by (5.6) and the fact that $\mu A_j = \mu(A_j \setminus A_{j+1}) + \mu A_{j+1}$, where μA_j is finite; this gives a contradiction.

Hence $\mu^+ A_n < \infty$ for some n and we apply the first part.

q.e.d.

We shall use Proposition 5.2 to prove the Hahn decomposition theorem. There are several proofs including one where it is a corollary of the R–N theorem. Hahn's proof of 1928 [48] and Sierpiński's [106] are quite efficient and do not use R–N. R. Frank [40] has given a proof using transfinite numbers. Hahn's original proof was given in 1921.

Theorem 5.3 (*Hahn decomposition*) *Let* $(X, \mathscr{A}, \mu)$ *be a signed measure space. There is a non-negative set* $P \in \mathscr{A}$ *such that* $N = X \setminus P$ *is non-positive.*

Proof. Assume $\mu^+ X < \infty$. The case that $\mu^+ X = \infty$ is Problem 5.2.

Set $\lambda = \sup \{\mu^+ A : A \in \mathscr{A} \text{ and } \mu^- A = 0\}$. The condition that $\mu^- A = 0$ means that if $E \subseteq A$ then $\mu E \geqslant 0$, and so A is non-negative. There is no problem about the existence of λ since $\mu^- \varnothing = 0$.

Since μ^+ is a measure and $\mu^+ X < \infty$ we have $0 \leqslant \lambda < \infty$.

Pick a sequence $\{A_n : n = 1, \ldots\} \subseteq \mathscr{A}$ such that $\mu^- A_n = 0$ and $\mu^+ A_n \to \lambda$. Define $P = \bigcup A_n$.

Note that $\mu^- P = 0$ since $\mu^- P \leqslant \sum \mu^- A_n$; hence P is a non-negative set.

Because $A_n \subseteq P$ we have $\mu^+ P \geqslant \mu^+ A_n$, and so $\mu^+ P \geqslant \lambda$.

On the other hand, $P \in \mathscr{A}$ and $\mu^- P = 0$ imply that $\lambda \geqslant \mu^+ P$. Thus $\lambda = \mu^+ P$.

Our final step is to prove that $N = X \setminus P$ is non-positive.

Assume no so that there is $B \subseteq N$ for which $\mu B > 0$.

By Proposition 5.2 there is $E \subseteq B$ such that $\mu^+ E = \mu E \geqslant \mu B > 0$.

Consequently, since $\mu^+ X < \infty$, we can apply Theorem 5.2d to ascertain that E is non-negative.

Because $E \cap P = \varnothing$,

$$\mu^+ (P \cup E) = \mu^+ P + \mu^+ E = \lambda + \mu^+ E > \lambda;$$

also, from the fact that $\mu^- P = 0$ (as we observed above),

$$\mu^- (P \cup E) = \mu^- P + \mu^- E = \mu^- E = 0,$$

where the last equality follows since E is non-negative (in fact, $F \subseteq E$ implies $\mu F \geqslant 0$ and

$$0 \leqslant \mu^- E = \sup \{-\mu F : F \subseteq E\} \leqslant 0).$$

Thus, we have a contradiction to the definition of λ by taking $A = P \cup E$. Hence N is non-positive.

q.e.d.

The sets P and N are an Hahn decomposition of the signed measure $(X, \mathscr{A}, \mu)$.

Remark 1 Given a signed measure $(X, \mathscr{A}, \mu)$. If $A \in \mathscr{A}$ satisfies

$$-\mu^- A = \mu A = \mu^+ A$$

then A is a null set. $A \in \mathscr{A}$ is null if and only if

$$\forall B \subseteq A, \, B \in \mathscr{A}, \qquad \mu B = 0.$$

Thus if A is null, $\mu A = 0$, but the converse is not true. It is a routine exercise to prove that the Hahn decomposition is not generally unique, whereas it is unique up to null sets (e.g. Problem 5.6).

Remark 2 We shall now give the Jordan decomposition of a signed measure $(X, \mathscr{A}, \mu)$. There are two points to be made. First, the Jordan decomposition is intimately related to the decomposition of a bounded variation function as a difference of increasing functions. Recall that increasing functions give rise to measures, and, in fact, bounded variation functions give rise to signed measures. The Riesz representation theorem gives a further relation between functions of bounded variation and signed measures. The second point is the relation between these two decompositions. Given an Hahn decomposition P, N of a signed measure $(X, \mathscr{A}, \mu)$, it is easy to compute that

$$\mu = v^+ - v^-,$$

where $\qquad \forall A \in \mathscr{A}, \qquad v^+ A = \mu(A \cap P) \quad$ and $\quad v^- A = -\mu(A \cap N).$

On the other hand we shall prove that

$$\mu = \mu^+ - \mu^-$$

independent of Hahn's result and then show that $\mu^\pm = v^\pm$.

Proposition 5.3 *Given a signed measure* $(X, \mathscr{A}, \mu)$. *If* $\mu^+ X = \infty$ *then there is* $A \in \mathscr{A}$ *such that* $\mu A = \infty$.

Proof. By the definition of μ^+ we can choose a sequence $\{A_n : n = 1, ...\} \subseteq \mathscr{A}$ such that $\mu A_n > n$ for each n. From Proposition 5.2 choose $E_n \subseteq A_n$ for which $\mu^+ E_n = \mu E_n \geqslant \mu A_n > n$. We shall prove that

$$\mu(\bigcup E_n) = \mu^+(\bigcup E_n);$$

this is sufficient since for $A = \bigcup E_n$, $\mu^+ A \geqslant \mu^+ E_n > n$, and so $\mu^+ A = \infty$.

Without loss of generality take $\infty > \mu E_n = \mu^+ E_n > n$ for each n.

From Theorem 5.2d, E_n is non-negative. Consequently, if $F \subseteq \bigcup E_n$ and $\bigcup E_n = \bigcup D_n$, where $\{D_n : n = 1, ...\}$ is a disjoint sequence and $D_n \subseteq E_n$, we have $\mu(F \cap D_n) \geqslant 0$ and

$$\mu F = \sum \mu(F \cap D_n) \geqslant 0.$$

Thus $\bigcup E_n$ is non-negative.

Finally, we apply Theorem 5.2c to conclude that $\mu A = \mu^+ A$.

$$\text{q.e.d.}$$

We could have proved Proposition 5.3 using the Hahn decomposition (cf. Problem 5.2): $\mu^+ X = \mu^+ P + \mu^+ N = \mu^+ P = \mu P$.

Theorem 5.4 (*Jordan decomposition*) *Let* $(X, \mathscr{A}, \mu)$ *be a signed measure. Then* $\mu = \mu^+ - \mu^-$; *in particular, either* μ^+ *or* μ^- *is bounded.*

Proof. Given $A \in \mathscr{A}$ for which $|\mu A| < \infty$. Take $B \subseteq A$. Then

$$\mu A = \mu B + \mu(A \setminus B) \tag{5.8}$$

and so $|\mu B|, |\mu(A \setminus B)| < \infty$. Therefore,

$$\mu B = \mu A - \mu(A \setminus B) \leqslant \mu A + \mu^- A, \tag{5.9}$$

and, consequently,

$$\mu^+ A \leqslant \mu A + \mu^- A$$

since (5.9) is true for all $B \subseteq A$.

To prove that

$$\mu A \geqslant \mu^+ A - \mu^- A \tag{5.10}$$

we must show that $\mu^- A < \infty$ when $|\mu A| < \infty$.

If $\mu^- A = \infty$, then from Proposition 5.3 there is $E \subseteq A$ for which $\mu E = -\infty$. Choose B above satisfying $E = A \setminus B$. Because $|\mu A| < \infty$ we conclude that $|\mu B|, |\mu E| < \infty$ from (5.8). Thus we have a contradiction and (5.10) follows.

Using (5.8) again we have $-\mu(A \setminus B) = \mu B - \mu A$ and therefore

$$\mu^- A \leqslant \mu^+ A - \mu A.$$

This fact combined with (5.10) yields a) for the case that $|\mu A| < \infty$.

If $\mu A = -\infty$, then $\mu^- A = \infty$, and so we must prove that $|\mu^+ A| < \infty$. If $\mu^+ A = \infty$ then there is $B \subseteq A$ for which $\mu B = \infty$ by Proposition 5.3, and this contradicts Proposition 5.1 since we assigned $\mu A = -\infty$.

The case $\mu A = \infty$ is treated similarly.

q.e.d.

The formula, $\mu = \mu^+ - \mu^-$, is the Jordan decomposition of μ.

Theorem 5.5 *Let* $(X, \mathscr{A}, \mu)$ *be a signed measure space.*
a) *If* P, N *is an Hahn decomposition of* X *then*

$$\forall A \in \mathscr{A}, \qquad \mu^+ A = \mu(A \cap P) \quad \text{and} \quad \mu^- A = -\mu(A \setminus P) = -\mu(A \cap N).$$

b) *If* $\mu = \mu_1 - \mu_2$, *where the* μ_j *are measures, then* $\mu_1 \geqslant \mu^+$ *and* $\mu_2 \geqslant \mu^-$.
c) *For each* $A \in \mathscr{A}$,

$$|\mu|(A) = \sup \left\{ \sum_{k=1}^{n} |\mu A_k| : \{A_1, ..., A_n\} \subseteq \mathscr{A} \text{ is a finite decomposition of } A \right\}$$

$$= \mu_A.$$

$(\{A_1, ..., A_n\} \subseteq \mathscr{A}$ *is a finite decomposition of A if* $A = \overset{n}{\underset{1}{\bigcup}} A$ *and* $A_j \cap A_k = \emptyset$ *if*
$j \neq k$).

d) *If for each* $A \in \mathscr{A}$, $|\mu A| < \infty$, *then* $K_\mu = \max\{\mu^+ X, \mu^- X\} < \infty$ *and* $\forall A \in \mathscr{A}$,
$|\mu A| \leqslant K_\mu$.

Proof. a) From the Jordan decomposition of μ,

$$\forall A \in \mathscr{A}, \qquad \mu(A \cap P) = \mu^+(A \cap P) - \mu^-(A \cap P).$$

Since P is non-negative, $A \cap P$ is non-negative and so $\mu(A \cap P) = \mu^+(A \cap P) \geqslant 0$
(and hence $\mu^-(A \cap P) = 0$). Also

$$0 \leqslant \mu^+(A \setminus P) \leqslant \mu^+(X \setminus P) = \mu^+ N = 0$$

from Theorem 5.4 and the fact that N is non-positive. Thus,

$$\mu^+ A = \mu^+(A \cap P) + \mu^+(A \setminus P) = \mu^+(A \cap P) = \mu(A \cap P).$$

b) If $\mu = \mu_1 - \mu_2$ then $\mu_1 \geqslant \mu$ and so

$$\forall A \in \mathscr{A}, \qquad \mu^+ A = \mu(A \cap P) \leqslant \mu_1(A \cap P) \leqslant \mu_1 A.$$

Hence $\mu_1 \geqslant \mu^+$.
If $|\mu A| < \infty$, $\mu^- A = \mu^+ A - \mu A \leqslant \mu_1 A - \mu A = \mu_2 A$.
The case $|\mu A| = \infty$ is Problem 5.5.
c) Given $A \in \mathscr{A}$ and a finite decomposition $\{A_1, ..., A_n\}$ of A. Since $|\mu|$ is a measure
and $\{A_j : j = 1, ..., n\}$ is disjoint,

$$\overset{n}{\underset{1}{\sum}} |\mu A_j| = \overset{n}{\underset{1}{\sum}} |\mu^+ A_j - \mu^- A_j| \leqslant \overset{n}{\underset{1}{\sum}} (\mu^+ A_j + \mu^- A_j) = \overset{n}{\underset{1}{\sum}} |\mu|(A_j) = |\mu|(A).$$

Thus $|\mu|(A) \geqslant \mu_A$.
For the opposite inequality take an Hahn decomposition P, N of X so that

$$A = (A \cap P) \cup (A \cap N).$$

Then $\qquad \mu_A \geqslant |\mu(A \cap P)| + |\mu(A \cap N)| = \mu^+ A + \mu^- A = |\mu|(A).$

d) If $\mu^+ X = \infty$ then we obtain a contradiction to our hypothesis since, by Proposition
5.3, there is $A \in \mathscr{A}$ for which $\mu A = \infty$.
Hence $\mu^+ X < \infty$. For each $A \in \mathscr{A}$ we choose E as in Proposition 5.2, and therefore

$$\mu A \leqslant \mu E = \mu^+ E \leqslant \mu^+ X < \infty.$$

Using part a) we compute

$$\forall A \in \mathscr{A}, \qquad \mu A = \mu^+ A - \mu^- A = \mu(A \cap P) + \mu(A \cap N) \geqslant \mu(A \cap N)$$
$$= -\mu^- A.$$

Consequently,

$$\mu A \geqslant -\mu^- A \geqslant -\mu^- X > -\infty$$

by Proposition 5.3 again.

Hence

$$\forall A \in \mathscr{A}, \qquad |\mu A| \leqslant \max \{\mu^+ X, \mu^- X\}.$$

q.e.d.

We expand on the first remark of this section concerning the construction of signed measures (and R–N as a converse) in the following

Proposition 5.4 *Given a measure space* $(X, \mathscr{A}, \mu)$. *Let*

$$f: X \to \mathbf{R}^*$$

be a μ-*measurable function and assume that*

$$\int_X f \, d\mu$$

exists. Define

$$\forall A \in \mathscr{A}, \; vA = \int_A f \, d\mu.$$

Then $(X, \mathscr{A}, v)$ *is a signed measure,*

$$v^{\pm} A = \int_A f^{\pm} \, d\mu, \qquad \text{and} \qquad |v|(A) = \int_A |f| \, d\mu. \tag{5.11}$$

Also, $P = \{x: f(x) > 0\}$, $N = X \setminus P$ *is an Hahn decomposition for* v; *and if neither* v^+ *nor* v^- *is identically zero then*

$$\exists A \in \mathscr{A} \quad \text{such that} \quad |v|(A) > |vA|.$$

$(f^+ (\text{resp. } f^-) = 0 \; (\text{resp. } f) \; \text{if } f \leqslant 0 \text{ and } f^+ \; (\text{resp. } f^-) = f \; (\text{resp. } 0) \; \text{if } f > 0).$

Let μ be a complex measure with real and imaginary parts μ_r and μ_i, respectively. Clearly, μ_r and μ_i are signed measures (with values in $\mathbf{R}$). From Theorem 5.4

$$\forall A \in \mathscr{A}, \qquad \mu A = \mu_r^+ A - \mu_r^- A + i\mu_i^+ A - i\mu_i^- A.$$

We would now like to define the total variation of μ; it is tempting to write it as

$$\mu_r^+ + \mu_r^- + \mu_i^+ + \mu_i^-.$$

In fact we shall be most interested in results that depend on certain decompositions of X; and so we take our cue from Theorem 5.5c and define the total variation $|\mu|$ of a complex measure μ to be

$$\forall A \in \mathscr{A},$$

$$|\mu|(A) = \sup \left\{ \sum_{j=1}^{n} |\mu A_j| : \{A_j : j = 1, ..., n\} \text{ is a finite decomposition of } A \right\}$$

$$= \mu_A.$$

Remark The boundedness of $|\mu|$ below can also be proved quite independent of our measure theoretic argument by using the following fact:

$$\forall z_1, ..., z_n \in \mathbf{C} \ \exists S \subseteq \{1, ..., n\} \quad \text{such that} \quad \left| \sum_{j \in S} z_j \right| \geqslant \frac{1}{\pi} \sum_{1}^{n} |z_j|.$$

Bourbaki has observed that the constant, $1/\pi$, is best possible (e.g. [12]). A deeper study of this inequality is found in [62] and there is an interesting refinement due to Grahame Bennett. We'll employ this sort of result in Chapter 6.

Theorem 5.6 Let $(X, \mathscr{A}, \mu)$ be a complex measure space.
a) $\forall A \in \mathscr{A}, |\mu|(A) \leqslant \mu_r^+ A + \mu_r^- A + \mu_i^+ A + \mu_i^- A.$
b) $(X, \mathscr{A}, |\mu|)$ is a finite measure space.
c) $\sup \{|\mu B| : B \in \mathscr{A}, B \subseteq A \in \mathscr{A}\} \leqslant |\mu|(A) \leqslant 4 \sup \{|\mu B| : B \in \mathscr{A}, B \subseteq A \in \mathscr{A}\}.$

Proof. a) Let $\{A_j : j = 1, ..., n\}$ be a finite decomposition of $A \in \mathscr{A}$. Then

$$\sum_{1}^{n} |\mu A_j| = \sum_{j=1}^{n} |\mu_r^+ A_j - \mu_r^- A_j + i\mu_i^+ A_j - i\mu_i^- A_j|$$

$$\leqslant \sum_{j=1}^{n} (\mu_r^+ A_j + \mu_r^- A_j + \mu_i^+ A_j + \mu_i^- A_j)$$

$$= \mu_r^+ A + \mu_r^- A + \mu_i^+ A + \mu_i^- A;$$

and we are done by taking the sup on the left-hand side.
b) From Theorem 5.5d, for each $A \in \mathscr{A}$,

$$|\mu|(A) \leqslant \mu_r^+ A + \mu_r^- A + \mu_i^+ A + \mu_i^- A \leqslant \mu_r^+ X + \mu_r^- X + \mu_i^+ X + \mu_i^- X$$
$$\leqslant 2(K_{\mu_r} + K_{\mu_i}) < \infty,$$

and so $|\mu|$ is bounded.

Now to prove that $|\mu|$ is a measure. Clearly, $|\mu|(\varnothing) = 0$ by the definition of $|\mu|$. Let $\{A_n : n = 1, ...\} \subseteq \mathscr{A}$ be a disjoint family and set $A = \bigcup_{1}^{\infty} A_n$.

Take $\alpha < |\mu|(A)$. Note that for any finite decomposition $\{B_j : j = 1, ..., k\} \subseteq \mathscr{A}$ of A, which satisfies $\alpha < \sum_1^k |\mu B_j|$ (which we can do by the definition of $|\mu|$), we have

$$\alpha < \sum_{n=1}^{\infty} \sum_{j=1}^{k} |\mu(B_j \cap A_n)|. \tag{5.12}$$

Then since $\sum_{j=1}^{k} |\mu(B_j \cap A_n)| \leqslant |\mu|(A_n)$ and $\alpha < |\mu|(A)$ is arbitrary,

$$|\mu|(A) \leqslant \sum_{n=1}^{\infty} |\mu|(A_n). \tag{5.13}$$

Take $\varepsilon > 0$. For each j let $\{B_{j,1}, ..., B_{j,n_j}\} \subseteq \mathscr{A}$ be a finite decomposition of A_j for which

$$\sum_{k=1}^{n_j} |\mu(B_{j,k})| > |\mu|(A_j) - \frac{\varepsilon}{2^j}.$$

For each m,

$$\sum_{j=1}^{m} |\mu|(A_j) < \sum_{j=1}^{m} \frac{\varepsilon}{2^j} + \sum_{j=1}^{m} \sum_{k=1}^{n_j} |\mu(B_{j,k})| < \varepsilon + |\mu|(A) \tag{5.14}$$

by the definition of $|\mu|$ and since $\bigcup \{B_{j,k} : k = 1, ..., n_j, j = 1, ..., m\} \subseteq A$ is a disjoint union.
Consequently,

$$\sum_{j=1}^{\infty} |\mu|(A_j) < \varepsilon + |\mu|(A),$$

and so

$$\sum_{j=1}^{\infty} |\mu|(A_j) \leqslant |\mu|(A).$$

This, coupled with (5.13), yields the desired countable additivity.
c) The first inequality is clear. For the second note that

$$|\mu_r^+ A| = |\mu_r(A \cap P_r)| \leqslant |\mu(A \cap P_r)|,$$

and apply a).

q.e.d.

Let $(X, \mathscr{A})$ be a measurable space and let $M(X)$ (resp. $SM(X)$) be the set of complex (resp. signed) measures. $M(X)$ is a complex vector space where

$$\forall \alpha, \beta \in \mathbf{C} \quad \text{and} \quad \forall A \in \mathscr{A}, \qquad (\alpha\mu + \beta\nu)(A) = \alpha\mu A + \beta\nu A;$$

and it is a normed space with norm

$$\|\mu\| = \|\mu\|_1 = |\mu|(X)$$

(cf. Problem 5.18). We shall frequently refer to the space "$M(X)$" without making special note of the corresponding measurable space $(X, \mathscr{A})$. If X is a locally compact space we say that a complex measure μ is regular if $|\mu|$ is regular. In this case $M(X)$ will denote the space of bounded regular complex measures (cf. Appendix III.2) and, because of the Jordan decomposition again, we refer to $M(X)$ as the space of bounded regular Borel measures.

Theorem 5.7 *Let $\mu \in M(X)$.*

a) *A function $f: X \to \mathbf{C}$ defined $|\mu|$-a.e. is in $L^1_{|\mu|}(X)$ if and only if*

$$f \in L^1_{\mu_r^+}(X) \cap L^1_{\mu_r^-}(X) \cap L^1_{\mu_i^+}(X) \cap L^1_{\mu_i^-}(X).$$

b) *For each $f \in L^1_{|\mu|}(X)$,*

$$\mu(f) = \int_X f \, d\mu = \int_X f \, d\mu_r^+ - \int_X f \, d\mu_r^- + i \int_X f \, d\mu_i^+ - i \int_X f \, d\mu_i^-$$

is well-defined, and $\mu: L^1_{|\mu|}(X) \to \mathbf{C}$ is a linear functional.

5.2 Discrete and continuous, absolutely continuous and singular measures

In this section we consider the problems for measures that were initiated for functions of bounded variation in Chapter 4.2. Let $(X, \mathscr{A})$ be a measurable space and take $\mu, \, \nu \in (S)M(X)$ (i.e. μ (resp. ν) is in either $M(X)$ or $SM(X)$). μ is **continuous** (denoted by $\mu \in (S)M_c(X)$) if

$$\forall x \in X, \qquad \mu(\{x\}) = 0;$$

and μ is **discrete** (denoted by $\mu \in M_d(X)$) if

$$\mu = \sum_{x \in D} a_x \delta_x,$$

where $D \subseteq X$ is countable, $\sum_{x \in D} |a_x| < \infty$, and

$$\forall A \in \mathscr{A}, \qquad \delta_x(A) = \begin{cases} 0, & \text{if } x \notin A \\ 1, & \text{if } x \in A. \end{cases}$$

Clearly, a non-zero continuous (resp. discrete) measure is not discrete (resp. continuous). μ is **absolutely continuous** with respect to ν (denoted by $\mu \ll \nu$) if

$$\forall A \in \mathscr{A}, \qquad |\nu|(A) = 0 \Rightarrow \mu A = 0.$$

Observe that $v \ll \mu$ in Proposition 5.4. To define a singular measure we say (as we did earlier after Example 4.9) that $\eta \in (S)M(X)$ is concentrated on $A \in \mathscr{A}$ if

$$\forall B \in \mathscr{A}, \qquad \eta B = \eta(A \cap B).$$

μ is singular with respect to v (denoted by $\mu \perp v$) if there are disjoint sets $C_\mu, C_v \in \mathscr{A}$ such that μ (resp. v) is concentrated on C_μ (resp. C_v). Obviously, $\mu \perp v$ if and only if $v \perp \mu$, and we say that μ and v are mutually singular.

Remark 1 Our project is to decompose a given μ in terms of the various types of measures defined above and to characterize the absolutely continuous μ with respect to a fixed v. The Radon–Nikodym theorem is the key result for this purpose.

Remark 2 Before proceeding in some generality let us recall the classical situation given in Chapter 4. Consider the measurable space $(\mathbf{R}, \mathscr{B})$, $F \in BV$, and the corresponding complex measure $\mu \in M(\mathbf{R})$ as in Theorem 4.3. $M(\mathbf{R})$ is determined by $(\mathbf{R}, \mathscr{B})$. Then

a) F is a continuous function $\Leftrightarrow$ for each $x \in \mathbf{R}$, $\mu(\{x\}) = 0$ (as defined after Theorem 4.5).

b) F is an absolutely continuous function $\Leftrightarrow \forall A \in \mathscr{M}$, for which $mA = 0$, we have $mF(A) = 0$ (Theorem 4.12).

c) F is a singular function $\Leftrightarrow$ there is $A \subseteq \mathbf{R}$, for which $m(\mathbf{R} \setminus A) = 0$, such that $mF(A) = 0$ (Theorem 4.14).

d) F is a discrete function $\Leftrightarrow$ there is $A \subseteq \mathbf{R}$ such that $\operatorname{card}(\mathbf{R} \setminus A) \leqslant \aleph_0$ and $mF(A) = 0$ (as in the remark after Proposition 4.5).

Clearly, $F(A)$ corresponds to C_μ in c) and d). Also,

$$F \text{ is absolutely continuous} \Rightarrow F \text{ is continuous,}$$

and F is discrete $\Rightarrow F$ is singular.

Since $m(\mathbf{R} \setminus C_m) = 0$, the notion of a "singular function" from Chapter 4 generalizes to the definition given above. The following was proved for functions of bounded variation in Theorem 4.5.

Theorem 5.8 *Given $\mu \in M(X)$. Then there is a unique decomposition*

$$\mu = \mu_c + \mu_d,$$

where $\mu_c \in M_c(X)$ and $\mu_d \in M_d(X)$.

Proof. Because of the Jordan decomposition theorem we take $\mu \geqslant 0$ without loss of generality. Also $\mu X < \infty$ since $\mu \in M(X)$. For each $\varepsilon > 0$ there are at most finitely many points $x \in X$ for which $\mu(\{x\}) > \varepsilon$.

Let Y_m be the finite collection corresponding to $\varepsilon_m = 1/m$.

Set $Y = \bigcup Y_m = \{x_1, ...\}$. For $A \in \mathscr{A}$ define

$$\mu_d A = \sum_j \mu(\{x_j\}) \delta_{x_j} A,$$

noting that $\mu(\{x_j\}) > 0$. Clearly,

$$\mu_d A \leqslant \sum_j \mu(\{x_j\}) \leqslant \mu X < \infty,$$

and hence $\mu_d \in M_d(X)$ (note that we use the fact that $\mu X < \infty$).

Define $\mu_c = \mu - \mu_d$. If $x \notin Y$ then $\mu(\{x\}) = 0 = \mu_d(\{x\})$ by the definition of Y; if $x \in Y$ we choose $A = \{x\}$ and compute $\mu_d A = \mu(\{x\})$ by the definition of μ_d. In either case $\mu_c(\{x\}) = 0$, and so $\mu_c \in M_c(X)$.

The uniqueness is trivial.

q.e.d.

Example 5.1 a) There are continuous measures μ such that $\mu \not\ll m^2$, where m^2 is Lebesgue measure on $X = [0, 1] \times [0, 1]$. In fact, for each Lebesgue measurable set $A \subseteq X$ define

$$\mu A = m^2(A \cap ([0, 1] \times \{0\})).$$

b) Let μ_c be the Cantor–Lebesgue continuous measure on $[0, 1]$ (e.g. Chapter 4.2). Then $mC = 0$ and $\mu_c C = 1$ so that μ_c is continuous (by its definition) but not absolutely continuous with respect to m (cf. Example 4.7).

Proposition 5.5 *Let $(X, \mathscr{A})$ be a measurable space, ν a measure, and $\mu \in (S)M(X)$. If $\mu \ll \nu$ and $\mu \perp \nu$ then $\mu = 0$.*

Proof. Since $\mu \perp \nu$ we have $\nu C_\mu = 0$.
$\mu \ll \nu$ implies $\mu A = 0$ for $A \in \mathscr{A}$ and $A \subseteq C_\mu$ because $\nu C_\mu = 0$. On the other hand $\mu A = 0$ if $A \cap C_\mu = \emptyset$ by the definition of C_μ.

q.e.d.

Proposition 5.6 *Let $\mu \in SM(X)$. $\mu^+ \perp \mu^-$, and if P, N is an Hahn decomposition for μ then we can take $C_{\mu^+} = P$ and $C_{\mu^-} = N$.*

Proof. From Theorem 5.5a, $\mu^+ A = \mu(A \cap P)$ and $\mu^- A = -\mu(A \cap N)$.
On the other hand, since P is non-negative, Theorem 5.2c implies that $\mu^+(A \cap P) = \mu(A \cap P)$.
A similar argument works for μ^- and we are done.

q.e.d.

We now prove the Lebesgue decomposition theorem. It can also be proved alongside R–N or as a corollary of R–N (cf. (5.20)).

Theorem 5.9 *Let* $(X, \mathscr{A}, v)$ *be a measure space, and assume that* μ *satisfies one of the following conditions:*

a) $(X, \mathscr{A}, \mu)$ *is a* σ-*finite measure space, or*

b) $\mu \in M(X)$.

Then there is a unique pair, μ_1 *and* μ_2, *of* σ-*finite measures in the case of* a) *or elements in* $M(X)$ *in the case of* b) *such that*

$$\mu = \mu_1 + \mu_2, \quad \mu_1 \perp v, \quad and \quad \mu_2 \ll v.$$

Proof. We shall prove the result for μ a bounded measure and refer to Problem 5.14 for the remaining cases.

Assume the result is false; thus suppose that whenever μ_1 and μ_2 are bounded positive measures for which $\mu = \mu_1 + \mu_2$ and $\mu_1 \perp v$ then there is $A \in \mathscr{A}$ for which $\mu_2 A > 0$ and $vA = 0$.

We shall first prove that

$$\forall B \in \mathscr{A} \quad \text{such that} \quad v(X \setminus B) = 0, \quad \exists F \subseteq B, \quad \text{where} \quad F \in \mathscr{A},$$

for which

$$vF = 0 \quad \text{and} \quad \mu F > 0.$$

In order to do this we first define the measures

$$\forall E \in \mathscr{A}, \quad \mu_B E = \mu(E \cap B)$$

and $\qquad \forall E \in \mathscr{A}, \quad \mu_{X \setminus B}(E) = \mu(E \cap (X \setminus B)).$

Then $\mu = \mu_B + \mu_{X \setminus B}$ and $v \perp \mu_{X \setminus B}$, so that by hypothesis there is $A \in \mathscr{A}$ for which $vA = 0$ and $\mu_B A > 0$.

Consequently, setting $F = A \cap B$, we have $vA = 0$ and compute

$$\mu F = \mu_B(A \cap B) + \mu_{X \setminus B}(A \cap B) = \mu_B(A \cap B) = \mu(A \cap B) = \mu_B A.$$

Let $\mathscr{G} = \{F \in \mathscr{A} : \mu F > 0 \text{ and } vF = 0\}$. $\mathscr{G} \neq \emptyset$ for we can take $B = X$. Define $S = \{\mu F : F \in \mathscr{G}\} \subseteq \mathbf{R}$. S is bounded below by 0 and above by $\mu X < \infty$.

It is easy to see that $\sup \{r \ \varepsilon \ S\} = \mu F$ for some $F \in \mathscr{G}$. In fact let $\mu F_n \to \sup \{r \in S\}$ and set $F = \bigcup F_n$. Clearly,

$$0 \leqslant vF \leqslant \sum vF_n = 0 \quad \text{and} \quad \mu F \geqslant \mu F_n \Rightarrow \mu F \geqslant \sup \{r \in S\}.$$

Fix this F. $v(X \setminus (X \setminus F)) = 0$ since $vF = 0$ so that by the above reasoning (for "$B = X \setminus F$") there is $D \subseteq X \setminus F$ satisfying $vD = 0$ and $\mu D > 0$.

Thus $F \cup D \in \mathscr{G}$ and $\mu(F \cup D) = \mu F + \mu D > \mu F$ since $D \subseteq X \setminus F$.

This contradicts the fact that $\mu F = \sup \{r \in S\}$.

q.e.d.

The following result tells us that absolutely continuous measures are "continuous" in a reasonable sense and Theorem 5.11 generalizes the fact that "indefinite integrals" (as in Proposition 5.4) are absolutely continuous as well as giving a converse to Theorem 5.10. We also refer to Example 5.4 in this regard.

Theorem 5.10 *Given a measure space* $(X, \mathcal{A}, v)$ *and* $\mu \in M(X)$. *If* $\mu \ll v$ *then for all* $\varepsilon > 0$ *there is* $\delta > 0$ *such that*

$$|vA| < \delta \Rightarrow |\mu A| < \varepsilon.$$

Proof. Let $\mu \geqslant 0$ and assume the result does not hold.

Then there is $\varepsilon > 0$ so that for all n we can find $A_n \in \mathcal{A}$ for which

$$|vA_n| < \frac{1}{2^n} \quad \text{and} \quad |\mu A_n| = \mu A_n \geqslant \varepsilon.$$

Setting $B_n = \bigcup_{k=n+1}^{\infty} A_k$ we have

$$vB_n \leqslant \sum_{n+1}^{\infty} vA_k < \frac{1}{2^n}.$$

Since $\mu \geqslant 0$, $\mu B_n \geqslant \mu A_{n+1} \geqslant \varepsilon$; consequently

$$\mu(\bigcap B_n) = \lim \mu B_n \geqslant \varepsilon,$$

noting that $\mu B_1 < \infty$.

Also $vB_1 < \infty$ and so

$$v(\bigcap B_n) = \lim vB_n \leqslant \lim 1/2^n = 0.$$

This last statement and the assumed absolute continuity contradict the fact that $\mu(\bigcap B_n) \geqslant \varepsilon$.

Now if $\mu = \mu^+ - \mu^-$ then the hypothesis that $\mu \ll v$ implies that $|\mu| \ll v$.

Therefore, from the first part,

$$\forall \varepsilon > 0 \quad \exists \delta > 0 \quad \text{such that} \quad |vA| < \delta \Rightarrow |\mu|(A) < \varepsilon.$$

This concludes the proof since $|\mu|(A) \geqslant |\mu A|$.

q.e.d.

Theorem 5.11 *Given a finite measure space* $(X, \mathcal{A}, v)$ *and let*

$$\mu \colon \mathcal{A} \to \mathbf{C}$$

be a finitely additive measure. Assume that

$$\forall \varepsilon > 0 \quad \exists \delta > 0 \quad \text{such that} \quad |vA| < \delta \Rightarrow |\mu A| < \varepsilon.$$

Then $\mu \in M(X)$ *and* $\mu \ll v$.

Proof. First we prove the countable additivity. Given a disjoint family $\{A_n: n = 1,...\} \subseteq \mathscr{A}$. From the finite additivity,

$$\mu(\bigcup A_n) = \sum_1^k \mu A_j + \mu\left(\bigcup_{k+1}^\infty A_j\right),$$

and so it is sufficient to prove that

$$\lim_{k \to \infty} \left| \mu\left(\bigcup_{k+1}^\infty A_j\right) \right| = 0. \tag{5.15}$$

Since $v(\bigcup_k A_j) = \sum_k v A_j$ is valid and finite for each k, and since $\lim_k \sum_k v A_j = 0$, we conclude (5.15) by our hypothesis. Clearly, $\mu \ll v$, for if $vA = 0$ then for each $\varepsilon > 0$, $|\mu A| < \varepsilon$.

This fact also tells us that $\mu\varnothing = 0$. Thus $\mu \in M(X)$.

q.e.d.

We can drop the hypothesis that v is finite in Theorem 5.11 if we assume that $\mu \in M(X)$.

Example 5.2 Let X be a compact space and assume that there is a non-zero Borel measure μ on X (e.g. m on $[0, 1]$). We'll show that there is a perfect set (i.e. closed non-$\varnothing$ and without isolated points) $P \subseteq X$—and so card $P > \aleph_0$—such that $\mu P = 0$ (e.g. $mC = 0$). To do this we follow Darst [29] and use the following two results by Rudin [93]: let Q be a compact space without any perfect subsets (Q need not be countable!)—

R1.   If $f \in C(Q)$ then card $f(Q) \leqslant \aleph_0$;

R2.   If $\mu \in M_c(Q)$ is a measure then $\mu Q = 0$.

R2 implies that there is a perfect set $A \subseteq X$ since otherwise μ is trivial. From Urysohn's lemma (Appendix I.1) there is a continuous surjection $f: A \to [0, 1]$. Let $\{C_\lambda\}$ be an uncountable collection of pairwise disjoint perfect subsets of $[0, 1]$. Set $A_\lambda = f^{-1}C_\lambda$ so that $\{A_\lambda\}$ is an uncountable disjoint family of closed sets in A. From R1 each A_λ contains a perfect subset P_λ. Now, if each $\mu P_\lambda \neq 0$ we have $\mu(\bigcup P_\lambda) = \infty$, and this is a contradiction. Consequently, let $P = P_\lambda$ for some P_λ for which $\mu P_\lambda = 0$.

Remark 1 The previous example is really not unexpected (in this subject nothing is). More spectacular is the following example due to Choquet [26]. Let $X = [0, 1] \times [0, 1]$ and let $\mathscr{B} \subseteq \mathscr{P}(X)$ be the Borel algebra. Then there is an uncountable set $N \subseteq X$ which is measurable for each measure on $\mathscr{B}$, such that

$$\forall \mu \in M_c(X), \qquad \mu N = 0$$

and for which card $K \leqslant \aleph_0$ when $K \subseteq N$ is compact.

Remark 2 For a given $\mu \in (S)M(X)$ we have defined C_μ. When X is locally compact we shall also be able to define the support of μ (denoted by supp μ). There is a relationship between these two notions which we shall discuss in Appendix III.5.

Example 5.3 a) Let $E \subseteq [0, 1]$ be of the form $\bigcap E_k$, $E_k = \bigcup\limits_{j=1}^{2^k} E_j^k$, where $\{E_j^k : j = 1, ..., 2^k\}$ is disjoint and each E_j^k is a closed interval. Let $\gamma_{j,k}$ be the midpoint of E_j^k and define the measure

$$\mu_k = \sum_{j=1}^{2^k} \frac{1}{2^k} \delta_{\gamma_{j,k}}.$$

Then each $\|\mu_k\|_1 = |\mu_k|([0, 1]) = 1$ so that by the Alaoglu theorem (e.g. Appendix I.9) $\{\mu_k : k = 1, ...\}$ has a weak $*$ limit μ (noting that $\mu_k \in C[0, 1]'$). Clearly, $\mu \in M_c[0, 1]$, $\|\mu\|_1 = 1$, and $\mu \geqslant 0$. This procedure can be extended to any perfect totally disconnected metric space E.

b) Suppose we adjust the above procedure in the following way. Let $F_n \subseteq [0, 1]$ consist of n points and define the measure

$$\mu_n = \sum_{\gamma \in F_n} \frac{1}{n} \delta_{\gamma}.$$

Then $\|\mu_n\|_1 = 1$ and $\mu_n \geqslant 0$. Set $F = \overline{\bigcup F_n}$. By the Alaoglu theorem there is a subsequence $\{\mu_{m_n} : n = 1, ...\} \subseteq \{\mu_n : n = 1, ...\}$ and $\mu \in M(F)$ such that $\mu_{m_n} \to \mu$ in the weak $*$ topology, $\mu \geqslant 0$, and $\|\mu\|_1 = 1$. Taking $F_n = \{1, \frac{1}{2}, ..., (1/n)\}$ we have $F = \{0, 1, \frac{1}{2}, ...\}$ and $\mu(\{\gamma\}) = 0$ for each $\gamma \in (0, 1]$. Also, $\mu(\{0\}) = 1$, and so $\mu = \delta_0$.

Example 5.4 We'll prove that if $\mu, \nu \in (S)M(X)$, where $\mu \ll \nu$ and ν is finite, then μ need not be finite. In the process we shall show the necessity of the hypothesis that μ be bounded in Theorem 5.10. Let ν be Lebesgue measure m on $[0, \infty)$ and set

$$\mu A = \int_A \frac{dx}{x}.$$

Clearly μ is unbounded and $\mu \ll \nu$. As far as Theorem 5.10 is concerned note that

$$\forall \varepsilon > 0 \quad \exists N \quad \text{such that} \quad \forall n > N, \quad \nu([0, 1/n]) = 1/n < \varepsilon,$$

whereas $\mu([0, 1/n]) = \infty$ for all n.

5.3 The Vitali–Lebesgue–Radon–Nikodym theorem

We have indicated the roles of Vitali and Lebesgue in the development of FTC in Chapter 4.3. Essentially, R–N can be considered as the natural generalization of FTC for more general spaces, and, in this context, we refer to the historical remarks in section A5.1 and the Epilogue in Hawkins' book [54]; the former studies the contribution of Vitali and Lebesgue whereas the latter examines Radon's paper and Stieltjes' influence (among other things).

We first prove R–N for the case that μ and v are bounded measures. We then discuss extensions of the result to other "measures". The proof of Theorem 5.12 actually works if μ is a real-valued signed measure.

Theorem 5.12 (R–N) *Let μ and v be bounded measures on a measurable space $(X, \mathscr{A})$. Assume that $\mu \ll v$. Then there is a unique element $f \in L^1_v(X)$ such that*

$$\forall A \in \mathscr{A}, \qquad \mu A = \int_A f \, dv. \tag{5.16}$$

Proof. For each $r \in \mathbf{R}$ consider the real-valued signed measure $\mu - rv$, and let P_r, N_r be an Hahn decomposition of X for $\mu - rv$.

a) Let $r < s$. We shall first check that

i) $\mu A \leqslant rvA \leqslant svA$ if $A \subseteq N_r$,

ii) $svA \leqslant \mu A$ if $A \subseteq P_s$,

iii) $vA = \mu A = 0$ if $A \subseteq N_r \setminus N_s$.

iii) follows from i) and ii) since $N_r \setminus N_s \subseteq X \setminus N_s = P_s$ and $\mu \ll v$. For the first inequality of i) we only need the definition of N_r. For the second inequality we use the facts that v is a measure and $r < s$.

ii) is immediate from the definition of P_s.

b) Set

$$E = \bigcup_{s \in \mathbf{Q}} \bigcup_{\substack{r < s \\ r \in \mathbf{Q}}} (N_r \setminus N_s),$$

$$R_r = N_r \setminus E,$$

and $G = X \setminus (\bigcup R_r \setminus \bigcap R_s).$

We shall prove that

i) $vE = 0$,

ii) $r < s \Rightarrow R_r \subseteq R_s$,

iii) $v\left(\bigcap_{r \in \mathbf{Q}} R_r\right) = 0$,

iv) $v\left(X \setminus \bigcup_{r \in \mathbf{Q}} R_r\right) = v(\bigcap R_r^{\sim}) = 0$,

v) $vG = 0$.

i) is clear from part a.iii) and the fact that E is a countable union.

ii) follows from an easy set theoretic manipulation using the definitions of E and R_r.

iii) Clearly,

$$\forall s \in \mathbf{Q}, \quad \bigcap_{r \in \mathbf{Q}} R_r \subseteq \bigcap N_r \subseteq N_s,$$

and so $(\mu - sv)\left(\bigcap_{r \in \mathbf{Q}} R_r\right) \leqslant 0.$

Consequently if $v\left(\bigcap_{r\in\mathbf{Q}} R_r\right) > 0$ and we let $s_n \to -\infty$, we conclude that $\mu(\bigcap R_r) = -\infty$ and this contradicts our hypothesis that μ is bounded.

iv) Compute (e.g. Fig. 6)

$$X \setminus \bigcup R_r = \bigcap (X \setminus (N_r \setminus E)) = \bigcap ((X \setminus N_r) \cup E) = E \cup (\bigcap N_r^\sim).$$

Thus it is sufficient to prove that $v(\bigcap P_r) = 0$.

We use the technique of iii): $\bigcap P_r \subseteq P_s$, assume $v(\bigcap P_r) > 0$, and obtain a contradiction.

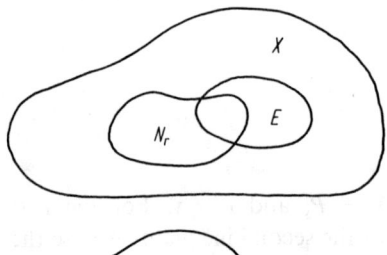

Fig. 6

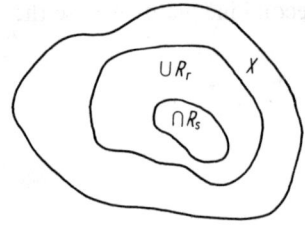

Fig. 7

v) Compute (e.g. Fig. 7)

$$\begin{aligned}
G = (\bigcup R_r \setminus \bigcap R_s)^\sim &= [(\bigcup R_r) \cap ((\bigcap R_s)^\sim)]^\sim \\
&= [(\bigcup R_r) \cap (\bigcup R_s^\sim)]^\sim \\
&= (\bigcap R_r^\sim) \cup (\bigcap R_s).
\end{aligned}$$

Thus $vG = 0$ by iii) and iv).

c) Define

$$f(x) = \begin{cases} 0, & \text{if } x \in G \\ \sup\{s \in \mathbf{Q}: x \notin R_s\}, & \text{if } x \notin G. \end{cases}$$

We'll show that f is v-measurable. In fact, if $\alpha \leq 0$, then

$$\{x: f(x) < \alpha\} = \bigcup_{\substack{r<\alpha \\ r\in\mathbf{Q}}} R_r,$$

and if $\alpha > 0$ then

$$\{x: f(x) < \alpha\} = G \cup (\bigcup_{\substack{r<\alpha \\ r\in\mathbf{Q}}} R_r).$$

These equalities are trivial to verify.

d) Let $r < s$ and take $A \subseteq R_s \setminus R_r$ (cf. b.ii).
We'll prove that

$$\left| \int_A f \, dv - \mu A \right| \leqslant (s - r) vA. \tag{5.17}$$

First observe that $R_s \setminus R_r \subseteq (\bigcup R_t) \setminus (\bigcap_u R_u)$ and so $A \cap G = \emptyset$.

Now, $x \notin R_t$ if and only if $f(x) \geqslant t$. Thus

$$r \leqslant f(A) \leqslant s,$$

and so

$$rvA \leqslant \int_A f(x) \, dv(x) \leqslant svA. \tag{5.18}$$

Further, $R_s \setminus R_r = (N_s \setminus E) \setminus (N_r \setminus E)$ and so $\mu A \leqslant svA$ since $A \subseteq N_s$. Also $A \subseteq P_r$, since $R_s \setminus R_r \subseteq (N_s \setminus E) \setminus N_r \subseteq X \setminus N_r$, and so $\mu A \geqslant rvA$.
Thus $-svA \leqslant -\mu A \leqslant -rvA$, which, when combined with (5.18), yields (5.17).

e) We next show that if $A \subseteq R_{n+1} \setminus R_n$ then

$$\mu A = \int_A f(x) \, dv(x). \tag{5.19}$$

For every integer p set

$$A_j = A \cap (R_{n+(j/p)} \setminus R_{n+[(j-1)/p]}), \qquad j = 1, \ldots, p.$$

Clearly, $\{A_j : j = 1, \ldots\}$ is a disjoint family and $A = \bigcup A_j$. Hence from d)

$$\left| \mu A - \int_A f \, dv \right| \leqslant \sum_{j=1}^{p} \left| \mu A_j - \int_{A_j} f \, dv \right| \leqslant \sum_{j=1}^{p} \frac{1}{p} vA_j = \frac{1}{p} vA,$$

and so (5.19) follows since p is arbitrary.

f) Given $A \in \mathscr{A}$ we write

$$A = (A \cap G) \cup (\bigcup_j A_j),$$

where $\{A_j : j = 1, \ldots\}$ is a disjoint family, and, for each j, there is an $n = n(j)$ for which $A_j \subseteq R_{n+1} \setminus R_n$.
Thus,

$$\mu A = \mu(A \cap G) + \mu(\bigcup_j A_j),$$

so that from (5.19)

$$\mu A = \mu(A \cap G) + \int_A f \, dv \tag{5.20}$$

(this is the Lebesgue decomposition of μ!).

The fact that $vG = 0$ (e.g. $b.v$)) and the hypothesis that $\mu \ll v$ yield

$$\mu A = \int_A f \, dv.$$

Because f is v-measurable and $\int_A f \, dv$ exists for each $A \in \mathscr{A}$ we conclude that $f \in L^1_v(X)$.

q.e.d.

Let $(X, \mathscr{A})$ be a measurable space and assume that μ and v satisfy the conditions that

$$\mu \in M(X) \quad \text{and } v \text{ is a } \sigma\text{-finite measure};$$

if $\mu \ll v$ then Theorem 5.12 can be extended in a routine way to assert the existence of a unique element $f \in L^1_v(X)$ such that

$$\forall A \in \mathscr{A}, \quad \mu A = \int_A f \, dv.$$

An analogous result can be formulated when $v \in M(X)$ (e.g. Problem 5.27b). Such statements are again denoted by R–N.

A necessary and sufficient condition that $f \in L^1_v(X)$ in any statement of R–N is that $\mu \in M(X)$. On the other hand, if, for example, v is a σ-finite measure, $\mu \in SM(X)$, and $\mu \ll v$, then there is a unique $\mathbf{C}$-valued v-measurable function f which satisfies (5.16). In this case $f \in L^1_v(B)$ whenever $vB < \infty$.

We rewrite Theorem 5.12 and the above remarks in the following way.

Define $M_{ac,v}(X) = \{\mu \in M(X): \mu \ll v\}$. Then

Theorem 5.13 *Let $(X, \mathscr{A}, v)$ be a σ-finite measure space. There is a natural bijective map*

$$L^1_v(X) \to M_{ac,v}(X)$$
$$f \mapsto \mu_f$$

where $\forall A \in \mathscr{A}, \quad \mu_f(A) = \int_A f \, dv.$

Remark Let X be a locally compact space and consider the sup-norm Banach space, $C_0(X)$ (e.g. Appendix III.2). R–N establishes an isomorphism between $M_{ac,v}(X)$ and those continuous linear functionals on $C_0(X)$ which are elements of $L^1_v(X)$; whereas the Riesz representation theorem (e.g. Appendix III) will yield an isomorphism between $(C_0(X))'$ and $M(X)$.

The isomorphism of Theorem 5.13 is also described in other terms. Given a statement of R–N with corresponding elements μ, v and f. f is the R–N derivative of μ with respect to v. As such, R–N becomes a natural generalization of FTC: FTC *characterizes absolutely continuous functions g as those for which*

$$g(x) - g(a) = \int_a^x g';$$

and R–N *characterizes absolutely continuous measures $\mu \ll v$ as those for which*

$$\mu A = \int_A f \, dv.$$

The fact that the R–N derivative can be viewed as a difference quotient is discussed in Problem 5.33.

Part b) of the following should be compared with Proposition 5.4.

Theorem 5.14 a) *Let $v \in M(X)$. Then there is a $|v|$-integrable function h, $|h| = 1$, such that*

$$\forall f \in L^1_{|v|}(X), \quad \int_X f \, dv = \int_X fh \, d|v|.$$

b) *Let v be a measure and let $g \in L^1_v(X)$. Define*

$$\forall A \in \mathscr{A}, \quad \mu A = \int_A g \, dv.$$

Then $\mu \in M(X)$ and

$$\forall A \in \mathscr{A}, \quad |\mu|(A) = \int_A |g| \, dv.$$

Proof. a) Clearly $v \ll |v|$ so that by R–N there is $h \in L^1_{|v|}(X)$ such that

$$\forall A \in \mathscr{A}, \quad vA = \int_A h \, d|v|.$$

Consequently, since we can approximate the elements of $L^1_{|v|}(X)$ by step functions it is sufficient to prove that $|h| = 1$, $|v|$-a.e.

Let $A_r = \{x : |h(x)| < r\}, r > 0$; and let $A_r = \bigcup A_j$, a finite disjoint union of elements from $\mathscr{A}$. Then

$$\sum |vA_j| = \sum \left| \int_{A_j} h \, d|v| \right| < r \sum |v|(A_j) = r|v|(A_r).$$

Therefore, $|v|(A_r) \leqslant r|v|(A_r)$, and so $|h| \geqslant 1$, $|v|$-a.e. (by considering $r \in (0, 1)$). We now prove that $|h| \leqslant 1$, $|v|$-a.e. Note that if $|v|(A) > 0$,

$$\left| \frac{1}{|v|(A)} \int_A h \, d|v| \right| = \left| \frac{vA}{|v|(A)} \right| \leqslant 1. \tag{5.21}$$

Take any closed disc $S \subseteq \{z \in \mathbf{C}: |z| \leqslant 1\}^{\sim}$ with center ζ and radius r; and set $B = h^{-1}S$. We must show that $|v|(B) = 0$.

If $|v|(B) > 0$ then

$$\left| \zeta - \frac{1}{|v|(B)} \int_B h \, d|v| \right| \leqslant \frac{1}{|v|(B)} \int_B |h(x) - \zeta| \, d|v|(x) \leqslant r$$

by the definition of S. This contradicts (5.21).

b) Clearly, from the definition of $|\mu|$,

$$\forall A \in \mathscr{A}, \qquad |\mu|(A) \leqslant \int_A |g| \, dv.$$

In particular, $\mu \in M(X)$.

The opposite inequality can be proved in an elementary way by approximating $|g|\chi_A$ by $g s_n$ where $\{s_n: n = 1, \dots\}$ is a sequence of simple functions satisfying

$$\left| \int_X g s_n \, dv \right| \leqslant |\mu|(A).$$

We shall give another proof of this (opposite) inequality which uses R–N.

Without loss of generality let v be σ-finite since $g \in L^1_v(X)$ (e.g. Problem 5.24).

By hypothesis we write $\mu = gv$ and since $\mu \in M(X)$ we have

$$\int f \, d\mu = \int f g \, dv.$$

We use part a) to assert that $\mu = h|\mu|$ where $|h| = 1$, $|\mu|$-a.e. It is at this point that we use the fact that v is σ-finite because part a) requires R–N which, in turn, we have only stated for σ-finite v.

Thus

$$|\mu| = \frac{g}{h} v.$$

Clearly, $g/h \geqslant 0$, v-a.e. since $|\mu| \geqslant 0$, and therefore $g/h = |g|$, v-a.e.

q.e.d.

Theorem 5.15 a) *If μ, $v \in M(X)$ and $\mu \perp v$ then*

$$|\mu + v| = |\mu| + |v|.$$

b) *If $\mu \in M(X)$ and v is a measure on X, then*

$$\mu = \mu_a + \mu_s + \mu_d,$$

where $\mu_d \in M_d(X)$, $\mu_a \ll v$, $\mu_s \perp v$, and $\mu_a, \mu_s \in M_c(X)$; also $(\mu_a + \mu_s) \perp \mu_d$ and

$$|\mu| = |\mu_a| + |\mu_s| + |\mu_d|.$$

Proof. a) Let $\mu = h|\mu|$, where $|h| = 1$ and $h \in L^1_{|\mu|}$ (resp. $v = g|v|$). Then

$$\mu + v = (h\chi_{C_\mu} + g\chi_{C_\mu^\sim})(|\mu| + |v|).$$

Observe that $|h\chi_{C_\mu} + g\chi_{C_\mu^\sim}| = 1$.

From Theorem 5.14b,

$$|\mu + v|(A) = \int_A |h\chi_{C_\mu} + g\chi_{C_\mu^\sim}|\,\mathrm{d}(|\mu| + |v|) = |\mu|(A) + |v|(A).$$

b) First, we know that $\mu = \mu_c + \mu_d$; and from the Lebesgue decomposition, Theorem 5.9, $\mu_c = \mu_a + \mu_s$, where $\mu_a \ll v$ and $\mu_s \perp v$.

It is easy to see that $\mu_c \perp \mu_d$, and so $|\mu| = |\mu_c| + |\mu_d|$.

Also in proving that $\mu \ll v$ and $\mu \perp v$ imply $\mu = 0$ we really showed that $\mu \ll v$ and $\lambda \perp v$ imply $\mu \perp \lambda$.

Thus $\mu_a \perp \mu_s$ and we apply a) again.

q.e.d.

I. Segal (1951) has characterized spaces for which R–N is valid as those which can be "localized"; and J. L. Kelley (Math. Ann. 163 (1966) 89–94) has streamlined Segal's work, making the characterization in terms of so-called Dedekind complete set functions on certain algebras of sets. These results show the equivalence of the Hahn decomposition theorem, the Lebesgue decomposition theorem, R–N, and (a form of) the Riesz representation theorem.

5.4 The relation between set and point functions

We restate Theorem 4.3 as

Theorem 5.16 *Given the measurable space* $(\mathbf{R}, \mathscr{B})$ *and a function* $f: \mathbf{R} \to \mathbf{R}$, *continuous from the right at each point, for which* $\lim\limits_{x \to -\infty} f(x) = 0$. $f \in BV_{\mathrm{loc}} \Leftrightarrow$

$$\mu_f(a, b] = f(b) - f(a) \tag{5.22}$$

can be uniquely extended to an element of $SM(\mathbf{R})$. *In this situation* f *is continuous at* $x_0 \Leftrightarrow \mu_f(\{x_0\}) = 0$.

To prove Theorem 5.16 consider the case of an increasing function f and define

$$\forall A \in \mathbf{R}, \qquad S(A) = \bigcup \{S(x): x \in A\},$$

where $S(x)$ is defined as follows:

$$S(x) = \begin{cases} \{f(x)\}, & \text{if } f \text{ is continuous at } x \\ [f(x), f(x+)], & \text{if } f(x+) > f(x). \end{cases}$$

Then it can be proved that

$$\mu A = m(S(A))$$

satisfies (5.22). The uniqueness assertion can be proved using the fact that Borel measures on **R** taking finite values on compact sets are regular.

We now complete Remark 2 prior to Theorem 5.8 and relate the results of Chapter 4 and the notions introduced in this chapter for the special case of **R**. A natural example to keep in mind is the Cantor–Lebesgue function and its corresponding measure.

Theorem 5.17 Given $f \in BV(\mathbf{R})$ and μ_f (as in (5.22)).

a) $\mu_f \ll m \Leftrightarrow f$ is absolutely continuous.

b) $\mu_f \perp m \Leftrightarrow f' = 0$, m-a.e.

c) If μ_f is discrete then $f' = 0$, m-a.e.

d) $f(x) = g(x) + \int\limits_{-\infty}^{x} f'$, where $\mu_g \perp m$; and $g = 0 \Leftrightarrow \mu_f \ll m$.

Proof. a) ($\Rightarrow$) $\mu_f \ll m$ implies that

$$\forall \varepsilon > 0 \quad \exists \delta \quad \text{such that} \quad A \in \mathcal{M} \quad \text{and} \quad mA < \delta \Rightarrow |\mu_f A| < \varepsilon.$$

Thus choose $\varepsilon > 0$ and let $A = \bigcup\limits_{k} (a_k, b_k)$ satisfy $mA < \delta$. Then

$$\varepsilon > |\mu_f A| = |\sum \mu_f(a_k, b_k)| = |\sum (f(b_k) - f(a_k))|.$$

($\Leftarrow$) Let $A \in \mathcal{M}$, $mA = 0$. Without loss of generality take $A \subseteq [-c, c]$ and let f be an increasing function. Given $\varepsilon > 0$, choose δ by the definition of absolute continuity (of f).

Now A is contained in a union $\bigcup\limits_{1}^{\infty} (a_k, b_k)$, for which $\sum (b_k - a_k) < \delta$, since $mA = 0$. By the choice of δ,

$$\left| \sum\limits_{1}^{\infty} (f(b_k) - f(a_k)) \right| \leqslant \varepsilon;$$

and so, since f is increasing,

$$\varepsilon \geqslant \left| \mu_f \left(\bigcup\limits_{1}^{\infty} (a_k, b_k) \right) \right| \geqslant |\mu_f A|.$$

d) is just a succinct way of writing some of our previous results. For c) we use b) and can take $C_m = \mathbf{R} \setminus C_{\mu_f}$ where card $C_{\mu_f} \leqslant \aleph_0$.

b) (cf. Theorem 4.14). ($\Leftarrow$) Without loss of generality let f be an increasing function. Assume that μ_f and m are not mutually singular. From the Lebesgue decomposition, $\mu_f = \mu_1 + \mu_2$, where $\mu_1 \ll m$, $\mu_2 \perp m$, and $\mu_1 \neq 0$. μ_1 and μ_2 are measures since μ is a measure.

As remarked after Theorem 5.16, μ_f is regular, and, for the same reason, so also are μ_1 and μ_2.

We define the function $f_1(x) = \mu_1(-\infty, x]$; thus, by a) and because $\mu_1 \ll m$, we conclude that f_1 is absolutely continuous. f_1 is an increasing function since μ_1 is a measure. From FTC

$$\mu_1((a, b]) = f_1(b) - f_1(a) = \int_a^b f_1'.$$

Consequently, since $\mu_1 \neq 0$ and μ_1 is regular, $f_1' > 0$ on some $A \in \mathcal{B}(\mathbf{R})$, where $mA > 0$.

If we set $f_2(x) = \mu_2(-\infty, x]$ then f_2 is characterized by μ_2 from Theorem 5.16. Hence $f = f_1 + f_2$, and f_2 is increasing since μ_2 is a measure.

Therefore,

$$f' = f_1' + f_2' \geqslant f_1',$$

and so $f' \neq 0$, m-a.e.

($\Rightarrow$) Assume $f' > 0$ on $A \in \mathcal{B}(\mathbf{R})$, where $mA > 0$. Again, without loss of generality, take f to be increasing and thus $f' \geqslant 0$, m-a.e.

Since f is m-measurable, f' is m-measurable, and we define

$$\forall B \in \mathcal{B}(\mathbf{R}), \qquad vB = \int_B f'.$$

Clearly $v \ll m$ and $vA > 0$.

Also

$$v((a, b]) = \int_a^b f' \leqslant f(b) - f(a) = \mu_f((a, b]),$$

where the inequality follows from Theorem 4.8.

It is easily verified that v is regular so that by a standard measure theoretic manipulation

$$\forall B \in \mathcal{B}(\mathbf{R}), \qquad vB \leqslant \mu_f B.$$

Consequently, $\mu_f A > 0$.

Note that $v \ll m$ and $\mu_f \perp m$ imply $v \perp \mu_f$ (e.g. Theorem 5.15b), and this contradicts the fact that $\mu_f A, vA > 0$.

Hence $f' = 0$, m-a.e.

$$\text{q.e.d.}$$

We wind up this peroration on the decomposition of measures with

Theorem 5.18 *Let* $(\mathbf{R}, \mathscr{B}, \mu)$ *be a regular measure. Then μ has the unique decomposition*

$$\mu = \mu_a + \mu_s + \mu_d$$

into regular Borel measures, where μ_d is discrete, $\mu_a \ll m$, $\mu_s \perp m$ and μ_s is continuous. Further, there are unique non-decreasing functions f, f_a, f_s, f_d (corresponding to μ, μ_a, μ_s, μ_d) where

$$f = f_a + f_s + f_d,$$

f_a *is absolutely continuous on every compact interval, f_s is continuous and $f_s' = 0$, m-a.e., $f_d' = 0$, m-a.e., and f_d has at most countably many discontinuities x_n, $n = 1, \ldots$, all of the form $f_d(x_n+) - f_d(x_n-)$. Finally*

$$\forall B \in \mathscr{B}(\mathbf{R}), \qquad \mu_a B = \int_B f_a(t)\, dt$$

and $\qquad \forall y \geqslant x, \qquad f_a(y) - f_a(x) = \int_x^y f'(t)\, dt.$

In Chapter 4 we gave several examples of increasing and strictly increasing continuous functions f for which $f' = 0$, m-a.e. An amusing application of some of our previous results including Theorem 5.15a and a characterization of this phenomenon is

Proposition 5.7 *Let f be an increasing continuous function on $[0, 1]$ with $f(0) = 0$ and $f(1) = \alpha$.*
a) *Let L be the length of the graph of f. Then $L = 1 + \alpha \Leftrightarrow f' = 0$, m-a.e.*
b) *Let f be strictly increasing with inverse function g. Then $f' = 0$, m-a.e., on $[0, 1] \Leftrightarrow g' = 0$, m-a.e., on $[0, \alpha]$.*

Proof. b) is Problem 5.38.
a) Take $\alpha > 0$ (the $\alpha = 0$ case is obvious).
Denote the graph of f by $G(x) = x + if(x)$ and so $L = V(G, [0, 1])$.
($\Leftarrow$) $\mu_f \perp m$ and $\mu_{if} \perp m$ by Theorem 5.17b and the hypothesis that $f' = 0$, m-a.e. Note that if $g(x) = x$ then $\mu_g = m$. Thus,

$$L = V(x + if(x), [0, 1]) = \|m + \mu_{if}\|_1 = \|m\|_1 + \|\mu_{if}\|_1$$
$$= 1 + \|\mu_f\|_1 = 1 + \alpha,$$

where $\|\mu_f\|_1 = \alpha$ since f is increasing.
($\Rightarrow$) Assume that $L = 1 + \alpha$ and let $f = f_1 + f_2$, where f_1 is absolutely continuous and $f_2' = 0$, m-a.e.

$\mu_{f_1} \perp \mu_{f_2}$ since $\mu_{f_1} \ll m$ and $\mu_{f_2} \perp m$; thus

$$\|\mu_f\|_1 = \|\mu_{f_1}\|_1 + \|\mu_{f_2}\|_1 = V(f_1, [0, 1]) + V(f_2, [0, 1]).$$

Consequently,

$$V(x + if_1(x), [0, 1]) = 1 + V(f_1, [0, 1]) \tag{5.23}$$

because

$$1 + \alpha = L = V(G, [0, 1]) \leqslant V(x + if_1, [0, 1]) + V(if_2, [0, 1])$$
$$\leqslant 1 + \|\mu_{f_1}\|_1 + \|\mu_{f_2}\|_1 = 1 + \|\mu_f\|_1 = 1 + \alpha,$$

where the fact that f is increasing yields the last equality.

Since f_1 is absolutely continuous we can use the integral formula for the arc length, $L_1 = V(x + if_1(x), [0, 1])$, of the graph of f_1. Hence

$$V(x + if_1(x), [0, 1]) = \int_0^1 (1 + (f_1')^2)^{1/2}. \tag{5.24}$$

On the other hand, $V(x + if_1(x), [0, 1]) = 1 + V(f_1, [0, 1])$ from (5.23), and so

$$V(x + if_1(x), [0, 1]) = 1 + \|\mu_{f_1}\|_1 = 1 + \mu_{f_1}[0, 1]$$

$$= 1 + f_1(1) - f_1(0) = 1 + \int_0^1 f_1'.$$

Combining this with (5.24) yields

$$\int_0^1 (1 + (f_1')^2)^{1/2} = \int_0^1 (1 + f_1'). \tag{5.25}$$

Now by squaring the integrands and noting that f_1 is increasing, we have $(1 + (f_1')^2)^{1/2} \leqslant 1 + f_1'$, m-a.e., so that we can actually conclude that

$$(1 + (f_1')^2)^{1/2} = 1 + f_1', \quad m\text{-a.e.}$$

from (5.25).

Therefore $f_1' = 0$, m-a.e., and consequently $f' = 0$, m-a.e., since $f_2' = 0$, m-a.e.

$$\text{q.e.d.}$$

Finally in this section we note the intimate relation between Steinhaus' theorem given in Problem 3.6b and the notion of absolute continuity: *take $\mu \in M(\mathbf{R})$; $\mu \ll m$ if and only if the implication*

$$|\mu|(K) > 0 \Rightarrow \text{int}\,(K - K) \neq \emptyset$$

is valid for each compact set $K \subseteq \mathbf{R}$ (this observation is due to S. M. Simmons).

5.5 $L_\mu^p(X)$, $1 \leqslant p \leqslant \infty$

For $p \geqslant 1$, q will designate $p/(p-1)$—in particular $p = 1$ if and only if $q = \infty$. We shall not consider L^p, $p < 1$, not because it is uninteresting but because it has a trivial duality theory. The $L_\mu^p(X)$ spaces are defined in Appendix I.2.

Theorem 5.19 *Let $(X, \mathscr{A}, \mu)$ be a measure space and take $1 \leqslant p \leqslant \infty$. (If $(X, \mathscr{A}, \mu)$ is not σ-finite, then the case, $L_\mu^\infty(X)$, requires an extra modification.)*

a) *$L_\mu^p(X)$ is a Banach space.*

b) *If $f \in L_\mu^p(X)$ and $g \in L_\mu^q(X)$, then $fg \in L_\mu^1(X)$ and*

$$\left| \int_X fg \, d\mu \right| \leqslant \int_X |fg| \, d\mu \leqslant \|f\|_p \|g\|_q.$$

c) *Let $\alpha_1, \ldots, \alpha_n > 0$ and $\sum \alpha_j = 1$. If $f_1, \ldots, f_n \in L_\mu^1(X)$ then*

$$\left| \int_X f_1^{\alpha_1} \cdots f_n^{\alpha_n} \, d\mu \right| \leqslant \int_X |f_1^{\alpha_1} \cdots f_n^{\alpha_n}| \, d\mu \leqslant \prod_1^n \|f_j\|_1^{\alpha_j}.$$

Parts a) and b) are discussed in Appendix I.2. Part b) is Hölder's inequality. Its verification is trivial for $p = 1$. If $p, q > 1$ it is proved by the following calculation. Set

$$h(t) = \frac{t^p}{p} + \frac{t^{-q}}{q}, \qquad t > 0,$$

and observe that

$$\forall t > 0, \qquad h(t) \geqslant h(1) = 1;$$

then take

$$t = \left(\frac{|f|}{\|f\|_p} \right)^{1/q} \Big/ \left(\frac{|g|}{\|g\|_q} \right)^{1/p}.$$

If $L_\mu^p(X)$ is considered as a Banach space then $(L_\mu^p(X))'$ is the space of continuous linear functions $L_\mu^p(X) \to \mathbf{C}$ (e.g. Appendix I.8); $(L_\mu^p(X))'$ is called the dual of $L_\mu^p(X)$. We shall now characterize the dual of $L_\mu^p(X)$. If $1 < p < \infty$ this can be done without R–N, but not without some effort except in the case $p = 2$ (cf. Problem 5.39). In any case we shall use R–N.

Proposition 5.8 *Let $(X, \mathscr{A}, \mu)$ be a bounded measure space and take $g \in L_\mu^1(X)$. Assume there is a constant $M > 0$ such that for all simple functions s,*

$$\left| \int sg \, d\mu \right| \leqslant M \|s\|_p.$$

Then $g \in L_\mu^q(X)$.

Proof. First assume that $p > 1$ and take a sequence $\{s_n : n = 1, \ldots\}$ of non-negative simple functions which increases pointwise to $|g|^q$.
Define

$$\varphi_n = s_n^{1/p} \overline{\operatorname{sgn} g},$$

where $\operatorname{sgn} g = \begin{cases} 0, & \text{if } g(x) = 0 \\ g(x)/|g(x)|, & \text{if } g(x) \neq 0. \end{cases}$

Clearly $\|\varphi_n\|_p = \left(\int_X s_n \, d\mu \right)^{1/p}.$ (5.26)

Observe that

$$0 \leqslant s_n = s_n^{1/p} s_n^{1/q} \leqslant s_n^{1/p} |g| = \varphi_n g,$$

and so when g is real-valued (and hence φ_n is simple)

$$.\ 0 \leqslant \int s_n \, d\mu \leqslant \int \varphi_n g \, d\mu \leqslant M \|\varphi_n\|_p.$$

Consequently from (5.26)

$$\int_X s_n \, d\mu \leqslant M^q,$$

and therefore

$$\int_X |g|^q \, d\mu \leqslant M^q$$

by Beppo Levi's theorem (i.e. LDC).
Now let $p = 1$ and assume, without loss of generality, that $g \geqslant 0$.
Assume $g \notin L_\mu^\infty(X)$ so that there is a sequence $\{A_n : n = 1, \ldots\} \subseteq \mathscr{A}$ for which $g \geqslant n$ on A_n and $\mu A_n > 0$. Then for $s_n = \chi_{A_n}$,

$$\frac{\displaystyle\int_X s_n g \, d\mu}{\|s_n\|_1} \geqslant \frac{n \mu A_n}{\mu A_n} = n.$$

This contradicts our hypothesis.   q.e.d.

If $1 < p < \infty$ the following is true for any measure space $(X, \mathscr{A}, \mu)$.

Theorem 5.20 *Let $(X, \mathscr{A}, \mu)$ be a σ-finite measure space and let $1 \leqslant p < \infty$. $F \in (L_\mu^p(X))' \Leftrightarrow$ there is a unique $g \in L_\mu^q(X)$ such that*

$$\forall f \in L_\mu^p(X), \qquad F(f) = \int fg \, d\mu. \qquad (5.27)$$

In this case, $\|F\| = \|g\|_q$ (cf. Problem 4.32).

Proof. ($\Leftarrow$) Assuming (5.37) we have $F \in (L_\mu^p(X))'$ by Theorem 5.19b.
With (5.27) we prove that $\|F\| = \|g\|_q$. The easier direction is to show

$$\|F\| = \sup_{\|f\|_p \leqslant 1} |F(f)| \leqslant \|g\|_q;$$

this is immediate by Theorem 19b.

For the opposite inequality we first consider the $1 < p < \infty$ case, and define

$$f = |g|^{q-1} \overline{\text{sgn } g}.$$

Then

$$f^p = |g|^{(q-1)p} (\overline{\text{sgn } g})^p = |g|^q (\overline{\text{sgn } g})^p,$$

and so $f \in L_\mu^p(X)$. Also

$$\|f\|_p = (\int |g|^q \, d\mu)^{1/p} = \|g\|_q^{q/p} = \|g\|_q^{q-1}. \tag{5.28}$$

For this f and using (5.28),

$$F(f) = \int fg \, d\mu = \int |g|^q \, d\mu = \|g\|_q^q = \|f\|_p \|g\|_q;$$

thus

$$\|F\| = \sup_{\substack{h \in L_\mu^p(X) \\ h \neq 0}} |F(h)|/\|h\|_p \geqslant \|g\|_q.$$

Now we must prove that $\|F\| \geqslant \|g\|_\infty$ when $p = 1$ and when we assume (5.27).
Take $\|g\|_\infty > \varepsilon > 0$ and choose a subset $A \subseteq \{x \in X : |g(x)| > \|g\|_\infty - \varepsilon\}$ for which
$0 < \mu A < \infty$.

Define

$$f = \frac{1}{\mu A} \chi_A \overline{\text{sgn } g}.$$

Clearly, $f \in L_\mu^1(X)$ since $\|f\|_1 = 1$. Further,

$$F(f) = \frac{1}{\mu A} \int_A g \, \overline{\text{sgn } g} \, d\mu = \frac{1}{\mu A} \int_A |g| \, d\mu \geqslant \|g\|_\infty - \varepsilon,$$

and so $\|F\| \geqslant \|g\|_\infty$.

($\Rightarrow$) Assume that $(X, \mathcal{A}, \mu)$ is a bounded measure space. The extension to the σ-finite
case is Problem 5.40. From the boundedness, $L_\mu^\infty(X) \subseteq L_\mu^p(X)$, and thus

$$\forall A \in \mathcal{A}, \qquad vA = F(\chi_A)$$

exists.

a) We use R–N to show that there is a unique element $g \in L_\mu^1(X)$ for which

$$\forall A \in \mathcal{A}, \qquad vA = \int_A g \, d\mu. \tag{5.29}$$

To do this we first prove that $v \in (S)M(X)$.

Let $\{A_n : n = 1, \ldots\} \subseteq \mathscr{A}$ be a disjoint family and set $A = \bigcup A_n$. Since $(X, \mathscr{A}, \mu)$ is bounded it is easy to check that

$$\lim_{n \to \infty} \left\| \sum_{n+1}^{\infty} \chi_{A_j} \right\|_p = 0,$$

so that by hypothesis (F is continuous)

$$\sum v A_n = \sum F(\chi_{A_n}) = F(\chi_A) = vA.$$

Thus $v \in (S)M(X)$.

Also $v \ll \mu$ from the definition of v, and hence (5.29) follows by R–N.

b) If s is simple then the linearity of F and (5.29) yield

$$F(s) = \int sg \, d\mu. \tag{5.30}$$

This and the continuity of F imply that

$$\left| \int sg \, d\mu \right| \leq \|s\|_p \|F\|.$$

Consequently, from Proposition 5.8, $g \in L_\mu^q(X)$.

c) Define

$$\forall f \in L_\mu^p(X), \qquad G(f) = \int fg \, d\mu,$$

so that $G \in (L_\mu^p(X))'$ by Theorem 5.19b. Therefore, $G - F \in (L_\mu^p(X))'$ and $G - F = 0$ on the set of simple functions.

Since these functions are dense in $L_\mu^p(X)$, $F = G$.

It remains to check that $g \in L_\mu^q(X)$ is unique. This is easy, for if $g_1 \in L_\mu^q(X)$ yielded $F(f) = \int fg_1 \, d\mu$ for each $f \in L_\mu^p(X)$ and if $\mu\{x : g_1(x) \neq g(x)\} > 0$ we construct $f \in L_\mu^p(X)$ in the obvious way to obtain

$$\int f(g - g_1) \, d\mu \neq 0.$$

q.e.d.

From the general properties of Banach spaces and Theorem 5.20 we know that $L_\mu^\infty(X)$ is complete (for X σ-finite). We shall calculate $(L_\mu^\infty(X))'$ for X σ-finite; it is possible to ease the condition of σ-finiteness and obtain a similar result.

Example 5.5 We'll show that $(L_m^\infty[0, 1])' \neq L_m^1[0, 1]$ (cf. Example I.1). Note that $C[0, 1]$ is a closed subspace of $L_m^\infty[0, 1]$. Define

$$F : C[0, 1] \to \mathbf{C}$$
$$g \mapsto g(0) \tag{5.31}$$

so that F is continuous on $C[0, 1]$ with $\|F\| = 1$. By the Hahn–Banach theorem F extends to an element of $(L_m^\infty[0, 1])'$. We now prove that there is no $f \in L_m^1[0, 1]$ such that

$$\forall g \in C[0, 1], \qquad F(g) = \int fg;$$

In fact, (5.31) defines the (discrete) Dirac measure, δ. If such an f existed choose a sequence $\{g_n : n = 1, \dots\} \subseteq C[0, 1]$ such that $g_n \to 0$ pointwise on $(0, 1]$, $g_n(0) = 1$, and $|g_n| \leqslant 1$. Then, from LDC, $F(g_n) \to 0$ whereas $F(g_n) = g_n(0) = 1$, a contradiction. On the other hand, $L_m^1[0, 1] \subseteq (L_m^\infty[0, 1])'$.

Let $(X, \mathscr{A}, \mu)$ be a σ-finite measure space and let $F(X)$ be the set of finitely additive set functions

$$v: \mathscr{A} \to \mathbf{C}$$

which vanish on the set $\{A \in \mathscr{A} : \mu A = 0\}$ and which have the property that

$$\sup\{|vA| : A \in \mathscr{A}\} < \infty.$$

We define

$$|v|(A) = \sup \sum_1^n |vA_j|,$$

where $A = \bigcup_1^n A_j$ is a disjoint union; it is easy to check that $|v| \in F(X)$. We set

$$\|v\| = |v|(X). \tag{5.32}$$

Further, if $f \in L_\mu^\infty(X)$ and $v \in F(X)$,

$$\int f \, dv$$

will denote

$$\lim \int s_n \, dv,$$

where s_n is simple, $\|s_n - f\|_\infty \to 0$, and $\int s \, dv = \sum_1^n a_j v A_j$ for $s = \sum a_j \chi_{A_j}$. $\int f \, dv$ is easily seen to be well defined and the following facts are checked in a routine fashion:

i) $\forall f, g \in L_\mu^\infty(X)$, $\forall \alpha, \beta \in \mathbf{C}$, and $\forall v \in F(X)$, $\int (\alpha f + \beta g) \, dv = \alpha \int f \, dv + \beta \int g \, dv$,

ii) $\forall f \in L_\mu^\infty(X)$ and $\forall v \in F(X)$, $|\int f \, dv| \leqslant \int |f| \, d|v|$,

iii) $F(X)$ is normed by (5.32),

iv) If $f \in L_\mu^\infty(X)$, $v \in F(X)$, and $\mu\{x : f(x) \neq 0\} = 0$, then

$$\int f \, dv = 0.$$

Theorem 5.21 *Let $(X, \mathcal{A}, \mu)$ be a σ-finite measure space. There is a surjective isometric isomorphism*

$$F(X) \to (L_\mu^\infty(X))'$$
$$v \mapsto F_v,$$

where $\forall f \in L_\mu^\infty(X), \qquad F_v(f) = \int f \, dv.$ (5.33)

Also $F(X)$ is a Banach space (clearly).

Proof. If $v \in F(X)$ is given and F_v is defined by (5.33) then property ii) above yields $\|F_v\| \leqslant \|v\|$ (F_v considered as an element of $(L_\mu^\infty(X))'$).
To prove that $\|v\| \leqslant \|F_v\|$ let $\varepsilon > 0$ and choose a finite decomposition $\{A_1, ..., A_n\} \subseteq \mathcal{A}$ of X such that

$$\sum_1^n |vA_j| > \|v\| - \varepsilon.$$

Define

$$g(x) = \sum a_j \chi_{A_j}, \qquad \text{where} \quad a_j = \overline{\text{sgn}\,(vA_j)}.$$

Clearly, $\|g\|_\infty \leqslant 1$ and

$$|F_v(g)| = \left| \int g \, dv \right| = |\sum a_j vA_j| = \sum |vA_j| > \|v\| - \varepsilon.$$

Consequently, $\|v\| = \|F_v\|$ and so the map $v \mapsto F_v$ is an isometry. It remains to prove the surjectivity.
Given $L \in (L_\mu^\infty(X))'$ and define v on $\mathcal{A}$ as

$$vA = L(\chi_A).$$

Note that

$$\sup\,\{|vA| : A \in \mathcal{A}\} \leqslant \sup\,\{|L(g)| : \|g\|_\infty \leqslant 1, g \in L_\mu^\infty(X)\} = \|L\|.$$

Also, if $\mu A = 0$ then $\chi_A = 0 \in L_\mu^\infty(X)$ and so $0 = L(0) = vA$. Thus to verify that $v \in F(X)$ we take $A, B \in \mathcal{A}$, for which $A \cap B = \emptyset$, and observe that

$$v(A \cup B) = L(\chi_{A \cup B}) = L(\chi_A) + L(\chi_B) = vA + vB.$$

Our final task then is to prove that if $g \in L_\mu^\infty(X)$ then

$$L(g) = \int g \, dv.$$

Let $\{s_n : n = 1, ...\}$ be a sequence of simple functions such that $\|s_n - g\|_\infty \to 0$, and note that

$$L(s_n) = \sum_j a_{j,n} L(\chi_{A_{j,n}}) = \sum_j a_{j,n} vA_{j,n} = \int s_n \, dv.$$

Since L is continuous

$$L(g) = \lim L(s_n) = \lim \int s_n \, dv = \int g \, dv,$$

where the last equality is the definition of the integral.

q.e.d.

We now note (e.g. [10, Chapter 2]) that generally the condition

$$\sup_{A \in \mathscr{A}} \{|vA|\} < \infty \tag{5.34}$$

for $v \in F(X)$ cannot be replaced by the condition that

$$\forall A \in \mathscr{A}, \quad |vA| < \infty. \tag{5.35}$$

Recall that if v is countably additive then (5.35) implies (5.34).

Problems for Chapter 5

Some of the more elementary problems in this set are Problems 5.1, 5.2, 5.5, 5.6, 5.7, 5.9, 5.11, 5.17, 5.20, 5.21, 5.22, 5.24, 5.26, 5.30, 5.36, 5.38, 5.40.

5.1 Prove Proposition 5.1.

5.2 Give the proof of Theorem 5.3 for the case that $\mu^+ X = \infty$. (*Hint.* Use Proposition 5.3.)

5.3 Let $(X, \mathscr{A}, \mu)$ be a measure space where $X = [1, \infty)$, $\mathscr{B} \subseteq \mathscr{A}$, and $\mu X < \infty$. Define $\varphi(t) = \int_X \cos xt \, d\mu(x)$. Prove that φ has a zero in $[0, \pi]$, and that π can't be replaced by a smaller number. (*Hint.* Set $I = \int_0^\pi \sin t\varphi(t) \, dt$ so that by Fubini's theorem (e.g. Appendix II)

$$I = \int_1^\infty (1 + \cos \pi x)/(1 - x^2) \, d\mu(x) \leqslant 0.$$

Since $\varphi(0) \geqslant 0$, φ must vanish on $[0, \pi]$. For the second part we construct μ with the property that φ's "first" zero is arbitrarily close to π. Take $\varepsilon > 0$ and $0 < \rho < 1$. Set

$$\mu = \sum_{k=0}^\infty \rho^k \delta_{k(2+\varepsilon)+1}$$

so that $$\varphi(t) = \frac{\cos t - \rho \cos (1 + \varepsilon)t}{1 + \rho^2 - 2\rho \cos (2 + \varepsilon)t} .)$$

5.4 Let f be an increasing function defined on $[0, 1]$. Consider the following condition:

$$\lim_n \int_0^1 f_n \, df = 0 \text{ for all sequences } \{f_n: n = 1, \ldots\} \subseteq C[0, 1] \text{ for which}$$

$$\forall g \in L_m^1[0, 1], \qquad \lim \int_0^1 f_n g = 0. \tag{P 5.1}$$

Prove that f is absolutely continuous if and only if (P 5.1) holds. Taking X to be a locally compact space and using the notation of Appendix III it is not difficult to prove the following generalization: *let* $v \in D^0(X)$, $\mu \in M(X)$; $\mu \ll v$ *if and only if for every sequence* $\{f_n: n = 1, \ldots\} \subseteq C_c(X)$ *such that* $\lim_n \int f_n g \, dv = 0$ *for all* $g \in L_v^1(X)$ *we have* $\lim_n \int f_n h \, d\mu = 0$ *for each* $h \in L_{|\mu|}^1(X)$.

5.5 Prove the case that $|\mu A| = \infty$ in Theorem 5.5b.

5.6 Prove by example that the Hahn decomposition of a signed measure μ need not be unique; but show that it is unique except for possibly a null set.

5.7 With regard to Theorem 5.2d find a signed measure space $(X, \mathcal{A}, \mu)$ and $A \in \mathcal{A}$ such that $\mu A = \infty$ and A is not non-negative.

5.8 Let $f \in L_m^1(\mathbf{R})$, $g \in L_m^\infty(\mathbf{R})$. Prove

a) $\forall r, \lim_{x \to \infty} \int_{-\infty}^r f(x - y)g(y) \, dy = 0.$

b) If $\lim_{x \to \infty} g(x) = G$ then $\lim_{x \to \infty} \int_{-\infty}^\infty f(x - y)g(y) \, dy = G \int_{-\infty}^\infty f(x) \, dx.$

Remark An important result in harmonic analysis is Wiener's Tauberian theorem: let $f \in L_m^1(\mathbf{R})$, $g \in L_m^\infty(\mathbf{R})$ and assume

$$\forall y \in \mathbf{R}, \qquad \int_{-\infty}^\infty f(x) e^{iyx} \, dx \neq 0;$$

if

$$\lim_{x \to \infty} \int_{-\infty}^\infty f(x - y)g(y) \, dy = G \int_{-\infty}^\infty f(x) \, dx$$

then

$$\forall h \in L_m^1(\mathbf{R}), \qquad \lim \int_{-\infty}^\infty h(x - y)g(y) \, dy = G \int_{-\infty}^\infty h(x) \, dx.$$

5.9 Given a signed measure space $(X, \mathcal{A}, \mu)$ where card $X \geqslant \aleph_0$, $\mathcal{A} = \mathscr{P}(X)$, and

$$\forall x \in X, \qquad \mu(\{x\}) = 1 \quad \text{or} \quad \mu(\{x\}) = -1. \tag{P 5.2}$$

If P, N is an Hahn decomposition of X prove that the cardinality of either P or N is finite.

5.10 Let $X = [-1, 1]$, let $\mathscr{A}$ be the algebra of all finite disjoint unions of intervals $[a, b) \subseteq X$, and define

$$f(x) = \begin{cases} 1/x, & \text{if} \quad x \neq 0 \\ 0, & \text{if} \quad x = 0 \end{cases}$$

Define $\mu \colon \mathscr{A} \to \mathbf{R}$ as

$$\mu\left(\bigcup_1^n [a_k, b_k)\right) = \sum_1^n [f(b_k) - f(a_k)].$$

Prove that μ is a well-defined finitely additive set function but that $\mu \neq \mu^+ - \mu^-$. What can you say about an Hahn decomposition for μ?

5.11 Let $(X, \mathscr{A}, \mu)$ be a signed measure space. Prove
a) $\forall A, B \in \mathscr{A}$, with $|\mu A| < \infty$ and $B \subseteq A$, $|\mu B| < \infty$.
b) $\forall \{A_n \colon n = 1, \ldots\} \subseteq \mathscr{A}$, increasing, $\lim_n \mu A_n = \mu(\bigcup A_n)$.
c) $\forall \{A_n \colon n = 1\} \subseteq \mathscr{A}$, decreasing and satisfying $|\mu A_1| < \infty$, $\lim \mu A_n = \mu(\bigcap A_n)$.

5.12 Let $([0, 1], \mathscr{A}, \mu_\alpha)$ be a family of measure spaces and let $f_n \colon [0, 1] \to \mathbf{R}^*$, $n = 1, \ldots$, have the property that for each α, $\{f_n \colon n = 1, \ldots\} \subseteq L^1_{\mu_\alpha}[0, 1]$ is $L^1_{\mu_\alpha}$-Cauchy.
Prove that there is a function $f \in \bigcap L^1_{\mu_\alpha}[0, 1]$ for which

$$\forall \alpha, \quad \|f - f_n\|_\alpha \to 0,$$

where $\| \ \|_\alpha$ is the $L^1_{\mu_\alpha}[0, 1]$ norm. (*Hint*. For the case "α" is not indexed by a sequence the notion of an absolute ultrafilter is used.)

5.13 Prove Proposition 5.4.

5.14 a) Prove the Lebesgue decomposition as a corollary to R–N.
b) Prove the uniqueness of the Lebesgue decomposition in Theorem 5.9.
c) Do the extension parts of Theorem 5.9: extend from μ a bounded measure to μ σ-finite (resp. $\mu \in M(X)$).
d) Show that if μ is counting measure on $(0, 1)$ and $\nu = m$ on $(0, 1)$ then μ has no Lebesgue decomposition with respect to m.

5.15 Let $(X, \mathscr{A})$ be a measurable space and let μ, ν be measures on $\mathscr{A}$. Prove that there are measures ν_1 and ν_2 such that

$$\nu_1 \ll \mu \quad \text{and} \quad \nu = \nu_1 + \nu_2,$$

where ν_2 satisfies the following singularity condition with respect to μ:

$$\forall A \in \mathscr{A}, \quad \exists B \subseteq \mathscr{A} \quad \text{such that} \quad \nu_2 A = \nu_2 B \quad \text{and} \quad \mu B = 0.$$

Also show that ν_2 is always unique and that ν_1 is unique if ν is σ-finite.

5.16 Given the measurable space $(\mathbf{T}, \mathscr{B})$, the Fourier coefficients of $\mu \in M(\mathbf{T})$ are defined by

$$\hat{\mu}(n) = \frac{1}{2\pi} \int\limits_0^{2\pi} e^{-inx} \, d\mu(x), \qquad n \in \mathbf{Z}.$$

a) If $k \geqslant 1$ is fixed, $k \in \mathbf{Z}$, and $\mu = \delta_{2\pi/k}$, verify that

$$\forall n \in \mathbf{Z}, \qquad \hat{\mu}(n + k) = \hat{\mu}(n). \tag{P 5.3}$$

Can you characterize the subclass of $M(\mathbf{T})$ which satisfies (P 5.3)?

b) Assume $\hat{\mu}(n) \to 0$ as $n \to \infty$. Prove that $\hat{\mu}(n) \to 0$ as $n \to -\infty$. (*Hint.* By R–N, there is a bounded μ-measurable function f, $|f| = 1$, such that for each n

$$|\hat{\mu}(n)| = \frac{1}{2\pi} \left| \int\limits_0^{2\pi} e^{-inx} f(x) \, d|\mu|(x) \right|.$$

Thus for $g(x) = \overline{f(x)}/f(x)$,

$$|\overline{\hat{\mu}(n)}| = \frac{1}{2\pi} \left| \int\limits_0^{2\pi} e^{inx} g(x) \, d\mu(x) \right|;$$

and so it is sufficient to prove that

$$\lim_{n \to \infty} \left| \int\limits_0^{2\pi} e^{-inx} g(x) \, d\mu(x) \right| = 0. \tag{P 5.4}$$

To do this observe first and trivially that if $g(x)$ has the form $t_N(x) = \sum\limits_{|k| \leqslant N} a_k \, e^{ikx}$ then (P 5.4) is immediate. By the Stone–Weierstrass theorem and the fact that each bounded μ-measurable function h on $\mathbf{T}$ is the pointwise (*m-a.e.*) limit of a bounded sequence of continuous functions we can compute

$$\lim_{n \to \infty} \int\limits_0^{2\pi} e^{-inx} h(x) \, d\mu(x) = 0$$

by LDC and Jordan decomposition.)

5.17 Let $(X, \mathscr{A}, \mu)$ be a signed measure space with values in $\mathbf{R}$. Prove that $\mu^{\pm} = \frac{1}{2}(|\mu| \pm \mu)$.

5.18 Given a measurable space $(X, \mathscr{A})$, assume X is locally compact, and take $M(X)$ with norm $\|\mu\| = |\mu|(X)$; prove that $M(X)$ is complete.

5.19 a) Prove Theorem 5.7.

b) For $\mu \in M(X)$, prove that

$$\left| \int f \, d\mu \right| \leqslant \int |f| \, d|\mu|.$$

5.20 Find a measure μ such that $\mu \neq \mu_c + \mu_d$. (*Hint*. Take μ to be counting measure.)

5.21 Find a discrete measure which is absolutely continuous with respect to ν for some ν.

5.22 Given an unbounded measure space $(X, \mathscr{A}, \mu)$. Prove that X contains measurable sets of arbitrarily large finite measure or else it contains a non-$\varnothing$ set A for which

$$\forall B \subseteq A, \qquad B \in \mathscr{A}, \qquad \mu B = 0 \quad \text{or} \quad \mu B = \infty.$$

5.23 Let $\mu \in M[0, 1]$, where $M[0, 1]$ is determined by the measurable space $([0, 1], \mathscr{A})$ and $\mathscr{B} \subseteq \mathscr{A}$. Prove that μ is discrete if and only if for each net $\{f_\alpha\}$ (e.g. [63, p. 65]) of bounded μ-measurable functions which decrease pointwise to 0 we can conclude that $\mu(f_\alpha) \to 0$ (cf. Appendix III for notation). (*Hint*. Assume μ has a continuous part ν. Set $f_\alpha = \chi_{E_\alpha}$, where $E_\alpha \subseteq [0, 1]$ and card $E_{\tilde{\alpha}} \leqslant \aleph_0$. The ordering $E > F$ for the net $\{f_\alpha\}$ is defined by $E \subseteq F$.

5.24 Let $\mu \in (S)M(X)$ and $f \in L^1_\mu(X)$. Prove that f vanishes outside of a σ-finite set.

5.25 Let $\mu, \nu \in (S)M(X)$. Find conditions on μ and ν so that

$$\mu \ll \nu \Rightarrow \mu \in M_c(X)$$

(cf. Theorem 5.10).

5.26 Let $\mu \in M(X)$ and assume $|\mu|(X) = \mu(X) = 1$. Prove that μ is a measure (cf. the proof of Theorem III.9). (*Hint*. Use Theorem 5.14.)

5.27 Given a measure space $(X, \mathscr{A}, \mu)$ and an $\mathbf{R}^*$ or $\mathbf{C}$-valued μ-measurable function f on X. Define $\nu A = \int_A f \, d\mu$ and let g be an $\mathbf{R}^*$ or $\mathbf{C}$-valued ν-measurable function on X.

Prove that

$$\forall A \in \mathscr{A}, \qquad \int_A g \, d\nu = \int_A fg \, d\mu,$$

and that $g \in L^1_{|\nu|}(X)$ if and only if $fg \in L^1_\mu(X)$. This result can be used to prove an extension of R–N from the case of bounded measures.

5.28 We give an alternative proof to Theorem 5.12. Let $\Phi = \{f \in L^1_\nu(X) : f \geqslant 0$ and $\forall A \in \mathscr{A}, \mu A \geqslant \int_A f \, d\nu\}$. $f = 0$ is in Φ and so $\Phi \neq \varnothing$. Set $r = \sup \left\{ \int_X f \, d\nu : f \in \Phi \right\}$ and check that $r = \int_X g\, dv$ for some $g \in \Phi$. Define

$$\forall A \in \mathscr{A}, \qquad \lambda A = \mu A - \int_A f\, dv$$

and obtain a contradiction by assuming that $\lambda A > 0$ for some A. The following lemma is used: let $a \ll b$ for bounded measures a and b, and assume that $aA \neq 0$; then there is a non-negative function $h \in L_b^1(X)$ such that $\int h\, db > 0$ and

$$aB \geqslant \int_B h\, db$$

for each b-measurable set B. To prove the lemma set $h = \varepsilon \chi_Q$ where $aA > \varepsilon bA$ and Q is a non-negative set with respect to $a - \varepsilon b$ for which $(a - \varepsilon b)(Q) > 0$. We then apply the lemma to the case $\lambda = a$, $v = b$, and $g + h \in \Phi$; and contradict the definition of r.

5.29 Give examples of $\mu \in M_c(\mathbf{T})$, resp. $\mu \in M_d(\mathbf{T})$, for which $\varlimsup_{|n| \to \infty} |\hat\mu(n)| > 0$ (cf. Problem 4.3).

5.30 Let $f \in L_m^p(0, \infty)$, $p \in [1, \infty)$, and assume that f is uniformly continuous on $(0, \infty)$. Prove that $\lim_{x \to \infty} f(x) = 0$.

5.31 Fix $\mu \in M(X)$ and let $Y \subseteq X$ be the intersection of all possible concentration sets C_μ (of μ). Prove that $Y = \{x : |\mu|(x) > 0\}$.

5.32 Given $([0, 1], \mathscr{M}, m)$. Is it true that for each uncountable set $\mathscr{A} \in \mathscr{M}$, $m\mathscr{A} = 0$, there is a measure $m_A \in M_c[0, 1]$ defined on $\mathscr{B}$, $m_A A = 1$, such that

$$D \subseteq B \subseteq A \Rightarrow m_A D = m_A B m_B D$$

(e.g. [116, pp. 77–78] gives a precise statement and discussion). Compare this with the problem of measure, mentioned in the Remark after the definition of "measure" in Chapter 2, which is answered by the statement that there are no continuous non-0 measures defined on $\mathscr{P}(X)$, card X = card $\mathbf{R}$ (cf. remark after Example 5.2).

5.33 We've indicated the equivalence of FTC and R–N. We develop this a bit more now. Suppose μ and v are finite measures and

$$\lim \frac{\mu A_\alpha}{v A_\alpha} = D_v\mu(x), \tag{P 5.5}$$

as $\{A_\alpha\} \subseteq \mathscr{A}$ ranges over some family for which $x \in A_\alpha$ and $x = \bigcap A_\alpha$. A good argument could be made for the equivalence of R–N and FTC if, in fact, (P 5.5) yielded

$$\mu A = \int_A D_v\mu\, dv.$$

Using a Vitali covering argument in $\mathbf{R}^n$ (with Lebesgue measure m^n) as well as R–N, it can be proved that

Theorem *Given* $\mu \in M(\mathbf{R}^n)$. *Then* a) $\mathrm{D}\mu \varepsilon L^1_{m^n}(\mathbf{R}^n)$; b) $\forall A \in \mathscr{B}(\mathbf{R}^n)$, $\mu A = \mu_s A +$
$\int\limits_A \mathrm{D}\mu \, \mathrm{d}x$, *where* $\mu_s \perp m^n$ *and* $\mathrm{D}\mu_s = 0$, m^n-*a.e.*

One has to be a bit careful in the definition of $\mathrm{D}\mu$, but in any case it is defined as in (P 5.5) (with $v = m^n$). For interesting expositions on these matters we refer to [22; 47; 94].

In the case of $\mathbf{R}$ an important way to view FTC, as we've seen in Chapter 4, is Theorem 4.10. An analogue of this latter result in $\mathbf{R}^n$ is

Theorem *Let* $f \in L^1_{m^n}(K)$ *for each compact set* $K \subseteq \mathbf{R}^n$. *Then for almost all* x

$$\lim_{m^n C_x \to 0} \frac{1}{m^n C_x} \int\limits_{C_x} |f(y) - f(x)| \, \mathrm{d}y = 0, \tag{P 5.6}$$

where C_x *is a cube centered at* x *with sides parallel to the axes.*

It is possible to generalize this result to general measure spaces; and we refer to [55] for such an extension whose proof uses ideas centered about the maximal theorem. The use of the maximal theorem to prove (P 5.6) is not new.

For our purposes now we just make the following introductory comments. In order to prove (P 5.6) we once again face the fundamental problem of analysis and wish to interchange the order of taking limits. Consequently, in light of LDC we'd like to dominate $(1/m^n C_x) \int_{C_x} |f|$; such a dominant, which happens to be useful, is the Hardy–Littlewood maximal function,

$$\overset{\cup}{f}(x) = \sup_{C_x} (1/m^n C_x) \int\limits_{C_x} |f|.$$

The original Hardy–Littlewood maximal theorem is

Theorem *If* $f \in L^p_{m^n}(\mathbf{R}^n)$, $1 < p \leq \infty$, *then* $\overset{\cup}{f} \in L^p_{m^n}(\mathbf{R}^n)$ *and* $\|\overset{\cup}{f}\|_p \leq K_{p,n}\|f\|_p$.

Prove that if $\overset{\cup}{f} \in L^1_{m^n}(\mathbf{R}^n)$ then $f = 0$ m^n-a.e.

5.34 Let $(X, \mathscr{A}, \mu)$ be a σ-finite measure space; and assume that $L^p_\mu(X)$ has a Hausdorff locally convex topology $\mathscr{T}$ so that with $\mathscr{T}$ (on $L^p_\mu(X)$), $(L^p_\mu(X)))' = L^{p/(p-1)}_\mu(X)$. Prove that $L^p_\mu(X)$ is $\mathscr{T}$-sequentially complete. (*Hint.* First prove the result for the topology $\mathscr{T}_w = \sigma(L^p_\mu(X), L^{p/(p-1)}_\mu(X))$.) For an arbitrary $\mathscr{T}$ choose a $\mathscr{T}$-Cauchy sequence $\{f_n: n = 1, ...\}$ so that, à fortiori, $\{f_n: n = 1, ...\}$ is $\mathscr{T}_w$-Cauchy. Thus $f_n \to f$ in $\mathscr{T}_w$. Take a convex neighborhood V of 0 so that by the Hahn–Banach theorem V is $\mathscr{T}_w$ closed if it is $\mathscr{T}$ closed. Hence, eventually, $f_n \in f + V$.) The notation "$(L^p_\mu(X))'$" requires some explanation in this exercise; and the problem should only be attempted if you've had some exposure to the theory of topological vector spaces.

5.35 Given $\alpha, \beta > 0$, $\alpha\beta < 1$. If $f \in L_\mu^{1+\alpha}(X)$, $g \in L_\mu^{1+\beta}(X)$, and $f^{1+\alpha}g^{1+\beta} \in L_\mu^1(X)$ then prove that

$$\left| \int_X fg \, d\mu \right|^{(1+\alpha)(1+\beta)/(1-\alpha\beta)} \leq$$

$$\left(\int_X |f|^{1+\alpha}|g|^{1+\beta} \, d\mu \right) \left(\int_X |f|^{1+\alpha} \, d\mu \right)^{\beta(1+\alpha)/(1-\alpha\beta)} \left(\int_X |g|^{1+\beta} \, d\mu \right)^{\alpha(1+\beta)/(1-\alpha\beta)}$$

5.36 Let $g \in L_m^1[0, 1]$ be non-negative and take $r \geq 1$. Prove that

$$\int_0^1 g^r \geq \left(\int_0^1 g \right)^r.$$

5.37 S. Bochner [15] proved a version of R–N for the case that μ and v are finitely additive. His original proof used the countably additive version of R–N as well as Lebesgue's theorem on the differentiation of monotone functions. There have been simplifications of proof during the years; we mention those due to L. Dubins [33] and C. Fefferman [39]. A statement of the result is: *let $\mathscr{A} \subseteq \mathscr{P}(X)$ be an algebra and let*

$$\mu, v \colon \mathscr{A} \to \mathbf{C}$$

be finitely additive; assume

$$\|v\| = \sup_{A \in \mathscr{A}} |vA| < \infty \tag{P 5.7}$$

and

$$\forall \varepsilon > 0 \,\, \exists \delta > 0 \text{ such that } |vA| < \delta \Rightarrow |\mu A| < \varepsilon; \tag{P 5.8}$$

then there is a sequence $\{s_n \colon n = 1, \dots\}$ of simple functions such that

$$\forall A \in \mathscr{A}, \quad \lim_n \int_A s_n \, dv = \mu A.$$

(P 5.7) is compared to absolute continuity in Theorem 5.10 and Theorem 5.11. The above result can be strengthened to read; if $\|v\| < \infty$ then the set $\{sv \colon s \text{ simple and } \langle sv, g \rangle = \int gs \, dv\}$ of finitely additive set functions is dense under the norm $\| \, \|$ of (P 5.7) in the space of finitely additive set functions μ which satisfy (P 5.8). The importance of something like (P 5.8) has already been made clear in Theorem 5.12. For the finitely additive case (P 5.7) is crucial.

5.38 Prove Proposition 5.7b.

5.39 In 1940 von Neumann proved R–N for $\mu \ll v$, where μ and v are bounded measures on $(X, \mathscr{A})$, by using "only" the fact that $(L_{\mu+v}^2(X))' = L_{\mu+v}^2(X)$. The proof is extremely slick and, as we've seen in Theorem 5.20, the converse also holds. The proof that $(L^2)' = L^2$ depends on the completeness of L^2 and is sometimes called the Riesz–Fischer theorem (cf. Appendix I.2); as mentioned in the text it can be proved elegantly and relatively simply in the context of Hilbert spaces. von Neumann's proof is given in several standard texts, e.g. [49, p. 178; 56, pp. 313ff.; 92, p. 242; 94, pp. 123–124].

We outline the proof:

a) set $\lambda = \mu + \nu$ and prove that

$$\exists g \in L^2_\lambda(X) \quad \text{such that} \quad \forall f \in L^2_\lambda(X), \quad \int_X f \, d\mu = \int_X fg \, d\lambda; \qquad \text{(P 5.9)}$$

b) show $0 \leqslant g \leqslant 1$, λ-a.e., by taking $f = \chi_A$, $\lambda A > 0$, in (P 5.9);

c) compute

$$\forall A \in \mathscr{A}, \quad \int_A (1 - g^{n+1}) d\mu = \int_A g(1 + g + \ldots + g^n) \, d\nu; \qquad \text{(P 5.10)}$$

d) prove that the left-hand side of (P 5.10) converges to μA as $n \to \infty$.

5.40 a) Prove the part of Theorem 5.20 which extends the result from the finite to σ-finite case, and which was omitted in the text.

b) Prove Theorem 5.20 for any measure space $(X, \mathscr{A}, \mu)$ in the $1 < p < \infty$ case.

5.41 Let $\{\mu_n, \mu: n = 1, \ldots\} \subseteq M(X)$ and assume

$$\forall A \in \mathscr{A}, \quad \mu_n A \to \mu A.$$

Can you find non-trivial situations to conclude that $\|\mu_n\|_1 \to \|\mu\|_1$ (cf. Chapter 6 and Theorem I.9)? (*Hint.* a) Take each μ_n to be a measure. b) Assume that $\mu_n \ll \nu$, where ν is a bounded measure, and that f_n converges in measure, where $f_n \in L^1_\nu(X)$ corresponds to μ_n.)

5.42 a) Given ([0, 1], $\mathscr{B}$, m) and assume $\mu \ll m$. We define $f(x)$ in the manner of Theorem 5.16 and so $\mu_f = \mu$. Prove that f' is the R–N derivative of μ with respect to m (cf. Problem 5.33).

b) Prove that if g is absolutely continuous in [0, 1] and $f \in C[0, 1]$ then

$$\int_0^1 f \, dg = \int_0^1 fg'$$

(where $\int_0^1 f \, dg$ is the usual Riemann–Stieltjes integral).

5.43 Let $\nu \in (S)M(X)$. Prove that $M_d(X)$ (resp. $M_c(X)$, $M_{ac,\nu}(X)$, $M_{s,\nu}(X)$) is closed in $M(X)$, where $M(X)$ is considered as a normed space with the total variation norm $\| \ \|_1$ (s = singular).

5.44 Given $(\mathbf{R}^n, \mathcal{M}^n, m^n)$, where m^n is Lebesgue measure on $\mathbf{R}^n$ and $\mathcal{M}^n$ is the family of Lebesgue measurable sets in $\mathbf{R}^n$.

If $f\colon \mathbf{R}^n \to \mathbf{C}$ is m^n-measurable and $A \in \mathcal{M}^n$ is fixed, define

$$\forall y > 0, \qquad w(y) = m^n\{x \in A\colon |f(x)| > y\}.$$

Prove

$$\forall f \in L^1_{m^n}(A), \qquad \int_A |f| = \int_0^\infty w(y)\, dy$$

(so that *the Lebesgue integral in $\mathbf{R}^n$ can be viewed as a one-dimensional improper integral*). (*Hint*. Note that if $m^n A < \infty$ then

$$w(y) \leqslant \frac{1}{y} \int_A |f|.)$$

5.45 Prove Wiener's characterization of continuous measures

$$\forall \mu \in M(\mathbf{T}), \qquad \sum_{x \in \mathbf{T}} |\mu(\{x\})|^2 = \lim_N \frac{1}{2N+1} \sum_{-N}^{N} |\hat{\mu}(n)|^2;$$

and so $\mu \in M_c(\mathbf{T})$ if and only if

$$\lim_{N \to \infty} \frac{1}{2N+1} \sum_{-N}^{N} |\hat{\mu}(n)|^2 = 0.$$

In particular, $\mu \notin M_c(\mathbf{T})$ if $\mu \in M(\mathbf{T})$ and $|\hat{\mu}| = 1$.

5.46 Let $\mathcal{A} \subseteq \mathcal{P}(X)$, card $\mathcal{A} \geqslant \aleph_0$, be an algebra. Prove that there is a finitely additive function $v\colon \mathcal{A} \to \mathbf{R}$ such that

$$\sup_{A \in \mathcal{A}} |vA| = \infty.$$

(*Hint*. First prove that there is an infinite disjoint family $\{A_n\colon n = 1, \dots\} \subseteq \mathcal{A}$. Choose $x_n \in A_n$ and define

$$\mu = \sum_1^\infty \frac{\pi^n}{4^n} \delta_{x_n}.$$

Since $\{\pi^n/4^n\colon n = 1, \dots\}$ is linearly independent over $\mathbf{Q}$ it can be imbedded in an Hamel basis H for $\mathbf{R}$. Define the appropriate function $f\colon H \to \mathbf{R}$, extend linearly, and set

$$\forall A \in \mathcal{A}, \qquad vA = f(\mu A).)$$

Remark By using a set theoretic argument it can be shown that there is a non-trivial finitely additive set function $v: \mathcal{P}(\mathbf{Z}^+) \to [0, 1]$ such that $vF = 0$ if $F \subseteq \mathbf{Z}^+$ is a finite set.

5.47 Decompose μ_E, the Cantor–Lebesgue measure for the perfect symmetric set E, as in Theorem 5.15 (for $v = m$). Don't forget the case $mE > 0$. With regard to this and Problem 4.21c we note the recent paper by J. Lipiński [74] which completely characterizes the set of points in which a singular function can have an infinite derivative.

5.48 Find a non-negative function $f \in L^2_m[1, \infty)$ such that

$$\int_1^\infty \frac{f(x)}{\sqrt{x}}\, dx$$

diverges. (*Hint.* Let $f = \sum_1^\infty k_n \chi_{[n, n+1)}$ and set

$$\frac{1}{n \log n} = k_n \int_n^{n+1} \frac{dx}{\sqrt{x}}\; .)$$

Appendix A 5.1 The Radon–Nikodym theorem: historical notes on Lusin's problem and Vitali

A 5.1.1 Lusin's problem

In his dissertation of 1915 (actually he published an annoucement at Paris in 1913 on the relevant material), L u s i n gave necessary and sufficient conditions that $f \in L^2_m(\mathbf{T})$ have a Fourier series convergent *m-a.e.*; and at the same time essentially posed the problem as to whether every $f \in L^2_m(\mathbf{T})$ has its Fourier series converging *m-a.e.*. Actually, men such as F a t o u (1906), J e r o s c h and W e y l (1908), W e y l (1909), W. H. Y o u n g (1912), H o b s o n (1913), P l a n c h e r e l (1913), and H a r d y (1913) had worked specifically on such issues. Refined "log-estimates" by K o l m o g o r o v–S e l i v e r s t o v (1925), P l e s s n e r (1926), and L i t t l e w o o d–P a l e y (1931) kept interest in the problem at a fine pitch. Finally, in 1966 (Acta Math. **116** (1966) 135–157), C a r l e s o n proved that if $f \in L^2_m(\mathbf{T})$ *then its Fourier series converges m-a.e. (to f)*; and in 1968, using the method of C a r l e s o n's proof and the theory of interpolation of operators, R. A. H u n t extended C a r l e s o n's result to the $L^p_m(\mathbf{T})$, $p > 1$, case (recall K o l m o g o r o v's example mentioned in Appendix A3.1.3). C. F e f f e r m a n has recently proved C a r l e s o n's theorem along the lines initiated by K o l m o g o r o v–S e l i v e r s t o v.

An important lemma for C a r l e s o n, and that aspect of C a r l e s o n's proof which leads us to FTC and R–N, is (his Lemma 5, p. 140):

Let $\{I_k : k = 1, \ldots\}$ be a disjoint cover of $(0, 1)$ by open intervals, where $mI_k = d_k$ and the center of I_k is t_k. Define

$$D(x) = \sum_k \frac{d_k^2}{(x - t_k)^2 + d_k^2}, \qquad x \in (0, 1),$$

and

$$U_M = \{x \in (0, 1): D(x) > M\}.$$

Then there are k, $K > 0$ such that for all M

$$mU_M \leqslant k\, e^{-KM}.$$

Zygmund [129] has used some important techniques due to Józef Marcinkiewicz to prove the above lemma in a way that puts it in an interesting perspective with regard to FTC and R–N. (Marcinkiewicz died in a prison camp in 1940 at the age of 30.) We give just a hint at Marcinkiewicz's approach. Recall from Chapter 4 that if $f \in L_m^1(\mathbf{R})$ then

$$\lim \frac{1}{mI} \int_I |f(t) - f(x)|\, dt = 0, \qquad m\text{-a.e.}, \tag{A 5.1.1}$$

where I is an interval containing x and the limit indicates that we let the mI's tend to 0. (A 5.1.1) occurs for all $x \in L(f)$ the set of Lebesgue points. Now there are certain classical theorems in Fourier analysis whose conclusions (c) hold for subspaces $X \subseteq L_m^1(T)$; and by the nature of their proofs (c) is valid for each $x \in L(f)$ of the given $f \in X$. During the 1920's and 1930's when it was popular to try to extend such results to all of $L_m^1(T)$, the difficulties that arose frequently culminated in some ingenious counterexample that showed the existence of $f \in L_m^1(T)$ such that (c) failed for some $x \in L(f)$. This situation did not preclude the possibility that (c) might hold m-a.e. for each $f \in L_m^1(\mathbf{T})$; and, in fact, this is the type of result that Marcinkiewicz obtained and that Zygmund used to prove Carleson's lemma.

A 5.1.2 Vitali and R–N

In [68, 1st edition, p. 94], Lebesgue considered the following definition:

(D–R)   A bounded function f is integrable if there is a function F with bounded derived numbers such that $F' = f$, m-a.e. The integral of f in (a, b) is $F(b) - F(a)$.

Such a definition generalizes the integrals of Riemann and Duhamel. Lebesgue introduced (D–R) by saying: "Je ne m'occuperai pas, pour le moment du moins, de la suivante." And, he keeps his word (and reticence) until on the very last page of text (p. 129), in a footnote no less, we read: "Pour qu'une fonction soit intègrable indéfinie, il faut de plus que sa variation totale dans une infinité dénombrable d'intervalles de longuer totale L tende vers zéro avec L. Si, dans l'énoncé de la page 94 (i.e. (D–R) above), on n'assiyettit pas f à ètre borneé, ni F à ètre à nombres dérivés bornés, mais seulement à la condition précédente, on a une definition de l'intégrale équivalente à celle développée dans ce Chapitre et applicable à toutes les fonctions sommables, bornées ou non." Thus, in an obtuse presentation and as a footnote and without proof, we are handed the fundamental theorem of calculus!

As we've seen V i t a l i defined the notion of absolute continuity in 1904 and went on to state and prove FTC. His next step in this business is [122]. He begins by proving the Vitali covering theorem and uses the covering theorem to prove an FTC in $\mathbf{R}^2$. He also deduces FTC (on $\mathbf{R}$) with the covering theorem. Because of the importance of set functions in the development of integration theory we note that in section 5 of [122] V i t a l i considers families of rectangles and the formula (R–N) for A, a "rettangolo coordinato."

Essentially, in his major work of 1910 [70], L e b e s g u e relies on the Vitali covering theorem and "les travaux de M. V o l t e r r a, à définir la dérivee de la fonction $F(A)$ en un point P comme la limite du rapport $F(A)/m(A)$, A étant un ensemble contenant P et dont on fait tendre toutes les dimensions vers zéro" [70, p. 361].

In [70], L e b e s g u e begins by quoting V i t a l i's FTC in $\mathbf{R}^2$; and notes that an "inadvertance" by V i t a l i in his proof is corrected by considering a r e g u l a r family of rectangles in $\mathbf{R}^2$. This constraint to define Radon–Nikodym derivatives by taking limits over regular families is necessary. Further, by proving FTC in $\mathbf{R}^2$ in terms of set functions (not depending on rectangular coordinates as V i t a l i had done), L e b e s g u e set the stage for the synthesis and generalization of R a d o n in 1913, where he (R a d o n) incorporated Stieltjes integrals into the scheme of things.

Concerning the above-mentioned footnote in [68, 1st edition], L e b e s g u e [70, p. 365] writes: "J'avais, dans mes Leçons, tout à fait incidemment et sans demonstration, fait connaître" the FTC.

A perusal of [70] indicates the crucial dependence of L e b e s g u e on the Vitali covering theorem ("un théorème capital" [70, p. 390]) and V i t a l i's original proof for L e b e s gue's setting ("La démonstration qu'on lira plus loin est presque copiée sur celle de M. V i t a l i" [70, p. 390]).

6 Weak convergence of measures

6.1 Vitali's theorems

We shall now prove Vitali's Theorem 3.10 and Theorem 3,11. As we noted in the remark after the statement of Theorem 3.11, Vitali's results give non-trivial necessary and sufficient conditions in order that

$$\lim_n \int_X |f_n - f| \, d\mu = 0, \qquad (6.1)$$

where $(X, \mathscr{A}, \mu)$ is a measure space and $\{f_n : n = 1, \ldots\} \subseteq L^1_\mu(X)$.

The following result was mentioned after the statement of Theorem 3.11 and is used to deduce LDC from Vitali's theorem.

Proposition 6.1 (Lebesgue) *Given a measure space $(X, \mathscr{A}, \mu)$ and μ-measurable functions f_n which converge μ-a.e. to a μ-measurable function f. If there is $g \in L^1_\mu(X)$ such that*

$$\forall n, \qquad |f_n| \leqslant g, \quad \mu\text{-a.e.},$$

then $f_n \to f$ in measure.

Proof. Without loss of generality assume that

$$\forall x \in X, \qquad |f_n(x)|, |f(x)| \leqslant g(x).$$

Then for each $\varepsilon > 0$,

$$A_{n,\varepsilon} = \bigcup_{j=n}^{\infty} \{x : |f_j(x) - f(x)| \geqslant \varepsilon\} \subseteq \{x : |g(x)| \geqslant \varepsilon/2\};$$

so that by the integrability of g, $\mu A_{n,\varepsilon} < \infty$ for $n = 1, \ldots$.
Since $f_n \to f$, μ-a.e. we have

$$\mu\left(\bigcup_{n=1}^{\infty} A_{n,\varepsilon}\right) = 0. \qquad (6.2)$$

Thus,

$$\overline{\lim_n} \, \mu\{x : |f_n(x) - f(x)| \geqslant \varepsilon\} \leqslant \overline{\lim_n} \, \mu A_{n,\varepsilon} = 0,$$

where the equality follows by (6.2) and because $\mu A_{n,\varepsilon} < \infty$.

q.e.d.

Theorem 6.1 (Theorem 3.10) *Given a finite measure space $(X, \mathscr{A}, \mu)$ and $\{f_n : n = 1, \ldots\} \subseteq$
$L^1_\mu(X)$. (6.1) is valid for some $f \in L^1_\mu(X)$* $\Leftrightarrow$
a) $\{f_n : n = 1, \ldots\}$ *converges in measure to a μ-measurable function f, and*
b) $\{f_n : n = 1, \ldots\}$ *is uniformly absolutely continuous.*

Proof. ($\Leftarrow$) Without loss of generality take each f_n to be real-valued.
i) Given b) we prove that $\{|f_n| : n = 1, \ldots\}$ is uniformly absolutely continuous.
Let $\varepsilon > 0$. Choose $\delta > 0$ such that

$$\forall A \in \mathscr{A}, \text{ for which } \mu A < \delta, \text{ and } \forall n, \left| \int_A f_n \, d\mu \right| < \varepsilon/2. \tag{6.3}$$

For each such A we define

$$A_n^+ = \{x \in A : f_n(x) \geqslant 0\} \quad \text{and} \quad A_n^- = \{x \in A : f_n(x) < 0\}.$$

A_n^+ and A_n^- are measurable since f_n is measurable; and

$$\forall n, \quad \mu A_n^+ < \delta \quad \text{and} \quad \mu A_n^- < \delta,$$

since $\mu A < \delta$.
By definition,

$$\int_{A_n^+} |f_n| \, d\mu = \left| \int_{A_n^+} f_n \, d\mu \right| < \varepsilon/2, \qquad \int_{A_n^-} |f_n| \, d\mu = \left| \int_{A_n^-} f_n \, d\mu \right| < \varepsilon/2.$$

Consequently, for each n,

$$\int_A |f_n| \, d\mu = \int_{A_n^+ \cup A_n^-} |f_n| \, d\mu = \int_{A_n^+} |f_n| \, d\mu + \int_{A_n^-} |f_n| \, d\mu < \varepsilon;$$

and so $\{|f_n| : n = 1, \ldots\}$ is uniformly absolutely continuous.
ii) We now prove that $\{f_n : n = 1, \ldots\}$ is L^1_μ-Cauchy.
This yields (6.1) because $L^1_\mu(X)$ is complete. The fact that $g = f$ μ-a.e., where
$\|f_n - g\|_1 \to 0$, follows from a).
For each $\sigma > 0$, define

$$A_{mn}(\sigma) = \{x : |f_m(x) - f_n(x)| \geqslant \sigma\} \quad \text{and} \quad B_{mn}(\sigma) = \{x : |f_m(x) - f_n(x)| < \sigma\}.$$

Then

$$\int_X |f_m - f_n| \, d\mu \leqslant \int_{A_{mn}(\sigma)} + \int_{B_{mn}(\sigma)} |f_m - f_n| \, d\mu$$

$$\leqslant \int_{A_{mn}(\sigma)} |f_m - f_n| \, d\mu + \sigma \mu B_{mn}(\sigma) \tag{6.4}$$

$$\leqslant \int_{A_{mn}(\sigma)} |f_m| \, d\mu + \int_{A_{mn}(\sigma)} |f_n| \, d\mu + \sigma \mu X.$$

Given $\varepsilon > 0$ there is $\sigma_0 > 0$ such that

$$\forall \sigma \leqslant \sigma_0, \qquad \sigma \mu X < \varepsilon/3 \tag{6.5}$$

(here we use the fact that $\mu X < \infty$).
From i) there is $\delta > 0$ for which

$$\int_A |f_n| \, d\mu < \varepsilon/3 \tag{6.6}$$

if $A \in \mathscr{A}$ satisfies $\mu A < \delta$.
From a) we can choose N where

$$\forall m, n > N, \qquad \mu A_{mn}(\sigma_0) < \delta. \tag{6.7}$$

Thus given $\varepsilon > 0$ we have picked σ_0 and δ as above (i.e. (6.5) and (6.6)) and then use a) to find N as in (6.7).
We apply (6.5) and (6.6) to (6.4) to obtain

$$\forall m, n > N, \qquad \int_X |f_m - f_n| \, d\mu < \varepsilon.$$

($\Rightarrow$) Fix $\varepsilon > 0$.
i) Let $A_n = \{x : |f_n(x) - f(x)| \geqslant \varepsilon\}$.
Then

$$\int_X |f_n - f| \, d\mu \geqslant \int_{A_n} |f_n - f| \, d\mu \geqslant \varepsilon \mu A_n$$

so that $\mu A_n \to 0$ from (6.1).
ii) By (6.1) choose N for which

$$\forall n, m \geqslant N, \qquad \int_X |f_n - f_m| \, d\mu < \varepsilon/3.$$

Since each $f_n \in L^1_\mu(X)$ we can find $\delta > 0$ such that if $\mu A < \delta$ for $A \in \mathscr{A}$ then

$$\forall n = 1, \cdots, N, \qquad \int_A |f_n| \, d\mu < \varepsilon/3.$$

Consequently for $n > N$ and $\mu A < \delta$, $A \in \mathscr{A}$,

$$\left| \int_A f_n \, d\mu \right| \leqslant \int_A |f_n - f_N| \, d\mu + \int_A |f_N| \, d\mu < \varepsilon.$$

Hence, for each n, $\left| \int_A f_n \, d\mu \right| < \varepsilon$ if $\mu A < \delta$.

q.e.d.

Remark One is tempted to prove directly that $f \in L^1_\mu(X)$ using Fatou's lemma, the uniform absolute continuity of $\{|f_n|: n = 1, \ldots\}$, and the fact (from a)) that $f_{n_k} \to f$, μ-a.e. for some subsequence of $\{f_n: n = 1, \ldots\}$. By this method,

$$\text{"} \int_A |f| \, d\mu \leqslant \varliminf \int_A |f_{n_k}| \, d\mu \leqslant \sup_n \int_A |f_n| \, d\mu < \varepsilon \text{"}$$

for $\mu A < \delta$; thus $f \in L^1_\mu(A)$, so that since μX is finite we write $X = \bigcup_1^n A_j$, $\mu A_j < \delta$, and have $f \in L^1_\mu(X)$. This procedure works except when there are not enough sets of small enough measure to guarantee that $X = \bigcup_1^n A_j$.

The notions of uniform absolute continuity and Vitali equicontinuity are obviously closely related.

Proposition 6.2 *Given a measure space* $(X, \mathcal{A}, \mu)$, *take* $M(X)$ *for the measurable space* $(X, \mathcal{A})$, *and let* $\{v_n: n = 1, \ldots\} \subseteq M(X)$.

a) *Assume that for each* n, v_n *is a measure and* $v_n \ll \mu$. *If* $\{v_n: n = 1, \ldots\}$ *is Vitali equicontinuous then it is uniformly absolutely continuous.*

b) *Assume* $\mu X < \infty$. *If* $\{v_n: n = 1, \ldots\}$ *is uniformly absolutely continuous then it is Vitali equicontinuous.*

Proof. We use the argument of Theorem 5.10.

Assume $\{v_n: n = 1, \ldots\}$ is not uniformly absolutely continuous.

Then choose $\varepsilon > 0$, $\{B_k: k = 1, \ldots\} \subseteq \mathcal{A}$, and integers $\{n_k: k = 1, \ldots\}$ such that

$$\mu B_k > 1/2^{k+1} \quad \text{and} \quad v_{n_k} B_k \geqslant \varepsilon. \tag{6.8}$$

Set $E_k = \bigcup_{j=k}^{\infty} B_j$ so that $E_{k+1} \subseteq E_k$. Also

$$\mu E_k \leqslant \sum_{j=k}^{\infty} 1/2^{j+1} = 1/2^k.$$

We write $A_k = E_k \setminus E_0$ where $E_0 = \bigcap_{k=1}^{\infty} E_k$; thus $A_{k+1} \subseteq A_k$ and $\bigcap A_k = \varnothing$.

Since $v_{n_k} \ll \mu$, $v_{n_k} E_0 = 0$; and by hypothesis $\lim_k v_{n_k} A_k = 0$.

Thus, $\lim_k v_{n_k} E_k = 0$, whereas, from (6.8), $v_{n_k} E_k \geqslant v_{n_k} B_k \geqslant \varepsilon$; this is the desired contradiction.

b) Given $\varepsilon > 0$ and $\{A_\alpha: n = 1, \ldots\} \subseteq \mathcal{A}$ decreasing to $\varnothing$. By hypothesis there is $\delta > 0$ so that if $A \in \mathcal{A}$, $\mu A < \delta$, then $|v_m A| < \varepsilon$ for each m.

Since $\mu X < \infty$, we have $\mu A_n \to 0$, and so there is N for which $\mu A_n < \delta$ if $n > N$.

Consequently $|v_m A_n| < \varepsilon$ for all $n > N$ and for all m.

$$\text{q.e.d.}$$

In light of Proposition 6.2 and part i) of the proof of the sufficient conditions in Theorem 6.1 we have: $\{v_n : n = 1, ...\}$ *is Vitali equicontinuous if and only if* $\{f_n : n = 1, ...\}$ *is uniformly absolutely continuous, where* $v_n A = \int_A |f_n| \, d\mu$ *and* $\mu X < \infty$.

The "only if" part does not require that $\mu X < \infty$.

Theorem 6.2 (Theorem 3.11) *Given a measure space* $(X, \mathscr{A}, \mu)$ *and* $\{f_n : n = 1, ...\} \subseteq L^1_\mu(X)$. (6.1) *is valid for some* $f \in L^1_\mu(X) \Leftrightarrow$

a) $\{f_n : n = 1, ...\}$ *converges in measure (to a μ-measurable function f), and*

b) $\{v_n : \forall A \in \mathscr{A}, \ v_n A = \int_A |f_n| \, d\mu\}$ *is Vitali equicontinuous.*

Proof. ($\Leftarrow$) If $\mu X < \infty$ the result follows by Theorem 6.1 and the above remarks. Note (e.g. Problem 5.24) that $\bigcup_n \{x : f_n(x) \neq 0\}$ is σ-finite so that without loss of generality we take X σ-finite.

Let $\{B_m : m = 1, ...\} \subseteq \mathscr{A}$ increase to cover X and assume that $\mu B_m < \infty$; if we set $A_m = B_{\tilde{m}}$ then $A_{m+1} \subseteq A_m$ and $\bigcap A_k = \emptyset$.

We prove that $\{f_n : n = 1, ...\}$ is L^1_μ-Cauchy. Take $\varepsilon > 0$.

From b) there is k such that

$$\forall m, \qquad \int_{A_k} |f_m| \, d\mu < \varepsilon/4$$

and so for each m and n

$$\int_{A_k} |f_m - f_n| \, d\mu < \varepsilon/2. \tag{6.9}$$

Using Theorem 6.1 and the restriction measure space $(B_k, \mathscr{A}, \mu)$, we see that there is N for which

$$\int_{B_k} |f_m - f_n| \, d\mu < \varepsilon/2 \tag{6.10}$$

if $m, n > N$.

(6.9) and (6.10) yield the result.

($\Rightarrow$) The proof of a) from (6.1) is the same as in Theorem 6.1. To prove b) pick any $\varepsilon > 0$ and take N such that

$$\forall n \geqslant N, \qquad \int_X |f_n - f| \, d\mu < \varepsilon/2. \tag{6.11}$$

Let $\{A_n: n = 1, \ldots\} \subseteq \mathscr{A}$ decrease to $\emptyset$. Each v_m is obviously Vitali continuous, and so there is $M > 0$ for which

$$\int\limits_{A_m} |f| \, d\mu < \varepsilon/2 \quad \text{and} \quad \int\limits_{A_m} |f - f_n| \, d\mu < \varepsilon/2 \tag{6.12}$$

if $m > M$ and $n = 1, \ldots, N$.
Consequently, for each $m > M$ and for all n

$$\int\limits_{A_m} |f_n| \, d\mu \leqslant \int\limits_{A_m} |f_n - f| \, d\mu + \int\limits_{A_m} |f| \, d\mu < \varepsilon,$$

where the last inequality follows from (6.11) or the second part of (6.12) depending on whether $n > N$ or $n \leqslant N$.

q.e.d.

As indicated after Theorem 3.11, LDC follows from Theorem 6.2. The hypotheses of LDC and Proposition 6.1 entail Theorem 6.2a, and the Vitali equicontinuity follows since "$|f_n| \leqslant g \in L_\mu^1(X)$."

In sections 6.2 and 6.3 we shall see that Vitali's results are intimately related to subsequent important work by Hahn, Nikodym, Saks, Dieudonné, and Grothendieck in the area of weak convergence of measures.

A point of information and a corresponding word of caution are in order. We use "weak convergence" in the topological sense introduced in Appendix I. Modern probabilists also have a notion of weak convergence of measures on complete separable metric spaces X. They say that $\mu_n \to \mu$ weakly if for every bounded continuous function f on X, $\langle \mu_n, f \rangle \to \langle \mu, f \rangle$. In this setting our weak sequential convergence obviously generates a stronger topology than "probabilistic weak convergence."

6.2 The Nikodym and Hahn–Saks theorems

In this section the work of Hahn dates from 1922 and that of Saks from 1933 [95; 96]; Nikodym announced his results in 1931 and published them in 1933. Our presentation is due to Brooks [20].
We use the following version of Schur's lemma [21] (cf. Theorem I.10).

Proposition 6.3 Let $\{c_{ij}: i, j \in \mathbf{Z}^+\} \subseteq \mathbf{C}$ have the following properties:

a) $\forall i, \sum\limits_{j=0}^{\infty} |c_{ij}| < \infty$,

b) $\forall S \subseteq \mathbf{Z}^+, \exists \lim\limits_{i \to \infty} \sum\limits_{j \in S} c_{ij}$.

Then there is $\{c_j : j \in \mathbf{Z}^+\} \subseteq \mathbf{C}$ such that $\sum |c_j| < \infty$ and

$$\lim_{i \to \infty} \sum_{j=0}^{\infty} |c_{ij} - c_j| = 0. \tag{6.13}$$

Proof. i) We first prove that if we are given $\{z_1, ..., z_n\} \subseteq \mathbf{C}$, then there is $S \subseteq \{j : 1 \leqslant j \leqslant n\}$ such that

$$\sum_{j=1}^{n} |z_j| \leqslant 4\sqrt{2} \left| \sum_{j \in S} z_j \right|. \tag{6.14}$$

This is a standard inequality and we refer to [94, p. 119] as well as our remarks prior to Theorem 5.6. Divide the complex plane into its four "diagonal" quadrants Q_1, Q_2, Q_3, Q_4 (e.g. Fig. 8).

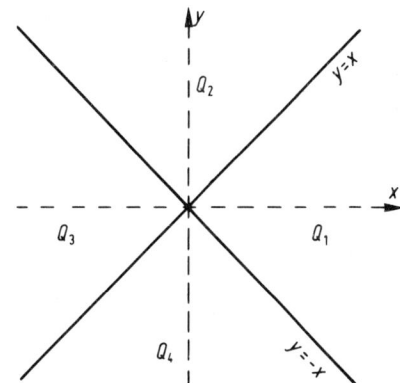

Fig. 8

Assume that $\sum_{z_j \in Q_1} |z_j| \geqslant \frac{1}{4} \sum_{1}^{n} |z_j|$, and let $S = \{j : z_j \in Q_1\}$.
Noting that $Q_1 = \{z = x + iy : |y| \leqslant x, x \geqslant 0\}$, we have

$$|z|/\sqrt{2} \leqslant \operatorname{Re} z$$

since, for $w = a + ib \in Q_1$,

$$\sqrt{2} \operatorname{Re} w = a\sqrt{2} = |a|\sqrt{2} = |a + ia| \geqslant |a + ib| = |w|.$$

Thus

$$\left| \sum_{j \in S} z_j \right| \geqslant \sum_{j \in S} \operatorname{Re} z_j \geqslant \frac{1}{\sqrt{2}} \sum_{j \in S} |z_j| \geqslant \frac{1}{4\sqrt{2}} \sum_{1}^{n} |z_j|,$$

which is (6.14).

ii) We now prove (6.13) using the additional hypothesis that the limit in b) is 0 for any S.

We assume (6.13) is false and obtain a contradiction. Thus there is an $\varepsilon > 0$ and a subsequence $\{i_k: k = 1, \ldots\}$ so that

$$\forall k \in \mathbf{Z}^+, \qquad \sum_{j=0}^{\infty} |c_{i_k j}| > \varepsilon > 0.$$

Observe that there are strictly increasing sequences $\{p_k: k = 1, \ldots\} \subseteq \mathbf{Z}^+$ and $\{n_k: k = 1, \ldots\} \subseteq \{i_k: k = 1, \ldots\}$ for which

$$\forall k, \qquad \sum_{j=0}^{p_k} |c_{n_k j}| < \varepsilon/r \tag{6.15}$$

$$\forall k, \qquad \sum_{j=p_{k+1}+1}^{\infty} |c_{n_k j}| < \varepsilon/r, \tag{6.16}$$

where $r > 2 + 8\sqrt{2}$.

To do this start by taking any p_1. Then, using (6.14), observe that for each n there is $Y_n \subseteq \{0, \ldots, p_1\}$ such that

$$\sum_{j=0}^{p_1} |c_{nj}| \leqslant 4\sqrt{2} \left| \sum_{j \in Y_n} c_{nj} \right|.$$

Since there are only finitely many possible subsets Y_n we obtain (6.15) for some n_1 from b). Then using a) choose p_2 for which (6.16) is true. Continue this process.

From (6.15) and (6.16) we compute

$$\sum_{j=p_k+1}^{p_{k+1}} |c_{n_k j}| = \sum_{j=0}^{\infty} - \sum_{j=0}^{p_k} - \sum_{j=p_{k+1}+1}^{\infty} > \varepsilon - \sum_{j=0}^{p_k} - \sum_{j=p_{k+1}+1}^{\infty} > \varepsilon \left(1 - \frac{2}{r}\right). \tag{6.17}$$

We now use (6.14) in the following way. For each k choose $S_k \subseteq \{j: p_k + 1 \leqslant j \leqslant p_{k+1}\}$ such that

$$4\sqrt{2} \left| \sum_{j \in S_k} c_{n_k j} \right| \geqslant \sum_{j=p_k+1}^{p_{k+1}} |c_{n_k j}|. \tag{6.18}$$

Letting $S = \bigcup S_k$ we combine (6.17) and (6.18), and obtain for each k that

$$\left| \sum_{j \in S} c_{n_k j} \right| \geqslant \left| \sum_{j \in S_k} c_{n_k j} \right| - \sum_{j=0}^{p_k} |c_{n_k j}| - \sum_{j=p_{k+1}+1}^{\infty} |c_{n_k j}|$$

$$\geqslant \varepsilon \left(\frac{1}{4\sqrt{2}} - \frac{1}{2r\sqrt{2}} - \frac{r}{2} \right) > 0.$$

This contradicts b).

iii) We reduce the general case to the setting of part ii).
Let $\{m_i : i = 1, ...\} \subseteq \mathbf{Z}^+$ increase to infinity and define

$$a_{ij} = c_{m_{i+1}j} - c_{m_ij}.$$

$\{a_{ij} : i = 1, ..., j = 1, ...\}$ satisfies the hypotheses of part ii) so that we can conclude

$$\lim_{i \to \infty} \sum_{j=0} |c_{m_{i+1}j} - c_{m_ij}| = 0. \tag{6.19}$$

Since $\{m_i : i = 1, ...\}$ is arbitrary, (6.19) tells us that $\{c_i : c_i = \{c_{i1}, c_{i2}, ...\}\}$ is a Cauchy sequence in $L^1_c(\mathbf{Z}^+)$.
The proof is finished by the completeness of $L^1_c(\mathbf{Z}^+)$.

q.e.d.

We can't replace hypothesis b) in Proposition 6.3 with the condition that

$$\exists \lim_{i \to \infty} \left| \sum_{j \in S} c_{ij} \right|;$$

in fact, let $c_{nm} = (1/2^m) \exp(in\pi/2)$.

Theorem I.10, which states the equivalence of weak and norm sequential convergence in $L^1_c(\mathbf{Z})$, is an immediate consequence of Proposition 6.3. In fact, if $\{c_i : i = 1, ...\} \subseteq L^1(\mathbf{Z})$, with $c_i = \{c_{i1}, ...\}$, converges weakly to 0 then

$$\forall S \subseteq \mathbf{Z}, \qquad \chi_S \in L^\infty_c(\mathbf{Z}) \quad \text{and} \quad \lim_i \int \chi_S c_i = 0;$$

thus the hypotheses of Proposition 6.3 are satisfied.

Theorem 6.3 (Nikodym) *Given a measurable space* $(X, \mathscr{A})$ *and* $\{\mu_n : n = 1, ...\} \subseteq M(X)$. *Assume*

$$\forall A \in \mathscr{A}, \qquad \lim_n \mu_n A = \mu A$$

exists. Then $\mu \in M(X)$ *and* $\{\mu_n : n = 1, ...\}$ *is Vitali equicontinuous.*

Proof. a) Clearly μ is finitely additive on $\mathscr{A}$.
Let $\{A_j : j = 1, ...\} \subseteq \mathscr{A}$ decrease to $\emptyset$. We'll prove

$$\lim_j \mu A_j = 0. \tag{6.20}$$

a.i) Set $E_k = A_k \setminus A_{k+1}$ so that $\{E_k : k = 1, ...\}$ is a disjoint family and

$$A_k = (A_k \setminus A_{k+1}) \cup (A_{k+1} \setminus A_{k+2}) \cup \cdots = \bigcup_k^\infty E_j.$$

Our immediate task is to verify (6.21) below.

Define

$$c_{ij} = \mu_i E_j.$$

We check that $\{c_{ij}: i = 1, ..., j = 1, ...\}$ satisfies the hypotheses of Proposition 6.3. For each i,

$$\sum_j |c_{ij}| = \sum_j |\mu_i E_j| \leq \sum_j |\mu_i|(E_j) = |\mu_i|\left(\bigcup_1^{\infty} E_j\right) = |\mu_i|(A_1) < \infty,$$

and condition a) of Proposition 6.3 is satisfied.

For condition b) let $S \subseteq \mathbf{Z}^+$ and note that

$$\lim_i \sum_{j \in S} c_{ij} = \lim_i \mu_i\left(\bigcup_{j \in S} E_j\right) = \mu\left(\bigcup_{j \in S} E_j\right) \in \mathbf{C}.$$

Consequently, since $\lim_i \mu_i E_j = \mu E_j$, we have

$$\lim_i \sum_{j=1}^{\infty} |\mu_i E_j - \mu E_j| = 0 \tag{6.21}$$

from Proposition 6.3.

a.ii) We now show that

$$\lim_{n \to \infty} \sum_{j=n}^{\infty} \mu E_j = 0. \tag{6.22}$$

To this end we calculate

$$\sum_{j=n}^{\infty} |\mu E_j| \leq \sum_{j=n}^{\infty} |\mu_i E_j - \mu E_j| + \sum_{j=n}^{\infty} |\mu_i E_j|$$

$$\leq \sum_{j=1}^{\infty} |\mu_i E_j - \mu E_j| + \sum_{j=n}^{\infty} |\mu_i E_j|.$$

Thus, $\quad \overline{\lim}_{n \to \infty} \sum_{j=n}^{\infty} |\mu E_j| \leq \sum_{j=1}^{\infty} |\mu_i E_j - \mu E_j| + \overline{\lim}_{n \to \infty} |\mu_i|\left(\bigcup_{j=n}^{\infty} E_j\right),$

so that by our hypothesis on $\{A_n: n = 1, ...\}$ and the fact that μ_i is bounded,

$$\overline{\lim}_{n \to \infty} \sum_{j=n}^{\infty} |\mu E_j| \leq \sum_{j=1}^{\infty} |\mu_i E_j - \mu E_j|.$$

Therefore, (6.22) follows from (6.21).

a.iii) Our next step is to prove

$$\lim_{j \to \infty} \left(\mu_i A_j - \sum_{k=j}^{\infty} \mu E_k\right) = 0, \quad \text{uniformly in } i. \tag{6.23}$$

(6.23) is certainly true for each i (not uniformly) by (6.22) and the hypothesis on $\{A_j : j = 1, \ldots\}$.

Given $\varepsilon > 0$, use (6.21) to choose $I > 0$ such that

$$\forall i > I, \qquad \sum_{j=1}^{\infty} |\mu_i E_j - \mu E_j| < \varepsilon.$$

Next, take $J > 0$ with the property that

$$\forall i = 1, \cdots, I, \qquad \sum_{j=J}^{\infty} |\mu_i E_j - \mu E_j| < \varepsilon.$$

Hence,

$$\forall j > J \quad \text{and} \quad \forall i, \quad \left| \mu_i A_j - \sum_{k=j}^{\infty} \mu E_k \right| = \left| \sum_{k=j}^{\infty} (\mu_i E_k - \mu E_k) \right|$$

$$\leqslant \sum_{k=J}^{\infty} |\mu_i E_k - \mu E_k| < \varepsilon,$$

and this is (6.23).

a.iv) To obtain finally (6.20) we first use the Moore–Smith theorem, in conjunction with a.iii), and the fact that

$$\lim_i \left(\mu_i A_j - \sum_{k=j}^{\infty} \mu E_k \right) = \mu A_j - \sum_{k=j}^{\infty} \mu E_k$$

to obtain

$$\lim_j \left(\mu A_j - \sum_{k=j}^{\infty} \mu E_k \right) = \lim_i \lim_j \left(\mu_i A_j - \sum_{k=j}^{\infty} \mu E_k \right). \tag{6.24}$$

The right-hand side of (6.24) is 0 as we observed after (6.23). Consequently we can apply (6.22) (again) to the left-hand side of (6.24), and (6.20) follows.

b) The desired Vitali equicontinuity is entailed by (6.22) and (6.23). To prove that μ is countably additive let $\{B_n : n = 1, \ldots\} \subseteq \mathscr{A}$ be disjoint and set $A_j = \bigcup_j^{\infty} B_n$. Thus $\{A_j : j = 1, \ldots\}$ decreases to $\emptyset$ and

$$\mu(\bigcup B_n) = \sum_1^j \mu B_n + \mu A_{j+1}.$$

The countable additivity follows by (6.20).

q.e.d.

Remark Given the measurable space $(\mathbf{R}, \mathscr{B})$ and define

$$\forall A \in \mathscr{B}, \qquad \mu_n A = m(A \cap [n, \infty)).$$

Then μ_n is a measure; and since $\mu_n \geqslant \mu_{n+1}$ on $\mathscr{B}$, we have that

$$\forall A \in \mathscr{B}, \qquad \lim \mu_n A = \mu A \tag{6.25}$$

exists with possibly infinite values. μ is not a measure. In fact $\mu(\bigcup[n, n+1)) = \infty$ and $\sum \mu[n, n+1) = 0$. It is interesting to note that if $\{\mu_n : n = 1, \ldots\}$ is a sequence of measures on a measurable space $(X, \mathscr{A})$ and $\mu_n \leqslant \mu_{n+1}$ then the limit μ in (6.25) is a measure.

We use Theorem 6.3 to prove

Theorem 6.4 (Hahn-Saks) *Given a measurable space* $(X, \mathscr{A})$, v *a measure, and* $\{\mu_n : n = 1, \ldots\} \subseteq M(X)$. *Assume*

a) $\forall A \in \mathscr{A}$, $\lim\limits_{n} \mu_n A = \mu A$ *exists,*

b) $\forall n, \mu_n \ll v$.

Then $\{\mu_n : n = 1, \ldots\}$ *is uniformly absolutely continuous.*

Proof. Assume that $\{\mu_n : n = 1, \ldots\}$ is not uniformly absolutely continuous.

Then there is $\varepsilon > 0$, a subsequence $\{n_m : m = 1, \ldots\}$, and $\{A_m : m = 1, \ldots\} \subseteq \mathscr{A}$ such that

$$v A_m < 1/2^m \tag{6.26}$$

and $\qquad |\mu_{n_m}(A_m)| \geqslant \varepsilon.$ \hfill (6.27′)

For simplification of notation write $n_m = m$ so that (6.27′) is replaced by

$$\forall m, \qquad |\mu_m A_m| \geqslant \varepsilon. \tag{6.27}$$

Using this we obtain a contradiction.

a) For each $k \geqslant 0$ we prove that there is a strictly increasing subsequence $\{n_i^k : i = 1, \ldots\} \subseteq \mathbf{Z}^+$ such that $\{n_i^{k+1} : i = 1, \ldots\}$ is a subsequence of $\{n_i^k : i = 1, \ldots\}$,

$$n_1^{k+1} > n_2^k, \quad n_2^{k+1} > n_3^k, \quad n_3^{k+1} > n_4^k, \ldots$$

and $\qquad \displaystyle\sum_{i=1}^{\infty} |\mu_{n_i^k}|(A_{n_i^{k+1}}) < \varepsilon/2,$ \hfill (6.28)

where the elements of the sum decrease to 0.

In order to do this first observe that (6.26) and the hypothesis $\mu_1 \ll v$ yield

$$\lim_{m} |\mu_1|(A_m) = 0.$$

Consequently, there is a sequence $\{n_i^1 : i = 1, \ldots\}$ for which $|\mu_1|(A_{n_i^1})$ is decreasing and

$$|\mu_1|(A_{n_i^1}) < \varepsilon/2^{i+1}.$$

Now since $|\mu_{n_i^1}| \ll v$ we use (6.26) again to obtain a sequence $\{n_i^2 : i = 1, ...\} \subseteq \{n_i^1 : i = 1, ...\}$ such that

$$\forall j, \quad n_j^2 > n_{j+1}^1 \qquad \text{and} \qquad \forall i, \quad |\mu_{n_i^1}|(A_{n_i^2}) < \varepsilon/2^{i+1},$$

and $\{|\mu_{n_i^1}|(A_{n_i^2})\}$ decreases.

Next we examine $\{|\mu_{n_i^2}|\}$ and construct $\{n_i^3 : i = 1, ...\}$ in the same way, etc.

b) Define $v_i = \mu_{n_1^i}$ and $B_i = A_{n_1^i}$.

Observe that when we chose $\{n_j^{k+1} : j = 1, ...\}$ from $\{n_j^k : j = 1, ...\}$ we had

$$n_1^{k+1} > n_2^k > n_1^k;$$

thus $n_1^{k+1} \geqslant n_1^k + 1$ and so $n_1^k \geqslant k$. Therefore,

$$vB_i < 1/2^i.$$

As such, we compute

$$\sum_{j=m+1}^{\infty} |v_m|(B_j) = \sum_{j=1}^{\infty} |v_m|(B_{m+j}) \leqslant \sum_{j=1}^{\infty} |v_m|(A_{n_j^{m+1}}) < \varepsilon/2, \tag{6.29}$$

where the last two inequalities follow from a).

The latter inequality is a consequence of (6.28).

For the former, note that for fixed m, $\{|v_m|(A_{n_j^{m+1}}) : j = 1, ...\}$ is decreasing and

$$n_j^{m+1} < ... < n_2^{m+j-1} < n_1^{m+j}.$$

c) Set $C_n = \bigcup_{j=n}^{\infty} B_j$ and $D = \bigcap_{n=1}^{\infty} \bigcup_{j=n}^{\infty} B_j = \bigcap_1^{\infty} C_n.$

c.i) Clearly $C_n \subseteq C_{n-1}$ so that $\{C_n : n = 1, ...\}$ decreases to D.

Consequently, $E_n = C_n \setminus D$ decreases to $\varnothing$.

Note that $vD = 0$ since

$$vD \leqslant vC_{n+1} \leqslant \sum_{n+1}^{\infty} vB_j < \sum_{n+1}^{\infty} 1/2^j = 1/2^n.$$

Thus by hypothesis b),

$$\forall i, \qquad v_i D = 0.$$

We therefore conclude that

$$\forall i \quad \text{and} \quad \forall n, \qquad v_i E_n = v_i C_n. \tag{6.30}$$

c.ii) From hypothesis a) and Theorem 6.3 we have

$$\lim_n v_m E_n = 0, \quad \text{uniformly in } m. \tag{6.31}$$

Applying (6.30) to (6.31),

$$\lim_n v_m C_n = 0, \quad \text{uniformly in } m.$$

Hence choose $N > 0$ such that

$$\forall n \geqslant N \quad \text{and} \quad \forall m, \qquad |v_m C_n| < \varepsilon/2. \tag{6.32}$$

c.iii) Note that

$$C_n = B_n \cup \left(B_n^{\sim} \cap \left(\bigcup_{n+1}^{\infty} B_j \right) \right),$$

where B_n and $\left(B_n^{\sim} \cap \left(\bigcup_{n+1}^{\infty} B_j \right) \right)$ are disjoint.

d) For $n \geqslant N$ and $m = n$ we use (6.27) and c.iii) to compute

$$\varepsilon \leqslant |v_n B_n| \leqslant |v_n C_n| + \left| v_n \left(B_n^{\sim} \cap \left(\bigcup_{n+1}^{\infty} B_j \right) \right) \right|;$$

so that from (6.32)

$$\varepsilon < \varepsilon/2 + |v_n| \left(\bigcup_{n+1}^{\infty} B_j \right) \leqslant \varepsilon/2 + \sum_{n+1}^{\infty} |v_n|(B_j). \tag{6.33}$$

We obtain a contradiction ($\varepsilon < \varepsilon$) using (6.29) in (6.33).

q.e.d.

Remark The Hahn–Saks theorem (Theorem 6.4) was originally proved by S a k s independent of Theorem 6.3. He then deduced Theorem 6.3 in the following way. Given $\{\mu_n : n = 1, \ldots\} \subseteq M(X)$ and the hypothesis of Theorem 6.3. Define

$$\forall A \in \mathscr{A}, \qquad vA = \sum_{1}^{\infty} \frac{1}{2^n \|\mu_n\|_1} |\mu_n|(A). \tag{6.34}$$

v is a bounded measure and $\mu_n \ll v$ for each n. Consequently we can apply Theorem 6.4 to obtain the uniform absolute continuity of $\{\mu_n : n = 1, \ldots\}$ with respect to v. Thus using Proposition 6.2b and the uniform absolute continuity we conclude that $\mu \in M(X)$ by a double limit argument; also, $\mu \ll v$ (this is the μ defined in the hypothesis of Theorem 6.3).

6.3 Weak convergence of measures

Outside of Theorem 6.7, Theorem 6.9b, and Theorem 6.11 the results of this section whose proofs we have omitted are a reasonable source of exercises.

Using the theorems in section 6.1 and 6.2 it is not difficult to prove [31, p. 89]—

Theorem 6.5 *Given a measure* $(X, \mathscr{A}, \mu)$ *and* $\{f_n : n = 1, \ldots\} \subseteq L^1_\mu(X)$. $\{f_n : n = 1, \ldots\}$ *converges weakly to some* $f \in L^1_\mu(X) \Leftrightarrow$

$$\forall A \in \mathscr{A}, \qquad \lim_{n \to \infty} \int_A f_n \, d\mu$$

exists (cf. Theorem I.9).

We can then employ the technique that Saks used to prove Nikodym's theorem (e.g. section 6.2) to rewrite Theorem 6.5 as—

Theorem 6.6 *Given a measurable space* $(X, \mathscr{A})$ *and* $\{\mu_n : n = 1, ...\} \subseteq M(X)$. $\{\mu_n : n = 1, ...\}$ *converges weakly to some* $\mu \in M(X) \Leftrightarrow$

$$\forall A \in \mathscr{A}, \qquad \lim_{n \to \infty} \mu_n A$$

exists.

A significant strengthening of Theorem 6.6 is due to Dieudonné [30, pp. 35–36] and Grothendieck [46, pp. 146–150].

We employ our convention (e.g. Appendix III) that if X is locally compact then $M(X)$ consists of the regular complex Borel measures.

Theorem 6.7 (Dieudonné–Grothendieck) *Let* X *be locally compact.* $\{\mu_n : n = 1, ...\}$ $\subseteq M(X)$ *converges weakly to some* $\mu \in M(X) \Leftrightarrow$

$$\forall U \subseteq X, \text{ open}, \quad \lim_{n \to \infty} \mu_n U$$

exists.

The following example should caution us from reading too much into Theorem 6.7.
Example 6.1 Let $f_n : [0, 1] \to \mathbf{R}^+$, $n = 1, ...$, have the form

$$f_n(x) = \sum_{0}^{2^n - 1} \Delta_k(x),$$

where Δ_k is an "isosceles triangle function" having its base of length $1/2^{2n+1}$ centered in $[k/2^n, (k + 1)/2^n]$, Δ_j is congruent to Δ_k, and $\int_0^1 f_n = 1$. It is easy to see that for all $(a, b) \subseteq [0, 1]$,

$$\lim_n \int_a^b f_n \, dx = \int_a^b dx.$$

In light of Theorem 6.5 we'd like to conclude that the sequence $\{f_n : n = 1, ...\}$ converges weakly to 1 (cf. Theorem 6.7 noting that, after all, open sets in $[0, 1]$ are just countable unions of open intervals). Such is not the case. Let $A = \{x : \forall n, f_n(x) = 0\}$. Since $m\{x : f_n(x) > 0\} = 1/2^{n+1}$, $m\{x : \exists n \text{ for which } f_n(x) > 0\} \leqslant \frac{1}{2}$; and thus $mA \geqslant \frac{1}{2}$. Consequently,

$$\lim_n \int_A f_n \, dx \neq \int_A dx \geqslant \frac{1}{2}$$

because $\int_A f_n = 0$.

An important problem with applications to harmonic analysis (e.g. [10, Chapter 2 and Chapter 7]) is to find other families of sets, besides the open sets, for which Theorem 6.7 is true. The best results to date are found in [125] (cf. Problem 3.4).

Results such as Theorem 6.7 are deduced from characterizations of the weakly compact sets in $M(X)$. The basic work was done by Dieudonné [30; 31], Dunford and Pettis [34, pp. 376–378], and Grothendieck [46]. Brooks, Darst, and H. H. Schaefer are some recent contributors to the area; the newer results generally consider finitely additive set functions which take values in locally convex spaces.

Because of RRT (e.g. Appendix III) and Theorem 6.7 it is interesting to investigate the relation between weak and weak * convergence in $M(X)$. Grothendieck proved: *if X is compact and the closure of every open set is open then any weak * convergent sequence in $M(X)$ is weak convergent*. This has been generalized by Seevers and Schaefer [99].

Example 6.2 Given the measurable space $([0, 1], \mathscr{B})$. Choose a sequence $\{\mu_n: n = 1, ...\} \subseteq M[0, 1]$ such that $\mu_n \to m$ in the weak * topology, card supp $\mu_n < \infty$, and supp $\mu_n \subseteq \mathbf{Q}$; for example we could take $\mu_n = \sum_{k=1}^{n} (1/n) \delta_{k/n}$. Let $A \subseteq [0, 1]$ be a closed set of irrationals with positive Lebesgue measure (e.g. Problem 1.4a). If $f = \chi_A$ then $\int f \, d\mu_n = 0$ and $\int f = mA > 0$. Obviously, $f \in M[0, 1]'$. Consequently, weak * convergence does not entail weak convergence in $M[0, 1]$. Also, we've proved that, generally, the following statement is false: *let X be a compact space $A \subseteq X$ closed, and assume that for $\mu_n, \mu \in M(X)$, $\mu_n \to \mu$ in the weak * topology; then*

$$\forall f \in C(X), \qquad \int_A f \, d\mu_n \to \int_A f \, d\mu.$$

In terms of weak convergence some of the results of sections 6.1 and 6.2 can be rephrased to read

Theorem 6.8 *Let $(X, \mathscr{A}, \mu)$ be a measure space and assume that the sequence $\{f_n: n = 1, ...\}$ converges weakly to f, where $f_n, f \in L^1_\mu(X)$. Then (6.1) is valid $\Leftrightarrow f_n \to f$ in measure on each $A \in \mathscr{A}$ satisfying $\mu A < \infty$* (cf. Example 3.9 and Example 3.13).

Proposition 6.4 *Let $(X, \mathscr{A}, \mu)$ be a measure space and take $\{f_n, f: n = 1, ...\} \subseteq L^1_\mu(X)$. Assume $f_n \to f$, μ-a.e., and $\|f_n\|_1 \to \|f\|_1$. Then for all $A \in \mathscr{A}$*

$$\lim_n \int_A |f_n| \, d\mu = \int_A |f| \, d\mu.$$

Proof. Take $A \in \mathscr{A}$. From Fatou's lemma

$$\varliminf_{n \to \infty} \int_A |f_n| \, d\mu \geqslant \int_A |f| \, d\mu \geqslant \int_X |f| \, d\mu - \varlimsup_{n \to \infty} \int_{A^\sim} |f_n| \, d\mu.$$

Since $\|f_n\|_1 \to \|f\|_1$ we have

$$\int_X |f| \, d\mu \geqslant \underline{\lim} \int_{A^\sim} |f_n| \, d\mu + \overline{\lim} \int_A |f_n| \, d\mu.$$

q.e.d.

Theorem 6.9b is due to Radon (1913) and F. Riesz (1928).

Theorem 6.9 *Given a measure space* $(X, \mathscr{A}, \mu)$, $\{f_n, f: n = 1, ...\} \subseteq L^p_\mu(X)$, $1 \leqslant p < \infty$, *and assume* $\|f_n\|_p \to \|f\|_p$.
a) *If* $1 \leqslant p < \infty$ *and* $f_n \to f$, μ-*a.e., then*

$$\lim_n \|f_n - f\|_p = 0.$$

b) *If* $1 < p < \infty$ *and* $f_n \to f$ *weakly, then*
$$\lim_n \|f_n - f\|_p = 0.$$

Proof. Different proofs of b) are given in [56, p. 233; 89, pp. 78–80], and we omit the proof here.

a) For any $a, b \geqslant 0$, $(a + b)^p \leqslant 2^p(a^p + b^p)$.
Let $|f_n| = a$, $|f| = b$ so that the non-negative functions,

$$g_n = 2^p(|f_n|^p + |f|^p) - |f_n - f|^p, \qquad n = 1, ...,$$

converge μ-a.e. to $2^{p+1}|f|^p$.

Using Fatou's lemma we have

$$2^{p+1} \int |f|^p \, d\mu \leqslant \underline{\lim} \int g_n \, d\mu = 2^{p+1} \int |f|^p \, d\mu - \overline{\lim} \int |f_n - f|^p \, d\mu,$$

where the equality follows since $\|f_n\|_p \to \|f\|_p$. Therefore,

$$\lim \int |f_n - f|^p \, d\mu = 0.$$

q.e.d.

Corollary 6.9.1 *Given a measure space* $(X, \mathscr{A}, \mu)$, $\{f_n, f: n = 1, ...\} \subseteq L^p_\mu(X)$, $1 \leqslant p < \infty$, *and assume* $f_n \to f$, μ-*a.e. Then* $\|f_n\|_p \to \|f\|_p \Leftrightarrow \|f_n - f\|_p \to 0$.

Theorem 6.10 *Given a measure space* $(X, \mathscr{A}, \mu)$, $1, < p < \infty$, *and* $\{f_n : n = 1, ...\} \subseteq L^p_\mu(X)$. *Assume* $f_n \to f$, μ-*a.e., and* $\sup_n \|f_n\|_p = K < \infty$. *Then* $f_n \to f$ *weakly (cf. Theorem I.17).*

Proof. From Fatou's lemma,

$$\|f\|_p^p \leqslant K^p,$$

and so $f \in L^p_\mu(X)$.

Take $\varepsilon > 0$ and $g \in L_\mu^q(X)$. We'll find N such that

$$\forall n \geqslant N, \qquad \left| \int (f - f_n) g \, d\mu \right| < \varepsilon.$$

Since $|g|^q \in L_\mu^1(X)$ there is $\delta > 0$ such that

$$\forall A \in \mathscr{A} \text{ satisfying } \mu A < \delta, \qquad \left(\int_A |g|^q \, d\mu \right)^{1/q} < \varepsilon/(6K).$$

Also from the integrability of $|g|^q$ there is $B \in \mathscr{A}$ such that $\mu B < \infty$ and

$$\left(\int_{B\sim} |g|^q \, d\mu \right)^{1/q} < \varepsilon/(6K).$$

Because of Egoroff's theorem there is $E \subseteq B$, $E \in \mathscr{A}$, for which

$$\mu(B \cap E^\sim) < \delta \quad \text{and} \quad f_n \to f \text{ uniformly on } E.$$

From the uniform convergence we take N so that

$$\forall n \geqslant N \quad \text{and} \quad \forall x \in E, \qquad |f(x) - f_n(x)| < \varepsilon/(3\|g\|_q (\mu E)^{1/p});$$

consequently,

$$\forall n \geqslant N, \qquad \left(\int_E |f - f_n|^p \, d\mu \right)^{1/p} \|g\|_q < \varepsilon/3.$$

Thus taking $A = B \cap E^\sim$ we estimate with Hölder's and Minkowski's inequalities as follows:

$$\int_X |f_n - f| |g| \, d\mu = \int_A + \int_{B\sim} + \int_E |f_n - f| |g| \, d\mu$$

$$< 2K \frac{\varepsilon}{6K} + 2K \frac{\varepsilon}{6K} + \frac{\varepsilon}{3} = \varepsilon.$$

q.e.d.

We can read Theorem 6.10 as follows: *let $f_n \to f$ μ-a.e.; then for $1 < p < \infty$, $\{\|f_n\|_p : n = 1, \ldots\}$ is bounded if and only if $f_n \to f$ weakly in $L_\mu^p(X)$.*

Example 6.3 Theorem 6.10 is false when $p = 1$. Let $f_n : [0, 1] \to \mathbf{R}^+$ be 0 on $[1/n, 1]$, n at $x = 0$, and linear on $[0, 1/n]$. Thus $f_n \to 0$, m-a.e., and $\|f_n\|_1 = \frac{1}{2}$, whereas $f_n \nrightarrow 0$ weakly.

Because of Theorem I.12 we know that if $f_n \to f$ weakly in $L_\mu^p(X)$ then certain linear combinations of the f_n's converge to f in the $L_\mu^p(X)$-topology. The Banach–Saks theorem [8; 89, pp. 78–81] which we now state is much finer in that instead of rather arbitrary linear combinations we can use arithmetic means.

Theorem 6.11 *Given a measure space* $(X, \mathscr{A}, \mu)$, $1 \leqslant p < \infty$, *and* $\{f_n, f : n = 1, ...\} \subseteq$ $L_\mu^p(X)$. *Assume* $f_n \to f$ *weakly. Then there is a subsequence* $\{f_{n_k} : k = 1, ...\}$ *whose arithmetic means* $(1/m) \sum_{k=1}^{m} f_{n_k}$ *converge in the* $L_\mu^p(X)$-*topology to* f.

Banach and Saks only proved Theorem 6.11 for the $1 < p < \infty$ cases; the result was proved for $L_\mu^1(X)$ in 1965 by Szlenk. Schreier (1930) showed that $C[0, 1]$ does not have the property of the Banach–Saks theorem.

Because of Problem 3.19d we see that if $\|f_n - f\|_1 \to 0$ then $f_{n_k} \to f$, μ-a.e., for some subsequence $\{f_{n_k} : k = 1, ...\} \subseteq \{f_n : n = 1, ...\}$. We then ask whether $f_n \to f$ weakly yields the same result. Generally the answer is negative as we've seen in Theorem 3.12, Example 3.9, Example 3.13, Problem 3.25, and Problem 3.26. Note that if we take $\{f_n : n = 1, ...\}$ as in Theorem 3.12 then $f_n \to \alpha$ weakly, where $\alpha = \int_0^1 f$ (cf. [127, pp. 87–88]).

We close this section by noting that $L_\mu^1(X)$ and $M(X)$ are weakly sequentially complete. This is proved using Schur's lemma and the results of sections 6.1 and 6.2 (e.g. [31]). On the other hand these spaces are never weakly complete (as uniform spaces); in fact, no infinite dimensional normed space is weakly complete.

Appendix A 6.1 Vitali

Giuseppe Vitali (August 26, 1875–February 29, 1932), the oldest of five children, was born in Ravenna. After graduating from the "liceo" in Ravenna, he studied mathematics at the University of Bologna in 1895. He then received a scholarship to the Scuola Normale Superiore (cf. Appendix A3.1.1). Dini and Bianchi were there at the time. He graduated in 1899 and in his thesis he extended a theorem of Mittag–Leffler to Riemann surfaces. His next work was devoted to abelian integrals. In 1901 he was Dini's assistant, and then he began his teaching career in a "liceo". After two brief appointments in Sassari and Voghera he taught at the Liceo Colombo in Genova from 1904 to 1922. (This, of course, does not match Dedekind's record of under-undergraduate teaching; also, Weierstrass' "defenders" would point out that his (Weierstrass) service in teaching penmanship and gymnastics (besides mathematics) should count for something extra.) Finally, in 1923, Vitali received a position at the University of Modena (a weak counterexample to one of the fundamental theorems of life that "you can keep a good person down"). In 1924 he went to the University of Padova where, at the end of 1926, he was struck with hemiplegia (a paralysis resulting from injury to the motor center of the brain). His intellectual powers were unaffected. He left Padova in 1930 for the University of Bologna.

Appendices

I Metric spaces and Banach spaces

I.1 Definitions of spaces

A topological space X is a pair $(X, \mathcal{T})$, where X is a non-empty set, $\mathcal{T} \subseteq \mathcal{P}(X)$, and $\mathcal{T}$ satisfies the conditions:

a) $\qquad \emptyset \in \mathcal{T}, X \in \mathcal{T}$

b) $\qquad \{U_\alpha : \alpha \in I, \text{ an index set}\} \subseteq \mathcal{T} \Rightarrow \bigcup_{\alpha \in I} U_\alpha \in \mathcal{T}$

c) $\qquad \{U_1, ..., U_n\} \subseteq \mathcal{T} \Rightarrow \bigcap_1^n U_j \in \mathcal{T}$.

The elements of $\mathcal{T}$ are called open sets and $\mathcal{T}$ is a topology for the set X. The interior of $S \subseteq X$, int S, is the largest open set contained in S. The complement of an open set is a closed set. $\mathcal{S} \subseteq \mathcal{T}$ is a basis for the topological space $(X, \mathcal{T})$ if for each $x \in X$ and $U \in \mathcal{T}$ for which $x \in U$ there is $V \in \mathcal{S}$ such that $x \in V \subseteq U$. We shall assume that all of our topological spaces X are Hausdorff:

$$\forall x, y \in X, x \neq y, \exists U_x, U_y \in \mathcal{T} \text{ such that } x \in U_x, y \in U_y, \text{ and } U_x \cap U_y = \emptyset.$$

A set $Y \subseteq X$ is dense in X if for each $x \in X$ and each open set U containing x there is a point $y \in Y \cap U$.

$K \subseteq X$ is compact if every covering of K by open sets contains a finite subcovering; and $K \subseteq X$ is relatively compact if its closure (the smallest closed set containing it) is compact. A topological space X is locally compact if

$$\forall x \in X \quad \text{and} \quad \forall U \in \mathcal{T} \text{ for which } x \in U, \exists K \subseteq X, \text{ compact, and}$$
$$\exists V \in \mathcal{T} \text{ such that } x \in V \subseteq K \subseteq U.$$

Two topological spaces $(X_i, \mathcal{T}_i)$, $i = 1, 2$, are homeomorphic if there is a bijection $f : X_1 \to X_2$ such that

$$\forall U \in \mathcal{T}_1, \ f(U) \in \mathcal{T}_2 \qquad \text{and} \qquad \forall U \in \mathcal{T}_2, \ f^{-1}(U) \in \mathcal{T}_1;$$

in this case, f is an homeomorphism. These two conditions characterize the continuity of f and f^{-1}, respectively, cf., the definition of continuity in Appendix I.4. We shall use the following result about continuous functions

Urysohn's lemma *Let X be a locally compact space (and therefore Hausdorff by our convention). If $K \subseteq X$ is compact and $U \subseteq X$ is an open set containing K, then there is a continuous function $f: X \to \mathbf{R}$ such that $f = 1$ on K and $f = 0$ on $U^\sim$.*

A metric space X is a pair (X, p), where X is a nonempty set and $p: X \times X \to \mathbf{R}^+$ satisfies:

$$\forall x, y \in X, \quad p(x, y) \geqslant 0,$$
$$\forall x, y \in X, \quad p(x, y) = 0 \quad \text{if and only if} \quad x = y,$$
$$\forall x, y \in X, \quad p(x, y) = p(y, x),$$
$$\forall x, y, z \in X, \quad p(x, z) \leqslant p(x, y) + p(y, z);$$

p is a metric. A ball $B(x, r)$ with center x and radius r in a metric space X is

$$B(x, r) = \{y \in X : p(x, y) < r\}.$$

A metric space is a topological space and U is defined to be open if

$$\forall x \in U, \quad \exists B(x, r) \subseteq U.$$

A sequence $\{x_n : n = 1, \ldots\} \subseteq X$, where X is a metric space, is Cauchy if

$$\forall \varepsilon > 0 \quad \exists N \quad \text{such that} \quad \forall n, m > N, p(x_n, x_m) < \varepsilon.$$

If X is a metric space in which every Cauchy sequence $\{x_n : n = 1, \ldots\}$ converges to some element x (i.e. $p(x_n, x) \to 0$) then X is complete. Two metric spaces (X_i, p_i), $i = 1, 2$, are isometric if there is a bijection $f: X_1 \to X_2$ such that

$$\forall x, y \in X_1, \quad p_1(x, y) = p_2(f(x), f(y));$$

in this case, f is an isometry.

An excellent reference for metric spaces is [44].

Let X be a vector space over F, $F = \mathbf{R}$ or $F = \mathbf{C}$.

X is a normed vector space if there is a function $\| \ \|: X \to \mathbf{R}^+$ such that

$$\forall x \in X, \|x\| = 0 \quad \text{if and only if} \quad x = 0,$$
$$\forall x, y \in X, \|x + y\| \leqslant \|x\| + \|y\|,$$
$$\forall a \in F, \forall x \in X, \|ax\| = |a| \, \|x\|;$$

$\| \ \|$ is a norm. A normed vector space is a metric space with metric $p(x, y) = \|x - y\|$. A complete normed vector space is a Banach space. It is sometimes necessary to consider topological vector spaces where the topology cannot be described by a metric (e.g. in the theory of distributions). In such cases we would still like to have a notion of completeness and this is accomplished through the theory of uniform spaces (e.g. [63]).

Let X be a normed vector space. $\sum x_n$ converges to $x \in X$, for $x_n \in X$, if $\|x - \sum_1^N x_n\| \to 0$; $\sum x_n$ is absolutely convergent if $\sum \|x_n\| < \infty$. We have the following important result:

Proposition I.1 *A normed vector space X is a Banach space $\Leftrightarrow$ every absolutely convergent series is convergent.*

Proof. ($\Rightarrow$) Take $x_n \in X$ for which $\sum \|x_n\| < \infty$ and choose $\varepsilon > 0$. If $\sum\limits_N^\infty \|x_n\| < \varepsilon/2$ then for each $n > m \geqslant N$

$$\left\| \sum_m^n x_j \right\| < \varepsilon;$$

thus $\sum x_j$ converges to some $x \in X$ since X is complete.

($\Leftarrow$) Let $\{x_n : n = 1, ...\} \subseteq X$ be a Cauchy sequence in X.
Hence, for each k there is $n_k \in \mathbf{Z}^+$ such that

$$\forall n, m \geqslant n_k, \qquad \|x_n - x_m\| < 1/2^k;$$

we can also choose $n_{k+1} > n_k$.

Set $y_k = x_{n_k} - x_{n_{k-1}}$ for $k = 1, ...$, where $x_{n_0} = 0$. Therefore $\sum x_k$ is absolutely convergent, so that by hypothesis and the fact that

$$\sum_{k=1}^m y_k = x_{n_m},$$

$\{x_{n_m} : m = 1, ...\}$ converges to some $x \in X$.
It is easy to check that $\lim \|x - x_n\| = 0$.

q.e.d.

An **Hilbert space** X is a **Banach space** with a function $(,): X \times X \to F$ which satisfies

$$\begin{aligned}
\forall x, y \in X, & \qquad \overline{(x, y)} = (y, x), \\
\forall x, y, z \in X, & \qquad (x + y, z) = (x, z) + (y, z), \\
\forall a \in F, \forall x, y \in X, & \qquad (ax, y) = a(x, y), \\
\forall x \in X, & \qquad \|x\| = (x, x)^{1/2};
\end{aligned}$$

$(,)$ is an **inner product**.

I.2 Examples

1. Given a measure space $(X, \mathscr{A}, \mu)$. If $0 < p < \infty$, $\mathscr{L}_\mu^p(X)$ is the set of all μ-measurable functions $f: X \to \mathbf{C}$ such that

$$\int_X |f|^p \, d\mu < \infty.$$

Set $L_\mu^p(X) = \mathscr{L}_\mu^p(X)/\sim$ where $\sim$ is the equivalence relation: $f \sim g$ if $f = g$, μ-a.e.

Thus $f \in L_\mu^p(X)$ is the collection, i.e. equivalence class, of all functions $g: X \to \mathbf{C}$ which are equal μ-a.e. For $p \geqslant 1$ we define

$$\|f\|_p = \left(\int_X |g|^p \, d\mu\right)^{1/p},$$

where g is an element of the equivalence class $f \in L_\mu^p(X)$; it is customary to write

$$\forall f \in L_\mu^p(X), \qquad \|f\|_p = \left(\int_X |f|^p \, d\mu\right)^{1/p}.$$

(An analogous remark holds for $L_\mu^\infty(X)$, which is discussed below). As such it is not difficult to verify that $L_\mu^p(X)$, $1 \leqslant p < \infty$, is a Banach space. The triangle inequality in this case is called **Minkowski's inequality**. This inequality is trivial for $L_\mu^1(X)$, and is verified in the following way, using Hölder's inequality (Theorem 5.19b), for $p, q > 1$:

$$\int |f + g|^p \, d\mu \leqslant \int |f + g|^{p-1} |f| \, d\mu + \int |f + g|^{p-1} |g| \, d\mu$$
$$\leqslant (\|f\|_p + \|g\|_p) \|f + g\|_p^{p/q}.$$

The completeness of $L_\mu^p(X)$, $1 \leqslant p < \infty$, is sometimes referred to as the **Riesz–Fischer theorem** and can be proved using Proposition I.1. In fact, letting $\{f_n : n = 1, \ldots\} \subseteq L_\mu^p(X)$ be absolutely convergent and $g_n = \sum_1^n |f_k|$, we see that g_n increases to an element $g \in L_\mu^p(X)$ by using Minkowski's inequality and Fatou's lemma; it is then straightforward to check, using LDC, that

$$\lim_n \left\| \sum_n^\infty f_k \right\|_p = 0;$$

consequently, $\sum_1^\infty f_k$ is convergent in $L_\mu^p(X)$ and we apply Proposition I.1.

2. Let $(X, \mathscr{A}, \mu)$ be a σ-finite measure space. $\mathscr{L}_\mu^\infty(X)$ is the set of all μ-measurable functions f with $\|f\|_\infty < \infty$ ("$\|f\|_\infty$" was defined after Example 2.14). Using the format of the previous example we define $L_\mu^\infty(X) = \mathscr{L}_\mu^\infty(X)/\sim$ with norm

$$\forall f \in L_\mu^\infty(X), \qquad \|f\|_\infty = \|g\|_\infty,$$

where $g = f$, μ-a.e. It is easy to prove that $L_\mu^\infty(X)$ is a Banach space.

3. If the σ-finite measure space $(X, \mathscr{A}, \mu)$ is also a topological space $(X, \mathscr{T})$ then $C_b(X)$ denotes the space of bounded continuous complex-valued functions on X. In this case

$$\forall f \in C_b(X), \qquad \|f\|_\infty = \sup_{x \in X} |f(x)| < \infty.$$

Since the uniform limit of continuous functions is continuous, $C_b(X)$ can be regarded as a closed subspace of $L_\mu^\infty(X)$. If X is compact we write $C(X) = C_b(X)$.

4. For any measure space $(X, \mathcal{A}, \mu)$, $L_\mu^2(X)$ is an Hilbert space with inner product

$$(f, g) = \int_X f(x)\overline{g(x)}\,dx.$$

The fact that the integral is defined follows from Hölder's inequality (Theorem 5.19b). The structurally important converse is: *let H be a non-zero Hilbert space; then there is a set X and a linear bijection*

$$L: H \to L_c^2(X),$$

where c is counting measure on $(X, \mathcal{P}(X))$, such that

$$(x, y) = \sum_{t \in X} (Lx)(t)\overline{(Ly)(t)}.$$

The fundamental elementary results of Hilbert space theory are used to prove this fact and to determine uniquely card X in terms of the cardinality of so-called orthonormal sets.

5. If the measure space $(X, \mathcal{A}, \mu)$ is also a compact space and $\mathcal{A}$ contains the Borel algebra, then

$$C(X) \subseteq L_\mu^\infty(X) \subseteq \ldots \subseteq L_\mu^p(X) \subseteq L_\mu^r(X) \subseteq \ldots \subseteq L_\mu^1(X), \qquad 1 \leqslant r \leqslant p.$$

In $(X, \mathcal{P}(X), c)$ (where X is topologized with the metric $\rho(x, y) = 0$ if $x = y$ and $\rho(x, y) = 1$ if $x \neq y$)

$$L_c^1(X) \subseteq \ldots \subseteq L_c^p(X) \subseteq L_c^r(X) \subseteq \ldots \subseteq L_c^\infty(X) = C_b(X), \qquad 1 \leqslant p \leqslant r.$$

In both cases we have the inequality $\|\ \|_p \geqslant \|\ \|_r$ so that the corresponding injection is continuous ("continuous" functions are defined in Appendix I.4).

6. A Banach space is an Hilbert space if and only if the parallelogram law, $\|x + y\|^2 + \|x - y\|^2 = 2(\|x\|^2 + \|y\|^2)$, is valid. Using this fact we see that there are Banach spaces which are not Hilbert spaces. We now give a standard example of a non-trivial complete metric vector space which is not a Banach space. Let X be the space of C^∞-functions on $[0, 1]$. Define the metric

$$\rho(f, g) = \sum_0^\infty \frac{\|f - g\|_{(k)}}{2^k(1 + \|f - g\|_{(k)})},$$

where $\|f\|_{(k)} = \sup_{0 \leqslant j \leqslant k} \|f^{(j)}\|_\infty;$

as such X is complete. If X is normed by $\| \ \|$ it is possible to show that

$$\forall n \ \exists C_n \quad \text{such that} \quad \forall f \in X, \ \text{for which} \quad \|f\| \leqslant 1, \qquad \|f^{(n)}\|_\infty \leqslant C_n.$$

It is then not difficult to find $f \in X$ such that

$$\forall n, \qquad \|f^{(n)}\|_\infty > nC_n,$$

from which we obtain the desired contradiction (to the hypothesis that X is normed).

7. If (X, ρ) is a metric space there is a complete metric space $(\tilde{X}, \tilde{\rho})$ such that $X \subseteq \tilde{X}$, $\tilde{\rho} = \rho$ on $X \times X$, and X is dense in $\tilde{X}$. $\tilde{X}$ is the set of equivalence classes of Cauchy sequences from X, where $\{x_n : n = 1, ...\}$ is defined to be equivalent to $\{y_n : n = 1, ...\}$ if $\rho(x_n, y_n) \to 0$; and $\tilde{\rho}(\{x_n\}, \{y_n\}) = \lim \rho(x_n, y_n)$. $(\tilde{X}, \tilde{\rho})$ is the completion of (X, ρ). A relevant example using this concept is the following: *define*

$$\forall f, \ g \in C[a, b], \qquad \rho(f, g) = R\int_a^b |f - g|;$$

then $\tilde{C}[a, b] = L_m^1[a, b]$ and $\tilde{\rho}(f, g) = \int_a^b |f - g|$ (cf. Appendix III).

For an alternative way to describe the completion of a metric space (X, ρ) let $B(X)$ be the Banach space of bounded real functions on X with metric $\sigma(f, g) = \sup \{|f(x) - g(x)| : x \in X\}$. Fix $x_0 \in X$ and define the function $F: X \to B(X)$, $x \mapsto f_x$, where

$$f_x(y) = \rho(x, y) - \rho(x_0, y).$$

Then F is an isometry $X \to F(X) \subseteq B(X)$ and $\tilde{X} = \overline{F(X)}$.

8. Let $p \in \mathbf{Z}$ be positive and for each $n \in \mathbf{Z} \setminus \{0\}$ let $\|n\|$ be the reciprocal of the highest power of p which divides n; set $\|0\| = 0$. Then $\|n + m\| \leqslant \|n\| + \|m\|$, $\|n\| = 0$ if and only if $n = 0$, and $\|nm\| \leqslant \|n\| \ \|m\|$ with equality if p is prime. $\rho(n, m) = \|n - m\|$ is the *p*-adic metric on $\mathbf{Z}$ and $(\mathbf{Z}, \rho)$ is not complete.

I.3 Separability

A topological space is *separable* if it contains a countable dense subset. It is not difficult to prove

Theorem I.1 *Given $(X, \mathcal{M}^n, m^n)$, $X \subseteq \mathbf{R}^n$, where m^n is Lebesgue measure on X. If $p \in [1, \infty)$, then $L_{m^n}^p(X)$ is separable.*

Example I.1 We'll prove that $L_m^\infty[0, 1]$ is not separable. Let $\{f_n: n = 1,...\}$ be an arbitrary sequence in $L_m^\infty[0, 1]$ and write

$$(0, 1] = \bigcup_1^\infty \left(\frac{1}{2^n}, \frac{1}{2^{n-1}} \right) = \bigcup_1^\infty E_n.$$

If

$$\operatorname{ess\,sup}_{x \in E_n} |f_n(x)| \leqslant \tfrac{1}{2},$$

define $g = 1$ on E_n; otherwise set $g = 0$ on E_n. Consequently, $g \in L_m^\infty[0, 1]$ and

$$\forall n, \qquad \|f_n - g\|_\infty \geqslant \tfrac{1}{2}.$$

Thus $\{f_n: n = 1, ...\}$ is not dense in $L_m^\infty[0, 1]$; since $\{f_n: n = 1, ...\}$ is arbitrary, $L_m^\infty[0, 1]$ cannot be separable.

Example I.2 The fact that a given base space X is separable has no bearing on the separability of $L_\mu^p(X)$, $1 \leqslant p < \infty$. Take $([0, 1], \mathscr{P}[0, 1], c)$, where c is counting measure. If $f \in L_c^1[0, 1]$ then $f = 0$ outside of a countable set. Thus if $\{f_n: n = 1, ...\} \subseteq L_c^1[0, 1]$ there is $y \in [0, 1]$ such that $f_n(y) = 0$ for each n. Define $g = \chi_{\{y\}} \in L_c^1[0, 1]$ so that

$$\forall n, \qquad \|f_n - g\|_1 \geqslant 1.$$

Theorem I.2 *Let $(X, \mathscr{A}, \mu)$ be a measure space and Y a complete separable metric space. If $f: X \to Y$ is measurable then there is a sequence $\{g_k: k = 1, ...\}$ of simple functions $X \to Y$ such that $\{g_k: k = 1, ...\}$ converges pointwise to f.*

I.4 Moore–Smith and Arzelà–Ascoli theorems

Given a metric space (X, ρ). $\{x_{m,n}\} \to x$ (i.e. $\lim\limits_{m,n} x_{m,n} = x$) if

$$\forall \varepsilon > 0 \; \exists N \quad \text{such that} \quad \forall m, n > N, \; \rho(x_{m,n}, x) < \varepsilon.$$

The following result can be generalized to uniform spaces with essentially the same proof.

Theorem I.3 (Moore–Smith) *Given $\{x_{m,n}: m = 1, ..., n = 1, ...\}$ in a complete metric space (X, ρ). Assume*

i) $\quad \exists \lim\limits_{n \to \infty} x_{m,n} = y_m$ *uniformly in m*

ii) $\quad \forall n \; \exists \lim\limits_{m \to \infty} x_{m,n} = z_n.$

Then $\lim\limits_n \lim\limits_m x_{m,n}$, $\lim\limits_m \lim\limits_n x_{m,n}$, and $\lim\limits_{m,n} x_{m,n}$ all exist and are equal.

Proof. i) means

$$\forall \varepsilon > 0 \ \exists K \quad \text{such that} \quad \forall n > K, \quad \text{and} \quad \forall m, \rho(y_m, x_{m,n}) < \varepsilon.$$

Using i) and ii) we show that $\{y_m : m = 1, \ldots\}$ is Cauchy by computing

$$\rho(y_m, x_{m,n}) < \frac{\varepsilon}{4}, \qquad \rho(z_k, x_{p,k}) < \frac{\varepsilon}{4}.$$

Since X is complete, $y_m \to w \in X$; and it is easy to check that

$$\lim_{m,n} x_{m,n} = w. \tag{I.1}$$

Thus, $\lim_{m} \lim_{n} x_{m,n} = \lim_{m,n} x_{m,n} = w.$

Finally, in order to prove that $\lim_{n} z_n = w$ take $\varepsilon > 0$ and write

$$\rho(z_n, w) \leqslant \rho(z_n, x_{m,n}) + \rho(x_{m,n}, w).$$

Since (I.1) holds,

$$\exists N \quad \text{such that} \quad \forall m, n > N, \qquad \rho(x_{m,n}, w) < \varepsilon;$$

and so $\quad \forall n > N, \qquad \rho(z_n, w) \leqslant \overline{\lim_{m}} \, \rho(z_n, x_{m,n}) + \varepsilon = \varepsilon.$

q.e.d.

Let (X, ρ) and (Y, d) be metric spaces. A function $f: X \to Y$ is **continuous at** $x \in X$ if

$$\forall \varepsilon > 0 \ \exists \delta > 0 \quad \text{such that} \quad \rho(z, x) < \delta \Rightarrow d(f(z), f(x)) < \varepsilon;$$

and f is **continuous on** X if it is continuous at each $x \in X$. $\{f_n : n = 1, \ldots\}$ is **equicontinuous at** $x \in X$ if

$$\forall \varepsilon > 0 \ \exists \delta > 0 \quad \text{such that} \quad \forall n, \rho(z, x) < \delta \Rightarrow d(f_n(z), f_n(x)) < \varepsilon.$$

$\{f_n : n = 1, \ldots\}$ is **equicontinuous on** X if it is equicontinuous at each $x \in X$.

Theorem I.4 (Arzelà–Ascoli) *Let X be a separable metric space, Y a compact metric space, and $\{f_n : n = 1, \ldots\}$ an equicontinuous family of functions $X \to Y$. Then there is a subsequence of $\{f_n : n = 1, \ldots\}$ which converges pointwise to a continuous function.*

Proof. Let $\{x_1, \ldots\} \subseteq X$ be dense. Since Y is compact,

$$\exists J_1 \subseteq \mathbf{Z}^+ \quad \text{such that} \quad \{f_n(x_1): n \in J_1\} \text{ is convergent.}$$

Pick $J_2 \subseteq J_1$ such that $\{f_n(x_2): n \in J_2\}$ is convergent, etc.

Consequently,

$$\forall j, \quad \exists \lim_k f_{n_k}(x_j) = g(x_j),$$

where $n_k \in J_k$ and $\lim_k n_k = \infty$.

Let $z \in X \setminus \{x_1, \ldots\}$ with $x_{q_p} \to z$ as $p \to \infty$. Then,

$$\lim_p f_{n_k}(x_{q_p}) = f_{n_k}(z), \quad \text{uniformly in } k,$$

by the equicontinuity hypothesis. Also

$$\forall p, \qquad \lim_k f_{n_k}(x_{q_p}) = g(x_{q_p}).$$

Consequently, by the Moore–Smith theorem,

$$\exists \lim_k f_{n_k}(z) = g(z).$$

The continuity of g is straightforward to check.   q.e.d.

Obviously the result is still true if, instead of assuming that Y is compact, we assume that the range of each f_n is compact in Y. It is also easy to prove that the convergence of $\{f_{n_k}: k = 1, \ldots\}$ is uniform on compact sets of X.

I.5 Uniformly continuous functions

Let (X, ρ) and (Y, d) be metric spaces. $f: X \to Y$ is uniformly continuous if

$$\forall \varepsilon > 0 \quad \exists \delta > 0 \quad \text{such that} \quad \rho(x, y) < \delta \Rightarrow d(f(x), f(y)) < \varepsilon.$$

If X is a compact metric space and $f: X \to \mathbf{R}$ is continuous then f is uniformly continuous. $f(x) = \sin(1/x)$ is a bounded continuous function $(0, 1] \to [-1, 1]$ which is not uniformly continuous. Observe that

$$f: [0, 1) \to [0, \infty)$$
$$x \mapsto x/(1 - x)$$

is bijective and bicontinuous (i.e. a homeomorphism), whereas the Cauchy sequence $\{1 - (1/n): n = 1, \ldots\}$ in $[0, 1)$ is transformed into the sequence $\{n - 1: n = 1, \ldots\}$, which is not Cauchy. In this case the range space is complete and $[0, 1)$ is not complete. Such a phenomenon leads us to distinguish between topological properties (dealing with homeomorphisms) and uniform properties (dealing with Cauchyness, uniform continuity, and completeness). Generally there are no relations between these two categories except the following: *let X be a metric space; X is compact $\Leftrightarrow$ it is complete and totally bounded* ((X, ρ) is totally bounded if

$$\forall \varepsilon > 0 \quad \exists x_1, \ldots, x_n \in X \quad \text{such that} \quad X \subseteq \bigcup B(x_j, \varepsilon)).$$

Theorem I.5 *Let Y be a metric space and let Z be a complete metric space. Assume that $X \subseteq Y$ and that $f: X \to Z$ is a uniformly continuous function. Then f has a unique uniformly continuous extension to $\bar{X}$.*

Given metric spaces (X, ρ) and (Y, d), and a continuous function $f: X \to Y$. f is absolutely continuous if

$$\forall \varepsilon > 0 \quad \exists \delta > 0 \quad \text{such that} \quad \forall \{x_1, ..., x_n\} \subseteq X,$$

$$\sum_1^{n-1} \rho(x_j, x_{j+1}) < \delta \Rightarrow \sum_1^{n-1} d(f(x_j), f(x_{j+1})) < \varepsilon.$$

If $\sigma: X \times X \to \mathbf{R}$ is defined by $\sigma(x, y) = \rho(x, y) + d(f(x), f(y))$ then (X, ρ) and (X, σ) have the same topologies and

$$f: (X, \sigma) \to (Y, d)$$

is absolutely continuous. For $X = Y = \mathbf{R}$ taken with the absolute value metric this definition of absolute continuity characterizes the class of Lipschitz functions which in turn is properly contained in the usual class of absolutely continuous functions as defined in Chapter 4.

Example I.3 Let (X, ρ) and (Y, d) be metric spaces and let $f: X \to Y$ be a continuous function. We'll show that it is not generally possible to find metrics σ and b on X and Y, respectively, so that $f: (X, \sigma) \to (Y, b)$ is absolutely and uniformly continuous. Take $f: (0, 1] \to [1, \infty)$, $f(x) = 1/x$, and the usual metrics. Assume we can find σ, b which yield both absolute and uniform continuity. Then from Theorem I.5 f has a unique uniformly continuous extension $[0, 1] \to [1, \infty)$, and this is obviously false.

I.6 Baire category theorem

An excellent reference for the Baire category theorem is [82]. A metric space is Baire if every countable intersection of open dense sets is dense. Since $\mathbf{R}$ is a complete metric space Theorem I.6 and Theorem I.7b yield (1.9).

Theorem I.6 (Baire I) *Every complete metric space X is Baire.*

Proof. a) We give Cantor's necessary conditions for the completeness of a metric space (as promised in Chapter 2.1). The converse is also true and easy. Take $\{A_n: n = 1, ...\} \subseteq X$ where each A_n is closed, non-empty, and $A_1 \supseteq A_2 \supseteq ...$. Assuming that $\lim_n \sup \{\rho(y, z): y, z \in A_n\} = 0$ we check that $\bigcap A_n = \{x\} \subseteq X$. For all n, let $x_n \in A_n$. $\{x_n: n = 1, ...\}$ is Cauchy, for if $m \geqslant n$,

$$\rho(x_m, x_n) \leqslant \sup \{\rho(y, x): y, z \in A_n\} = dA_n \to 0, \quad n \to \infty.$$

By the completeness of X there is a point $x \in X$ such that $\rho(x_n, x) \to 0$. Now, for each n, $x_m \in A_n$ when m is sufficiently large; consequently $x \in \bigcap A_n$ since A_n is closed. If $y \in \bigcap A_n$, then $\rho(x, y) \leqslant dA_n$ for each n so that by hypothesis, $\rho(x, y) = 0$. Thus $x = y$.

b) Given U_n, open and dense, so that $A_n = U_n^\sim$ is nowhere dense (i.e. int $A_n = \emptyset$). $(\cap U_n)^\sim = \cup A_n = A$ where each A_n is closed. We prove that if V is open then $V \cap (\cap U_n) \neq \emptyset$.

Choose an open set, V_1, such that $\bar{V}_1 \subseteq V$ and $d\bar{V}_1 < 1$. Since V_1 is not a subset of A_1,

$$V_1 \cap U_1 \neq \emptyset,$$

and $V_1 \cap U_1$ is open.

Choose an open set V_2 such that $\bar{V}_2 \subseteq V_1 \cap U_1$ and $d\bar{V}_2 < \frac{1}{2}$. Generally, then, we choose open sets V_n with $\bar{V}_n \subseteq V_{n-1} \cap U_{n-1}$ and $d\bar{V}_n < 1/n$.

The hypotheses of a) are satisfied for $\bar{V}_n$, and hence $\cap \bar{V}_n = \{x\}$.

Therefore

$$x \in \cap (V_n \cap U_n) \subseteq V_1 \cap (\cap U_n).$$

q.e.d.

Let X be a metric space. $A \subseteq X$ is a set of **first category** if it is the countable union of nowhere dense sets; any other subset of X is a set of **second category**. The following is straightforward to prove.

Theorem I.7 (**Baire II**) *The following are equivalent for a metric space X:*

a) *X is Baire.*

b) *Every countable union of closed nowhere dense sets has empty interior.*

c) *Every non-empty open set is of second category.*

d) *If $\cup A_n$, A_n closed, contains an open set then some A_j contains an open set.*

e) *The complement of every set of first category is dense in X.*

Example I.4 a) First category sets aren't necessarily nowhere dense; in fact, take $\mathbf{Q} \subseteq \mathbf{R}$ noting that $\mathbf{Q}$ is of first category and $\bar{\mathbf{Q}} = \mathbf{R}$.

b) Let $S \subseteq \mathbf{R}$. If $\{x - y : x, y \in S\}$ is a set of first category than S is a set of first category (and so if S is of second category then $\{x - y : x, y \in S\}$ is of second category).

c) It is easy to construct a first category set of Lebesgue measure 1 in $[0, 1]$. Let E_n be a perfect symmetric set with $mE_n \geqslant 1 - (1/n)$. Then $E = \cup E_n$ does the trick (cf. Problem 2.5 and Problem 2.6).

d) Clearly $[0, 1]$ does not contain a countable dense $\mathcal{G}_\delta$, $D = \cap_j U_j$. In fact if $D = \{d_j : j = 1, \ldots\}$ were such a set then $V_j = U_j \setminus \cup_1 d_n$ is open and dense, and $\cap V_j = \emptyset$; this contradicts Baire I.

e) Let $E \subseteq [0, 1]$ be any perfect symmetric set with a countable set A of accessible points ($a \in A$ is **accessible** if it is the endpoint of a contiguous open interval). Note that if $\{U_\alpha\}$ is an open covering of A it does not necessarily follow that $E \subseteq \cup U_\alpha$. For example, if $x \in E \setminus A$ consider $[0, x) \cup (x, 1]$. For each $a_n \in A$ let $\{I_{m,n} : m = 1, \ldots\}$ be a sequence of open intervals about a whose lengths tend to 0. Then $\{a_n : n = 1, \ldots\} =$

$\bigcap_m I_{m,n}$. Now let $V_m = \bigcup_n I_{m,n}$ so that $E \cap V_m$ is open and dense in E. Observe that $A \subseteq \bigcap_m (E \cap V_m)$, properly. To prove this note that $U_m = E \cap V_m \setminus \{a_1, ..., a_m\}$ is open and dense in E so that $\bigcap U_m$ is dense; but $A \cap (\bigcap U_m) = \emptyset$ and $\bigcap (E \cap V_m) = A \cup (\bigcap U_m)$.

Example I.5 Let $C_c(\mathbf{R})$ be the vector space of continuous functions $f: \mathbf{R} \to \mathbf{C}$ which vanish outside of some compact set (depending on f). We define sequential convergence in $C_c(\mathbf{R})$ as

$$f_n \to f \text{ in } C_c(\mathbf{R}), f_n, f \in C_c(\mathbf{R}), \text{ if } \|f_n - f\|_\infty \to 0 \text{ and}$$
$$\exists r > 0 \quad \text{such that} \quad \forall n, \quad f_n = 0 \quad \text{on } [-r, r]^\sim.$$

We'll prove that with this convergence $C_c(\mathbf{R})$ can't be a complete metric space $(C_c(\mathbf{R}), \rho)$. If such a metric ρ exists then $C_c(\mathbf{R})$ is a Baire space; we'll show that $(C_c(\mathbf{R}), \rho)$ is of first category to obtain the contradiction. First note that

$$C_c(\mathbf{R}) = \bigcup_n C_{c,n}, \qquad C_{c,n} = \{f \in C_c(\mathbf{R}): f = 0 \text{ on } [-n, n]^\sim\}.$$

Clearly $C_{c,n} = \bar{C}_{c,n}$ and it is sufficient to check that int $C_{c,n} = \emptyset$. Assume not and let $V \subseteq C_{c,n}$ be an open neighborhood of 0 in $C_c(\mathbf{R})$. Choose $f_k \in C_{c,n+1} \setminus C_{c,n}$ such that $\rho(f_k, 0) \to 0$; consequently $f_k \in V \subseteq C_{c,n}$ and this contradicts the definition of f_k. There is, in fact, a (completely regular) topology on $C_c(\mathbf{R})$ whose uniform structure renders $C_c(\mathbf{R})$ complete and whose sequential convergence is that given above (e.g. Appendix III).

Example I.6 In Chapter 1 we discussed everywhere continuous nowhere differentiable functions. The soft analysis proof of their existence uses Baire I. Take $C[0, 1]$ with the $\| \ \|_\infty$-norm so that $C[0, 1]$ is complete with the canonical metric $\rho(f, g) = \|f - g\|_\infty$.
Define

$$F_n = \left\{ f \in C[0, 1]: \exists x \in [0, 1] \text{ such that } \forall h > 0, \left| \frac{f(x + h) - f(x)}{h} \right| < n \right\}.$$

Each F_n is closed and nowhere dense, and so $C[0, 1] \neq \bigcup F_n$. Consequently, the set of continuous nowhere differentiable functions is dense in $C[0, 1]$.

Example I.7 The proof of Problem 2.22b is easy: let $A_1 \subseteq [0, 1]$ be a perfect symmetric set of measure $\frac{1}{4}$; let $A_2 = \bigcup_j A_{j,2}$ where $A_2 = \frac{1}{8}$ and $A_{j,2}$ is a perfect symmetric set of positive measure in the jth contiguous interval of A_1; let $A = \bigcup_1^\infty A_j$. A generalization of this result is due to Kirk [65]: *let $(X, \mathscr{A}, \mu)$ be a separable metric space, assume μ is continuous, and suppose $\mathscr{B}(X) \subseteq \mathscr{A}$; then there is $A \in \mathscr{B}(X)$ such that for each open set I of positive measure*

$$0 < \mu(A \cap I) < \mu I.$$

I.7 Uniform boundedness principle

Theorem I.8 (Uniform boundedness principle) *Let* (X, ρ) *be a complete metric space and let* $\mathscr{F}$ *be a set of continuous functions* $X \to \mathbf{C}$. *Assume*

$$\forall x \in X \quad \exists M_x > 0 \quad \text{such that} \quad \forall f \in \mathscr{F}, \quad |f(x)| \leqslant M_x.$$

Then there is a non-empty open set $U \subseteq X$ *and a constant* M *such that*

$$\forall x \in U \quad \text{and} \quad \forall f \in \mathscr{F}, \quad |f(x)| \leqslant M.$$

Proof. For each $f \in \mathscr{F}$ and m, define

$$A_{m, f} = \{x : |f(x)| \leqslant m\} \quad \text{and} \quad A_m = \bigcap_f A_{m, f}.$$

Since f is continuous, A_m is closed. We show that $X = \bigcup A_m$; in fact, if $x \in X$ choose $m = M_x$, so that $x \in A_m$.

Consequently from Baire I, II $U = \text{int } A_n \neq \varnothing$ for some n and we take $M = n$.

$$\text{q.e.d.}$$

Given $f_n, f \in L_m^1[0, 1]$. $f_n \to f$ weakly if

$$\forall g \in L_m^\infty[0, 1], \qquad \int_0^1 (f - f_n)g \to 0.$$

As we pointed out in Example 3.13, weak convergence does not necessarily imply norm convergence (cf. Theorem I.10). In fact, weak convergence does not imply pointwise a.e. convergence, uniform convergence, or convergence in measure (e.g. Chapter 6.3).

We have the following characterization of weak sequential convergence.

Theorem I.9 *Given* $f_n, f \in L_m^1[0, 1]$. $f_n \to f$ *weakly* $\Leftrightarrow$

i) $$\forall A \in \mathscr{M}, \int_A (f_n - f) \to 0,$$

ii) $$\sup_n \|f_n\|_1 = k < \infty.$$

Proof. ($\Rightarrow$) Define $F_n : L_m^\infty[0, 1] \to \mathbf{C}$ as

$$\forall g \in L_m^\infty[0, 1], \qquad F_n(g) = \int (f_n - f)g.$$

Since $F_n(g) \to 0$ for each $g \in L_m^\infty[0, 1]$ by hypothesis, we know that

$$\forall g \in L_m^\infty[0, 1], \qquad \exists N_g \quad \text{such that} \quad \forall n \geqslant N_g, \quad |F_n(g)| \leqslant 1;$$

and so

$$\forall g \in L_m^\infty[0, 1], \qquad \exists M_g = \max\{1, |F_1(g)|, \dots, |F_{N_g - 1}(g)|\}$$

such that

$$\forall n, \qquad |F_n(g)| \leqslant M_g.$$

We can apply Theorem I.8 and conclude that

$$\exists M \text{ such that } \forall n \quad \text{and} \quad \forall g \in L_m^\infty[0, 1], \qquad |F_n(g)| \leqslant M\|g\|_\infty.$$

Clearly,

$$\|f_n\|_1 = \int f_n g_n, \quad \text{where} \quad g_n \in L_m^\infty[0, 1] \quad \text{and} \quad |g_n| = 1;$$

Therefore ii) follows.

($\Leftarrow$) Take any $\varepsilon > 0$. We prove that for any fixed $g \in L_m^\infty[0, 1]$,

$$\overline{\lim_n} \left| \int (f_n - f)g \right| \leqslant \varepsilon. \tag{I.2}$$

Choose a simple function h such that $\|g - h\|_\infty \leqslant \varepsilon/2k$. Then

$$\left| \int (f_n - f)g \right| \leqslant \|g - h\|_\infty \|f_n - f\|_1 + \left| \int (f_n - f)h \right|,$$

and this gives (I.2).

q.e.d.

The incredible thing is that we can drop condition ii) in Theorem I.9 even in the setting of quite general measure spaces; this is proved in Chapter 6.

The weak sequential convergence defined above is actually sequential convergence for a certain topology (called the weak topology) on $L_m^1[0, 1]$. We'll discuss the weak topology generally in section I.9 but for now consider a special result for the case of $L_c^1(\mathbf{Z})$. This result (Schur's lemma) is studied in greater detail in Chapter 6. The proof we now outline uses the Baire category theorem.

Take $x = \{x_j : j = 1, \ldots\} \in L_c^\infty(\mathbf{Z})$ and $a = \{a_j : j = 1, \ldots\} \in L_c^1(\mathbf{Z})$. Define

$$N_{\varepsilon, x}(a) = \{b = \{b_j\} \in L_c^1(\mathbf{Z}) : |\textstyle\sum x_j(b_j - a_j)| < \varepsilon\}.$$

$N_{\varepsilon, x}(a)$ is a "neighborhood" of a and $\{N_{\varepsilon, x}(a)\}$, as ε, x, and a vary, is a "neighborhood basis" (cf. section I.1) for the weak topology $\mathscr{W}$ on $L_c^1(\mathbf{Z})$—this means that $W \in \mathscr{W}$ if and only if

$$\forall a \in W \quad \exists x \in L_c^\infty(\mathbf{Z}) \quad \text{and} \quad \exists \varepsilon > 0 \quad \text{such that} \quad N_{\varepsilon, x}(a) \subseteq W.$$

$\mathscr{W}$ is in fact a topology on $L_c^1(\mathbf{Z})$ and if U is open for the $\| \ \|_1$-topology on $L_c^1(\mathbf{Z})$ then $U \in \mathscr{W}$. It is not difficult to verify that

$$S = \{a \in L_c^1(\mathbf{Z}) : \textstyle\sum |a_j| = 1\} \quad \text{is} \quad \| \ \|_1\text{-closed}$$

and   the weak closure of S is the $\| \ \|_1$-closure of $B(0, 1)$

(where the ball $B(0, 1)$ is defined in terms of $\| \ \|_1$) (cf. Theorem I.11). Consequently, $\mathscr{W}$ is strictly weaker than the norm topology on $L_c^1(\mathbf{Z})$. Note that weak sequential convergence for $L_c^1(\mathbf{Z})$ is precisely the analogue (for $\mathbf{Z}$) of that defined (for $[0, 1]$) immediately after Theorem I.8. Schur's lemma tells us that $\mathscr{W}$ and $\|\ \|_1$ yield the same convergent sequences in $L_c^1(\mathbf{Z})$.

Theorem I.10 (Schur) *If* $\{a^{(n)}: n = 1, \ldots\} \subseteq L_c^1(\mathbf{Z})$ *converges to* 0 *in* $\mathscr{W}$ *then* $\|a^{(n)}\|_1 \to 0$.

Proof. Let $Y = \{x \in L_c^\infty(\mathbf{Z}): \sup |x_j| \leqslant 1\}$ and define

$$\forall x, \ y \in Y, \qquad d(x, y) = \sum |x_j - y_j|/2^j.$$

(Y, d) is complete and sets of the form

$$\forall x \in Y, \qquad S_{J, \delta}(x) = \{y: |x_j - y_j| < \delta, |j| \leqslant J\}$$

are a basis at x for the topology of (Y, d).
Next we define

$$\forall \varepsilon > 0 \quad \text{and} \quad \forall m, \qquad A_m = \{y \in Y: |\sum_j y_j a_j^{(n)}| \leqslant \varepsilon, \forall n \geqslant m\}.$$

It can be shown that A_m is closed in (Y, d) and $Y = \bigcup A_m$, so that by Baire I, II there is m such that int $A_m \neq \emptyset$.
From this point it is straightforward to prove that $\|a^{(n)}\|_1 \to 0$.

q.e.d.

I.8 Hahn–Banach theorem

Our presentation of the Hahn–Banach theorem is standard. There are basically three distinct parts to the proof. The first and crucial step is the lemma below which allows us to extend continuous linear functionals from a closed subspace Y to the closed subspace generated by Y and an element x (the setting here is necessarily with real normed spaces); second, an axiom of choice argument is used to "blow up" this finite procedure to extend maps in the infinite dimensional case; finally, an ingenious trick due to Bohnenblust and Sobczyk yields the result for the complex case. The general problem of extending linear functions

$$L: Y \to Z,$$

$Y \subseteq X$, and X and Y normed spaces, is generally intractable. Also, *for infinite dimensional spaces the axiom of choice is actually equivalent to the Hahn–Banach theorem.*

If X and Z are normed spaces, $Y \subseteq X$ is a subspace, and $L: Y \to Z$ is linear we define

$$\|L\| = \sup \{\|Ly\|: \|y\| \leqslant 1, y \in Y\}.$$

The space of continuous linear functions $X \to \mathbf{C}$ is denoted by X'.

Lemma *Let X be a real normed space, $Y \subseteq X$ a closed subspace, and Z the closed subspace of X generated by Y and some $z \in X \setminus Y$. If $L: Y \to \mathbf{R}$ is linear and continuous then there is a continuous linear functional $K: Z \to \mathbf{R}$ such that $K = L$ on Y and $\|K\| = \|L\|$.*

Proof. If $x, y \in Y$,

$$Lx - Ly \leqslant \|L\| \, \|x + z\| + \|L\| \, \|y + z\|;$$

and so

$$\sup_{u \in Y} \left(-\|L\| \, \|u + z\| - Lu \right) = a \leqslant b = \inf_{u \in Y} \left(\|L\| \, \|u + z\| - Lu \right).$$

For fixed $c \in [a, b]$ we define $K(y + rz) = Ly + rc$, where $r \in \mathbf{R}$, $y \in Y$, and $\{y + rz: r \in \mathbf{R}, y \in Y\} = Z$.

q.e.d.

Theorem I.11 (Hahn–Banach) a) *Let $Y \subseteq X$ be a subspace of the normed space X, and assume $L: Y \to \mathbf{C}$ is linear and continuous. Then there is $K \in X'$ such that $K = L$ on Y and $\|K\| = \|L\|$.*
b) *If $Y \subseteq X$ is a closed subspace of the normed space X and $z \notin Y$ then there is $L \in X'$ such that $Lz \neq 0$ and $L = 0$ on Y.*

Proof. i) a) $\Rightarrow$ b) Define $L_z(y + rz) = r$, where $y + rz$, for $y \in Y$ and $r \in \mathbf{R}$, is a typical element of the closed subspace generated by Y and z. Note that

$$a = \inf_{y \in Y} \|z + y\| > 0$$

and $|L_z(y + rz)| \leqslant (1/a)\|y + rz\|.$

Thus we apply part a) directly.

ii) For part a) choose Y closed without any loss of generality. In fact it is trivial to extend L to $\bar{Y}$ by Theorem I.5.

iii) We now prove a) for the real case, assuming that Y is closed and that $Y \subseteq X$ properly.

Let $\mathscr{L}$ be the family of all continuous linear functions $K: Z \to \mathbf{R}$ such that $Y \subseteq Z$, $K = L$ on Y, and $\|K\| = \|L\|$. From the Lemma, $\mathscr{L}$ is non-trivial.

We order $\mathscr{L}$ by setting $K \leqslant K_1$ if $Z \subseteq Z_1$ and $K_1 = K$ on Z.

From Zorn's lemma (i.e. the axiom of choice) there is a maximal element $K: Z \to \mathbf{R}$ and we easily check that $Z = X$.

iv) Let W be a complex vector space. If $K: W \to \mathbf{C}$ is real linear then K is complex linear if and only if $K(ix) = iKx$.

Given $L: Y \to \mathbf{C}$, complex linear (as in a)). Set $L_1 = \operatorname{Re} L$, $L_2 = \operatorname{Im} L$ and note that L is real linear. Thus $L(iy) = iLy$ on Y, and, using this fact, we compute that

$$\forall y \in Y, \qquad L_2(y) = -L_1(iy).$$

Because of iii) we can extend L_1 to K_1 on X, considered as a real vector space, such that $\|L_1\| = \|K_1\|$. Set $Kx = K_1x - iK_1(ix)$ on X.

Similar computations show that K has the desired properties.

q.e.d.

Example I.8 Let X be a normed vector space $x, y \in X$, $x \neq y$. By the Hahn–Banach theorem we see that there is $L \in X'$ such that $Lx \neq Ly$. In fact let Y be the subspace generated by $x - y$, define $K(r(x - y)) = r\|x - y\|$, observe that $\|K\| = 1$, and use Theorem I.11.

I.9 The weak and weak * topologies

Let X be a normed vector space. X' is a Banach space normed by

$$\forall x' \in X', \qquad \|x'\| = \sup \{|x'(x)|: \|x\| \leqslant 1\};$$

as such X' is the dual of X.

We then consider $(X')' = X''$ normed analogously noting that X'' is a Banach space and that X can be embedded isometrically and algebraically isomorphically onto a subspace of X''. The map defining this isomorphism is given by

$$\forall x \in X, \qquad x(x') = x'(x).$$

X is reflexive if $X = X''$ under this canonical map.

The weak topology on X, denoted by $\sigma(X, X')$, has a basis at $0 \in X$ given by sets of the form

$$\{x \in X: |x'_j(x)| < \varepsilon, j = 1, ..., n\},$$

where $\varepsilon > 0$ and $\{x'_1, ..., x'_n\}$ is an arbitrary finite subset of X'. Similarly, we define $\sigma(X', X'')$. The weak * topology on X' denoted by $\sigma(X', X)$, is defined analogously with corresponding sets

$$\{x' \in X': |x'(x_j)| < \varepsilon, j = 1, ..., n\}, \quad \varepsilon > 0, \quad x_j \in X.$$

Clearly, $\sigma(X', X)$ is generally weaker than $\sigma(X', X'')$, i.e. $\sigma(X', X) \subseteq \sigma(X', X'')$. $K \subseteq X$ is convex if for each $x, y \in K$ and $0 \leqslant r \leqslant 1$,

$$rx + (1 - r)y \in K.$$

An important application of Theorem I.11 is

Theorem I.12 *Let X be a normed vector space and let $K \subseteq X$ be convex. Then K has the same norm and $\sigma(X, X')$ closure.*

Theorem I.13 (Alaoglu) *Let X be a normed vector space and let*

$$B' = \{x' \in X': \|x'\| \leqslant 1\}.$$

*Then B' is weak * compact.*

Proof. For each $x \in X$ define

$$D_x = \{z \in \mathbf{C}: |z| \leqslant \|x\|\}.$$

Clearly, $B' \subseteq D = \prod_{x \in X} D_x$. Since *the product of compact spaces is compact* (this statement is equivalent to the axiom of choice and it is called Tychonoff's theorem) and since it is easy to check that B' is closed in D, B' is a compact subset of D.

It is immediate from definition that the induced product topology on B' is its weak * topology.

q.e.d.

In this regard note that—

Theorem I.14 *Let X be a Banach space. $Y \subseteq X'$ is weak * compact $\Leftrightarrow Y$ is weak * closed and norm bounded.*

Proof. The sufficient condition for weak * compactness follows from Theorem I.13.

For the necessary condition we must verify that weak * boundedness implies norm boundedness (since weak * compactness yields weak * boundedness).

This follows from Theorem I.8 noting that X is complete.

q.e.d.

Since weak * boundedness implies norm boundedness we see that every *weak * convergent sequence is norm bounded.*

Remark We need X to be complete in Theorem I.14. For a counter-example let X be the vector space of all finite sequences of complex numbers normed by $\|x\| = \sup_n \{|x_n|\}$, $x = \{x_n: n = 1, \ldots\}$. Set $x'_n(x) = n|x_n|$ and $Y = \{0\} \cup \{x'_n: n = 1, \ldots\} \subseteq X'$. $x'_n \to 0$ in $\sigma(X', X)$, whereas $\|x'_n\| = n$. The situation is corrected by the following result: *let X be a normed vector space and $Y \subseteq X'$ weak * compact; Y is norm bounded if and only if the weak * closure of the smallest convex set containing Y is weak * compact.*

A very useful result (e.g. [10, p. 141]) concerning weak * closures is the *Kreĭn–Šmul'yan theorem: let X be a Banach space and let $K \in X'$ be convex; by definition, a net [63, p. 65] $\{x'_n\} \subseteq X'$ converges to 0 in the ks topology if $x'_\alpha \to 0$ uniformly on compact sets of X; then the ks and weak * closures of K are identical.*

Note, of course, that finite sets are compact.

Example I.9 Define Y to be the space of functions $f: [0, 1] \to [0, 1]$ having the form $f = \chi_A$, where the subset $A \subseteq [0, 1]$ is a finite disjoint union of intervals. The weak * closure of Y, as a subset of $L_m^\infty[0, 1]$ is

$$\{f \in L_m^\infty[0, 1]: 0 \leqslant f \leqslant 1\}.$$

Theorem I.15 *Let X be a separable normed space. Then B' is sequentially compact in the $\sigma(X', X)$ topology (cf. Theorem I.13).*

Proof. Let $\{x_k': k = 1, \ldots\} \subseteq B'$ and let $\{x_n: n = 1, \ldots\}$ be a countable dense subset of X. By the expected diagonal argument there is a subsequence $\{x_{k_j}': j = 1, \ldots\} \subseteq \{x_k': k = 1, \ldots\}$ such that

$$\forall n, \quad \lim_j x_{k_j}'(x_n) = x'(x_n).$$

For $x \in X$, let $x_{n_p} \to x$, so that

$$\lim_p x_{k_j}'(x_{n_p})$$

exists uniformly in j. We complete the proof by the Moore–Smith theorem.

q.e.d.

Note that a compact topological space is metrizable if and only if it has a countable basis. Thus *if a normed space X is separable, B' with the weak * topology is metrizable* (by definition of the weak * topology).

An immediate corollary of the uniform boundedness principle is—

Theorem I.16 *Let X be a normed space and assume $x_n \to x$ in $\sigma(X, X')$. Then $\{\|x_n\|: n = 1, \ldots\}$ is bounded.*

Using Theorem I.15 we can prove the following "converse" to Theorem I.16.

Theorem I.17 *Let X be a reflexive Banach space and let $\{x_n: n = 1, \ldots\}$ be a norm bounded sequence in X. Then there is a subsequence which converges to some $x \in X$ in the $\sigma(X, X')$ topology (cf. Theorem 6.10).*

If "subsequence" is replaced by "subnet" in Theorem I.17 the result is immediate from the Alaoglu theorem. It is interesting to compare this result with Theorem I.9 noting that $L_m^1[0, 1]$ is not reflexive. Because of Theorem I.17 it is easy to check that reflexive Banach spaces are weakly sequentially complete.

From the definition of the weak * topology *the dual of X', taken with the weak * topology, is X.* Since Theorem I.11b is valid for locally convex topological vector spaces we can therefore conclude

Theorem I.18 *Let X be a normed space and let $Y \subseteq X'$ be a $\sigma(X', X)$ closed subspace. If $y' \notin Y$ there is $x \in X$ such that $y'(x) \neq 0$ and*

$$\forall x' \in Y, \quad x'(x) = 0.$$

I.10 Linear maps

If X and Y are Banach spaces, $L(X, Y)$ denotes the space of continuous linear functions $X \to Y$. Parts a) and b) of the following result are the open mapping theorem and closed graph theorem, respectively.

Theorem I.19 a) *Let $L \in L(X, Y)$ be bijective. Then $L^{-1} \in L(Y, X)$.*
b) *Let X and Y be Banach spaces and let $L: X \to Y$ be linear. Assume that*

$$\|x_n - x\| \to 0 \quad and \quad \|Lx_n - y\| \to 0 \tag{I.3}$$

imply $y = Lx$. Then $L \in L(X, Y)$.

The proof of a) depends on the Baire category theorem. Part b) is clear from a) by applying a) to the situation

$$X \times LX \to X$$
$$(x, Lx) \mapsto x,$$

where the norm on $X \times LX$ is given by $\|(x, Lx)\| = \|x\| + \|Lx\|$. (I.3) is used to check that $X \times LX$ is complete.

Example I.10 The closed graph theorem does not say that if $X \times LX$ is closed in $X \times Y$ then L is continuous; it asserts the continuity of L if each $(x, y) \in \overline{X \times LX} \subseteq X \times Y$ can be approximated by $\{(x_n, Lx_n): n = 1, \ldots\}$, $x_n \in X$. Assume $\{x_\alpha\} \subseteq X$ and $\{y_\alpha\} \subseteq Y$ are Hamel bases with $\|x_\alpha\| \leqslant 1$, $\sup \|y_\alpha\| = \infty$. Taking card $X = $ card Y we define $Lx_\alpha = y_\alpha$, and extend L linearly to all of X. Then L is a linear surjection and $X \times LX = X \times Y$, but $X \times LX$ does not satisfy (I.3); clearly L is not continuous.

Example I.11 We shall put two norms on $L_c^\infty(\mathbf{Z})$ so that $L_c^\infty(\mathbf{Z})$ is a Banach space for each norm but such that neither identity mapping $L_c^\infty(\mathbf{Z}) \to L_c^\infty(\mathbf{Z})$ is continuous. Choose $\| \ \|_\infty$ for the first norm. To define the second norm first observe that

$$\text{card } L_c^1(\mathbf{Z}) = \text{card } L_c^\infty(\mathbf{Z}). \tag{I.4}$$

To prove (I.4) consider the injection

$$L_c^\infty(\mathbf{Z}) \to L_c^1(\mathbf{Z})$$
$$\{x_n: n = 1, \ldots\} \mapsto \{x_n/2^n: n = 1, \ldots\}.$$

Thus card $L_c^\infty(\mathbf{Z}) \leqslant$ card $L_c^1(\mathbf{Z})$. On the other hand card $L_c^1(\mathbf{Z}) \leqslant$ card $L_c^\infty(\mathbf{Z})$ since $L_c^1(\mathbf{Z}) \subseteq L_c^\infty(\mathbf{Z})$. (I.4) follows from the Schroeder–Bernstein theorem (e.g. Problem 1.6). Consequently if H_p is an Hamel basis for $L_c^p(\mathbf{Z})$ then card $H_\infty = $ card H_1, and so we choose any bijection $b: H_\infty \to H_1$. We extend b by linearity to a bijection $L: L_c^\infty(\mathbf{Z}) \to L_c^1(\mathbf{Z})$. By Theorem I.19a, the non-separability of $L_c^\infty(\mathbf{Z})$, and the separability of $L_c^1(\mathbf{Z})$, we see that $L \notin L(L_c^\infty(\mathbf{Z}), L_c^1(\mathbf{Z}))$.

The second norm on $L_c^\infty(\mathbf{Z})$ is then defined by

$$\|x\| = \|Lx\|_1$$

(it is easy to check that $L_c^\infty(\mathbf{Z})$ with $\| \ \|$ is complete).

The following was given by L. Carleson with regard to an interpolation problem [24].

Theorem I.20 *Let X and Y be Banach spaces with norms $\| \ \|_X$ and $\| \ \|_Y$ respectively. Assume $Y \subseteq X$ and $\| \ \|_Y \geqslant \| \ \|_X$ on Y. If*

$$\exists\{x_n : n = 1,\ldots\} \subseteq Y, \|x_n\|_Y \leqslant 1, \text{ such that } \forall x' \in X', \|x'\| \leqslant M \sup_n |x'(x_n)|$$

then $X = Y$ and $\| \ \|_Y \leqslant M \| \ \|_X$.

If $L \in L(X, Y)$, then the transpose, L', of L is the element of $L(Y', X')$ defined by

$$\forall x \in X \quad \text{and} \quad \forall y' \in Y, \qquad L'y'(x) = y'(Lx).$$

Theorem I.21 *Let X and Y be Banach spaces and assume $L \in L(X, Y)$ is injective and $\overline{LX} = Y$. The following are equivalent:*

a) $LX = Y$,

b) L' *is open,*

c) $L'Y' = X'$.

II Fubini's theorem

We now present the Fubini–Tonelli (F–T) theorem; this is one of the most useful results in analysis.

Let $(X, \mathscr{A}, \mu)$ and $(Y, \mathscr{C}, v)$ be complete measure spaces. The F–T theorem gives conditions so that

$$\int_X \left(\int_Y f(x, y)\, dv(y) \right) d\mu(x) = \int_Y \left(\int_X f(x, y)\, d\mu(x) \right) dv(y) = \iint_{X \times Y} f\, d\mu \times v. \quad \text{(II.1)}$$

Obviously, just the formal expression in (II.1) requires some explanation.

Any proof of a general statement of F–T demands a good deal of detail which can be found in basic texts. Our purpose is to point out the two or three main steps for such a proof. Since (II.1) involves a switching of limits you can expect to find important use of LDC for several computations in the proof. We shall also indicate a simple proof of F–T for a special case.

If $A \in \mathscr{A}$ and $B \in \mathscr{C}$ then $A \times B$ is measurable rectangle; the space of such rectangles is denoted by $\mathscr{R}$. For any $A \times B \in \mathscr{R}$ we define

$$\rho(A \times B) = \mu A v B. \quad \text{(II.2)}$$

The first important step is to prove:

> There is an extension $\mu \times v$ of ρ as a complete measure on a σ-algebra $\mathscr{D}$ containing $\mathscr{R}$. (II.3)

The proof of (II.3) involves first extending ρ in a certain way to the algebra consisting of finite disjoint unions from $\mathscr{R}$, and then using Problem 3.27 (Caratheodory's theorem).

For the next step, if $Q \subseteq X \times Y$ define

$$\forall x \in X, \qquad Q_x = \{y \in Y : (x, y) \in Q\}.$$

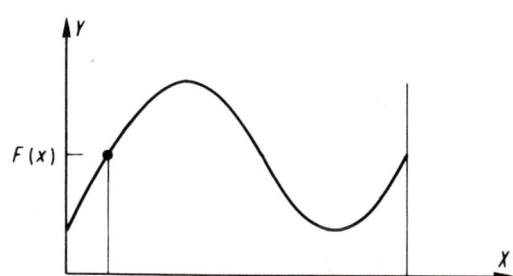

Fig. 9

When $Q \subseteq X \times Y$ and $Q_x \in \mathscr{C}$ we set (e.g. Fig. 9).

$$F(x) = \nu Q_x.$$

Consequently, as in the case of ordinary integration, we should have

$$\mu \times \nu(Q) = \int_X F(x) \, d\mu(x) \tag{II.4}$$

(cf. our remarks on Peano in Chapter 2.1). Precisely, the second step is:

> let $Q \in \mathscr{R}_{\sigma\delta}$ with $\mu \times \nu(Q) < \infty$; then F is μ-measurable and (II.4) is valid. $\tag{II.5}$

The third step extends (II.5) to any $Q \in \mathscr{Q}$ satisfying $\mu \times \nu(Q) < \infty$. In order to prove this we must use the fact, which is easy to show, that if $Q \in \mathscr{Q}$ then there is $P \in \mathscr{R}_{\sigma\delta}$ such that $Q \subseteq P$ and $\mu \times \nu(P) = \mu \times \nu(Q)$ (cf. Proposition 2.3). Note that this step clinches (II.1) for the case of

$$f(x, y) = \chi_Q(x, y).$$

The proof of F–T for "arbitrary" $f(x, y)$ involves standard approximation techniques.

Before stating F–T let us note that, in order to prove the third step from (II.5), $\mathscr{R}_{\sigma\delta}$ can't approximate $Q \in \mathscr{Q}$ from within by an element $P \in \mathscr{R}_\sigma$.

Example II.1 Let $(X, \mathscr{A}, \mu)$ and $(Y, \mathscr{C}, \nu)$ be $([0, 1], \mathscr{M}, m)$ and choose a perfect totally disconnected set $E \subseteq [0, 1]$ satisfying $mE > 0$. Set

$$Q = \{(x, y): x - y \in E, x \in X, y \in Y\},$$

so that $m \times m(Q) > 0$. Now if we take any $A \times B \in \mathscr{R}$ then $A - B \subseteq E$ if $A \times B \subseteq Q$. Note that if A and B are both of positive measure then $A - B$ contains an interval (e.g. Problem 3.6b); this contradicts the fact that $A - B \subseteq E$. Thus $m \times m(A \times B) = 0$.

Theorem II.1 (F–T) *Let* $(X, \mathscr{A}, \mu)$ *and* $(Y, \mathscr{C}, \nu)$ *be complete measure spaces and assume* f *is* $\mu \times \nu$-*measurable.*
a) *If* $f \in L^1_{\mu \times \nu}(X \times Y)$ *then* $f_x(y) = f(x, y) \in L^1_\nu(Y)$, μ-*a.e.,* $f_y(x) = f(x, y) \in L^1_\mu(X)$, ν-*a.e.,*

$$\int_Y f(x, y) \, d\nu(y) \in L^1_\mu(X) \quad \text{and} \quad \int_X f(x, y) \, d\mu(x) \in L^1_\nu(Y),$$

and (II.1) *holds.*
b) *If* $(X, \mathscr{A}, \mu)$ *and* $(Y, \mathscr{C}, \nu)$ *are* σ-*finite and one of the iterated integrals,*

$$\int_X \left(\int_Y |f| \, d\nu \right) d\mu \quad \text{or} \int_Y \left(\int_X |f| \, d\mu \right) d\nu,$$

is finite, then $f \in L^1_{\mu \times \nu}(X \times Y)$.

Remark If $\mu \in M(X)$ we have $\mu \ll |\mu|$ and $\mu = h|\mu|$, $|h| = 1$, by R–N; thus Theorem II.1 is applicable for elements of $M(X)$.

If a function does not satisfy the hypotheses of F–T (noting that a measure space can always be completed) there is probably no other version of F–T which will help— counter-examples abound; and if it is possible to switch limits at all, it will have to be done by ad hoc, and probably ingenious, methods.

In our sketch of the proof of Theorem II.1 we have avoided all details, and there are many (although our three step hint makes for a feasible exercise). We now give a simple proof of F–T for a special case.

The following is a standard result from advanced calculus and is easy to prove.

Lemma Let $g: [a, b] \to \mathbf{C}$ be bounded and assume $\lim_n S_{P_n}$ and $\lim_n s_{P_n}$ exist and are equal for every sequence $\{P_n : n = 1, \ldots\}$ of partitions for which card $P_n = n + 1$ and the norms $|P_n| \to 0$ (cf. Chapter 3 for notation). Then $R \int_a^b g$ exists.

Theorem II.2 Let $(X, \mathscr{A}, \mu)$ be a measure space and take $f(x, y): X \times [a, b] \to \mathbf{C}$. Assume $R \int_a^b f(x, y) \, dy$ exists μ-a.e., $f(x, y)$ is μ-measurable for all $y \in [a, b]$, and that there is $F \in L^1_\mu(X)$ for which

$$\forall (x, y) \in X \times [a, b], \qquad |f(x, y)| \leqslant F(x).$$

Then $\int_X f(x, y) \, d\mu(x)$ is Riemann integrable and

$$\int_X \left(\int_a^b f(x, y) \, dy \right) d\mu(x) = \int_a^b \left(\int_X f(x, y) \, d\mu(x) \right) dy.$$

Proof. Let $\{P_n : n = 1, \ldots\}$ be a sequence of partitions of $[a, b]$ for which the norms $|P_n| \to 0$ and card $P_n = n + 1$; and form

$$S_{P_n}(x) = \sum_1^n f(x, M_j)(y_j - y_{j-1}).$$

Then

$$|S_{P_n}(x)| \leqslant (b - a)F(x) \tag{II.6}$$

and

$$\lim_n S_{P_n}(x) = R \int_a^b f(x, y) \, dy, \qquad \mu\text{-a.e.}$$

Now
$$\int_X S_{P_n}(x)\, d\mu(x) = \sum_1^n \left(\int_X f(x, M_j)\, d\mu(x) \right) (y_j - y_{j-1}).$$

By LDC, which we can use by (II.6),

$$\lim_n \sum_1^n \left(\int_X f(x, M_j)\, d\mu(x) \right) (y_j - y_{j-1})$$

$$= \lim_n \int_X S_{P_n}(x)\, d\mu(x) = \int_X \left(R \int_a^b f(x, y)\, dy \right) d\mu(x). \tag{II.7}$$

This is true for all such $\{P_n : n = 1, \ldots\}$ and for $s_{P_n}(x)$; consequently, by the Lemma,

$$R \int_a^b \left(\int_X f(x, y)\, d\mu(x) \right) dy$$

exists and (II.7) completes the proof.

q.e.d.

III The Riesz representation theorem (RRT)

III.1 Riesz's representation theorem

Let $C[a, b]$ be the real Banach space of continuous functions $f: [a, b] \to \mathbf{R}$ with norm

$$\forall f \in C[a, b], \qquad \|f\|_\infty = \sup_{x \in [a, b]} |f(x)|$$

(e.g. Appendix I.2). From the theory of Riemann–Stieltjes integration we know that if $g \in BV[a, b]$, then $L_g: C[a, b] \to \mathbf{R}$, defined by

$$L_g(f) = \int_a^b f \, dg,$$

is an element of $C[a, b]'$; in fact,

$$|L_g(f)| \leq \|f\|_\infty V(g, [a, b]).$$

The converse is Riesz's representation theorem (cf. (4.10) and Proposition 4.3):

Theorem III.1 (F. Riesz, 1909) *If $L \in C[a, b]'$ then there is $g \in BV[a, b]$ such that $\|L\| = V(g, [a, b])$ and*

$$\forall f \in C[a, b], \qquad L(f) = \int f \, dg. \tag{III.1}$$

Proof. For the sake of simple subscripts let $[a, b] = [0, 1]$.

Since $L \in C[0, 1]'$ we use the Hahn–Banach theorem to conclude the existence of $K \in L_m^\infty[0, 1]'$ such that $L = K$ on $C[0, 1]$ and $\|L\| = \|K\|$.

Let $\chi_x = \chi_{[0, x]}$ and write

$$K(\chi_x) = g(x). \tag{III.2}$$

We shall prove that $g \in BV[0, 1]$. Given a partition $0 = x_0 \leq x_1 \leq \ldots \leq x_n = 1$, we set $t_i = \text{sgn}\,[g(x_i) - g(x_{i-1})]$, $i = 1, \ldots, n$.

Thus

$$\sum_1^n |g(x_i) - g(x_{i-1})| = \sum_1^n t_i(g(x_i) - g(x_{i-1}))$$

$$= \mathcal{K} \sum_1^n t_i(\chi_{x_i} - \chi_{x_{i-1}})) = K \sum_1^n t_i \chi_{(x_{i-1}, x_i]}).$$

Note that $|t_i|$ is 0 or 1 so that

$$\left\| \sum_1^n t_i \chi_{(x_{i-1}, x_i]} \right\|_\infty \leqslant 1.$$

Hence,

$$V(g, [0, 1]) \leqslant \|K\| = \|L\|. \tag{III.3}$$

Now, for $f \in C[0, 1]$, define

$$f_n(x) = \sum_{j=1}^n f(j/n)(\chi_{j/n}(x) - \chi_{(j-1)/n}(x));$$

therefore if $j = 1, \ldots, n$,

$$\forall x \in \left(\frac{j-1}{n}, \frac{j}{n} \right), \qquad f_n(x) = f\left(\frac{j}{n} \right).$$

Since f is continuous, $\|f - f_n\|_\infty \to 0$. Thus

$$\lim_n K(f_n) = K(f) = L(f). \tag{III.4}$$

Using (III.2) we see that

$$K(f_n) = \sum_{j=1}^n f\left(\frac{j}{n} \right) \left(g\left(\frac{j}{n} \right) - g\left(\frac{j-1}{n} \right) \right),$$

from which we conclude

$$\lim_n K(f_n) = \int_0^1 f \, dg \tag{III.5}$$

(III.4) and (III.5) yield (III.1).

Fix $\varepsilon > 0$, and letting $\|L\| \leqslant |L(f)| + \varepsilon$ for some $f \in C[0, 1]$ which satisfies $\|f\|_\infty \leqslant 1$, we apply (III.1) and compute

$$\|L\| \leqslant V(g, [0, 1]) + \varepsilon;$$

since ε is arbitrary we have $\|L\| \leqslant V(g, [0, 1])$, which, when combined with (III.3), yields $\|L\| = V(g, [0, 1])$.

q.e.d.

Part of the motivation that led to Theorem III.1 came from the moment problem: given a sequence $\{a_n : n = 1, ...\} \subseteq \mathbf{C}$ of moments of a distribution of electric charges and given $\{f_n : n = 1, ...\} \subseteq C[a, b]$ the problem is to find $\mu \in M[a, b]$ such that

$$\forall n, \quad \int f_n \, d\mu = a_n.$$

As an enticement to Schwartz's theory of generalized functions [35; 57] and its important relations with Fourier analysis and partial differential equations we restate Theorem III.1 in the following way:

> Let T be a Schwartz generalized function on $\mathbf{R}$; T is a complex regular Borel measure $\Leftrightarrow T = f'$, where $f \in BV$ is continuous from the right at each point and f' represents the first distributional derivative of f.   (III.6)

Observe that f' in (III.6) is not, in general, the usual derivative of f (which exists m-a.e. since $f \in BV$). In fact, the usual derivative of C_C is 0, m-a.e., whereas the distributional derivative is μ_C.

III.2 RRT

Let X be a locally compact space and let $C_0(X)$ be the complex vector space of continuous functions $f : X \to \mathbf{C}$ with the property that

$$\forall \varepsilon > 0 \qquad \exists K_{f, \varepsilon} \subseteq X, \text{ compact, such that } \quad \forall x \notin K_{f, \varepsilon}, \quad |f(x)| < \varepsilon,$$

i.e., f "vanishes at infinity". $C_0(X)$ is a Banach space with norm $\|f\|_\infty = \sup_{x \in X} |f(x)|$. For locally compact spaces X, $M(X)$ shall denote the space of regular complex measures on the measurable space $(X, \mathscr{B})$ (cf. the definition of $M(X)$ for arbitrary measure spaces given prior to Theorem 5.7). We know from Problem 5.18 that $M(X)$ is a Banach space with norm $\|\mu\|_1 = |\mu|(X)$.

A far reaching extension of Theorem III.1 is

Theorem III.2 (RRT) *Let X be a locally compact space. $M(X)$ and $C_0(X)'$ are isometrically isomorphic; in fact, for each $L \in C_0(X)'$ there is a unique $\mu \in M(X)$ such that*

$$\forall f \in C_0(X), \qquad L(f) = \int_X f \, d\mu. \tag{III.7}$$

There are several excellent expositions of the proof of Theorem III.2 (e.g. [56; 94]), and we omit it here.

Remark Let $\{K_N : N = 1, \dots\} \subseteq L_m^1(\mathbf{T})$ be an approximate identity so that, in particular, $\sup_N \|K_N\|_1 < \infty$. By Problem 5.43, $\{K_N : N = 1, \dots\} \subseteq M(\mathbf{T})$ is norm bounded in $M(\mathbf{T})$; and by choosing $r > 0$ properly in the estimate,

$$\forall f \in C(\mathbf{T}), \quad \left| f(0) - \int f(x) K_N(x)\, dx \right| \leqslant \int_{-r}^{r} |(f(x) - f(0)) K_N(x)|\, dx$$

$$+ \int_{\pi \geqslant |x| \geqslant r} |(f(x) - f(0)) K_N(x)|\, dx,$$

we see that $K_N \to \delta$ in the weak * topology on $M(\mathbf{T})$. As a result, it is easy to verify that

$$\overline{\lim_N} \|K_N\|_2 = \infty$$

(assuming, of course, that $\{K_N : N = 1, \dots\} \subseteq L_m^2(\mathbf{T})$). In fact, if we had the L^2-boundedness, then there is $K \in L_m^2(\mathbf{T})$ such that

$$\forall f \in C(\mathbf{T}), \qquad f(0) = \int f K;$$

at this point it is routine to argue to a contradiction.

III.3 Radon measures

We have seen that it is quite effective to study the space $M(X)$. Unfortunately, $M(X)$ is not quite Camelot. For example, $m \notin M(\mathbf{R})$. We now define the space of Radon measures which takes care of this particular flaw.

Let X be locally compact and let $C_c(X)$ be the complex vector space of continuous functions $f : X \to \mathbf{C}$ whose support, $\operatorname{supp} f = \overline{\{x : f(x) \neq 0\}}$, is compact. There is a Hausdorff locally convex topology $\mathcal{T}$ on $C_c(X)$ such that sequential convergence $f_n \to f$, where $\{f_n, f : n = 1, \dots\} \subseteq C_c(X)$, is defined by

$$\|f_n - f\|_\infty \to 0$$

and $\qquad \exists V \subseteq X$, open and relatively compact, such that $\forall n$, $f_n = 0$ on $\tilde{V}$

(cf. Example I.5). For each compact set $K \subseteq X$ let $C_c(K) = \{ f \in C_c(X) : \operatorname{supp} f \subseteq K \}$. $\mathcal{T}$ is characterized by the property that a linear functional $\mu : C_c(X) \to C$ is continuous if and only if μ is continuous on each sup normed space $C_c(K)$.

The space of continuous linear functionals $\mu : C_c(X) \to \mathbf{C}$ is the space of Radon measures and is denoted by $D^0(X)$ (this notation indicates that the Radon measures form a special subset of Schwartz's generalized functions).

Example III.1 Let $f_n \in C_c(\mathbf{R})$ take the values

$$f_n(x) = \begin{cases} \dfrac{1}{n}, & \text{if } |x| \leqslant n \\[2mm] 0, & \text{if } |x| > n + 1, \end{cases}$$

and assume that for all x, $|f_n(x)| \leqslant 1/n$; then $\|f_n\|_\infty \to 0$ whereas $f_n \nrightarrow 0$ in the topology of $C_c(\mathbf{R})$.

Clearly, then, $\mathscr{T}$ is a finer topology than the usual sup norm topology on $C_c(X)$ and so $M(X) \subseteq D^0(X)$ since $\overline{C_c(X)} = C_0(X)$. In this context $M(X)$ is the space **bounded Radon measures**. If we define a functional μ_m by

$$\forall f \in C_c(\mathbf{R}), \qquad \mu_m(f) = \int_{\mathbf{R}} f \, dm,$$

we see that $\mu_m \in D^0(\mathbf{R}) \setminus M(\mathbf{R})$; consequently, Lebesgue measure m is an unbounded Radon measure. As another type of example we have—

Example III.2 Define

$$\mu = \sum_{n=-\infty}^{\infty} a_n \delta_n, \tag{III.8}$$

where $\{a_n : n = 1, \ldots\} \subseteq \mathbf{C}$. Clearly, $\mu \in D^0(\mathbf{R})$. If we put $a_0 = 0$ and $a_n = 1/n$ for $n \neq 0$, and if we interpret the sum in (III.8) as

$$\mu = \sum_{1}^{\infty} \frac{1}{n} (\delta_n - \delta_{-n}),$$

then $\mu \in D^0(\mathbf{R}) \setminus M(\mathbf{R})$. To prove this, let f_k have the form of the function in Fig. 10, where each triangle has height $1/\log k$. Then $\|f_k\|_\infty \to 0$, whereas

$$\mu(f_k) \sim 2, \quad k \to \infty.$$

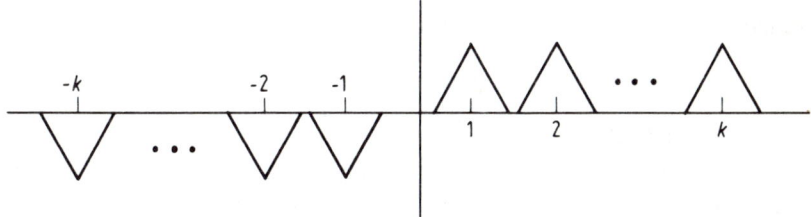

Fig. 10

Note that if X is compact then $D^0(X) = M(X)$. On the other hand if $X = (-\pi, \pi)$, so that $\bar{X}$ is compact, we have $D^0(X) \setminus M(X) \neq \emptyset$. For example, define

$$\forall A \in \mathscr{B}((-\pi, \pi)), \qquad \mu A = m(\tan A).$$

III.4 Radon measures and countably additive set functions

Let X be locally compact. If $\mu: C_c(X) \to \mathbf{C}$ is a linear functional for which $\mu(f) \geq 0$, whenever $f \in C_c(X)$ is non-negative, then μ is a positive linear functional. An essential ingredient in the proof of Theorem III.2 is to "reduce" the proof to showing (III.7) for positive linear functionals; this is generalized to $D^0(X)$ as follows [18, Chapitre III.1.5]:

Theorem III.3 a) *Let $\mu: C_c(X) \to \mathbf{C}$ be a positive linear functional. Then $\mu \in D^0(X)$.*
b) *Let $\mu \in D^0(X)$. There exist positive linear functionals μ_i, $i = 1, ..., 4$, such that*

$$\mu = (\mu_1 - \mu_2) + i(\mu_3 - \mu_4).$$

It is just a matter of checking definitions to verify part a); the proof of part b) is more involved.

We now discuss the extension problem. Given $\mu \in D^0(X)$ we wish to find a large super-space Y of $C_c(X)$ so that μ is defined on Y and the canonical properties of an integra-tion theory hold; the extension from $C_c(X)$ to Y is the Lebesgue extension. This extension is closely related to (III.7) (cf. (III.14)).

$f: X \to \mathbf{R}^*$ is *lower semicontinuous (lsc)* at $x_0 \in X$ if $f(x_0) = -\infty$ or

$$\forall \varepsilon > 0 \quad \exists V \subseteq X, \text{ an open set containing } x_0, \text{ such that } \forall x \in V,$$
$$f(x) > f(x_0) - \varepsilon.$$

Upper semicontinuity (usc) at $x_0 \in X$ and lower and upper semicontinuity of f on X are defined in the obvious ways.

$f: X \to R$ is continuous at x if and only if it is both *lsc* and *usc* at x. $f: X \to \mathbf{R}^*$ is *lsc* if and only if

$$f = \sup \{g: g \leq f \text{ and } g \text{ is continuous}\}.$$

We let $C_c^+(X)$ (resp. $J^+(X)$) be the non-negative elements of $C_c(X)$ (resp. the *lsc* functions). Thus a non-negative function $f: X \to \mathbf{R}^*$ has the form

$$f = \sup \{g: g \leq f \text{ and } g \in C_c^+(X)\} \tag{III.9}$$

if and only if $f \in J^+(X)$.

Motivated by the intuitive linearity property of measuring and the mathematical prejudice of integrating continuous functions on bounded intervals, one has a strong suspicion that a positive linear functional μ on $C_c(X)$ can be regarded in some way in the context of integration theory. More precise motivational remarks are found in [18, Introduction]. The fundamental idea of Lebesgue, in terms of functional analy-sis, was to complete $C_c(X)$ in terms of μ when the space has a suitable topology; this completion is the Lebesgue extension.

In order to define this "suitable topology" for a positive linear functional $\mu: C_c(X) \to$ $\mathbf{R}^*$ note that

$$\|f\|_\mu = \mu(|f|) \tag{III.10}$$

is a semi-norm on $C_c(X)$ (a semi-norm has all the properties of a norm except that $\|x\|$ can be 0 for some non-zero x).

Because of (III.9) we define

$$\forall f \in J^+(X), \qquad \mu^*(f) = \sup\{\mu(g): g \leqslant f \text{ and } g \in C_c^+(X)\}. \tag{III.11}$$

Then for any non-negative $\mathbf{R}^*$-valued function f on X we set

$$\mu^*(f) = \inf\{\mu^*(g): g \geqslant f \text{ and } g \in J^+(X)\}. \tag{III.12}$$

(III.12) is well defined since $g = +\infty$ is in $J^+(X)$; and (III.11) and (III.12) are compatible for $f \in J^+(X)$. Thus, with an eye to (III.10) we let $\mathscr{F}_\mu(X)$ be the set of $\mathbf{R}^*$-valued functions f on X for which

$$\|f\|_\mu = \mu^*(|f|) \tag{III.13}$$

is finite. $\mathscr{F}_\mu(X)$ is then a semi-normed space and $\mathscr{L}_\mu^1(X)$ (cf. Appendix I.2) is the closure of $C_c(X)$ in $\mathscr{F}_\mu(X)$. μ has a unique continuous linear extension, which we again designate by μ, to $\mathscr{L}_\mu^1(X)$ and $\mu(f)$ is the integral of $f \in \mathscr{L}_\mu^1(X)$. $\mathscr{L}_\mu^1(X)$ is the Lebesgue extension of μ. This process was first accomplished by Lebesgue when he extended integration theory from the Riemann integrable functions on $[a, b]$ to $\mathscr{L}_m^1[a, b]$, cf. Appendix I.2.7.

In order to establish the relation between the above procedure and (III.7) we consider the following situation. Let μ be a positive linear functional on $C_c(X)$. $B \subseteq X$ is a μ-integrable set if $\chi_B \in \mathscr{L}_\mu^1(X)$. Compact and open relatively compact sets are μ-integrable and the family of μ-integrable sets is an algebra $\mathscr{A}$. We define the following set function ν_μ on $\mathscr{A}$:

$$\forall B \in \mathscr{A}, \qquad \mu(\chi_B) = \nu_\mu(B) = \int_X \chi_B \, d\nu_\mu. \tag{III.14}$$

The integral sign in (III.14) is written only to establish the relationship with (III.7). On this matter and related to Theorem III.2, we have the following result [18, pp. 149–174]:

Theorem III.4 *Given a measure space* $(X, \mathscr{B}, \nu)$ *where the locally compact space* X *has a countable basis. Assume that* $\nu B < \infty$ *for every compact set* $B \in \mathscr{B}$. *Then there is a unique positive linear functional* $\mu \in D^0(X)$ *such that*

$$\forall B \in \mathscr{B}, \qquad \mu^*(\chi_B) = \nu(B).$$

III.5 Support and the approximation theorem

Let X be a locally compact space and let $\mu \in D^0(X)$. We shall define the support of μ, denoted by supp μ. Naturally we want the notion of supp μ to extend that of supp f, where $f \in C_c(X)$. To make this definition it is necessary to use the following partition of unity result [18, Chapitre III.2.1]:

Theorem III.5 *Let $\{V_\alpha\}$ be an open covering of X and let $\mu_\alpha \in D^0(V_\alpha)$ (noting that V_α is also locally compact). Assume that for each α and β and each $f \in C_c(X)$ for which* supp $f \subseteq V_\alpha \cap V_\beta$ *we have*

$$\mu_\alpha(f) = \mu_\beta(f).$$

Then there is a unique $\mu \in D^0(X)$ such that

$$\forall \alpha \quad and \quad \forall f \in C_c(V_\alpha), \qquad \mu(f) = \mu_\alpha(f).$$

Because of this we conclude that

Corollary III.5.1 *Let $\{V_\alpha\}$ be a family of open sets in X and let $\mu \in D^0(X)$. Assume that*

$$\forall \alpha \quad and \quad \forall f \in C_c(V_\alpha), \qquad \mu(f) = 0.$$

Then

$$\forall f \in C_c \left(\bigcup_\alpha V_\alpha \right), \qquad \mu(f) = 0.$$

We say that $\mu \in D^0(X)$ is z e r o on $V \subseteq X$, open, if

$$\forall f \in C_c(V), \qquad \mu(f) = 0.$$

For a given $\mu \in D^0(X)$, let $\mathcal{V}$ be the family of open sets $V_\alpha \subseteq X$ such that μ is 0 on V_α; thus if $V = \bigcup \{V_\alpha : V_\alpha \in \mathcal{V}\}$, then, by Corollary III.5.1, μ is 0 on V. Hence, we can well-define the s u p p o r t of $\mu \in D^0(X)$ as the complement of the union of the open sets $V \subseteq X$ for which μ is 0 on V.

Note that if supp $f \cap$ supp $\mu = \emptyset$, where $f \in C_c(X)$ and $\mu \in D^0(X)$, then $\mu(f) = 0$; in fact, just take an open set $V \supseteq$ supp f with $V \subseteq ($supp $\mu)^\sim$, so that μ is 0 on V and hence $\mu(f) = 0$. This does not tell us that if $f = 0$ on supp μ then $\mu(f) = 0$. In fact, this result is true for measures:

Theorem III.6 *Let $\mu \in D^0(X)$ and assume $f \in C_c(X)$ vanishes on* supp μ. *Then $\mu(f) = 0$.*

Proof. Let $K =$ supp f, $E =$ supp μ. By definition of the topology on $C_c(X)$ there is $M_K > 0$ such that for each $g \in C_c(K)$ with supp $g \subseteq K$ we have

$$|\mu(g)| < M_k \|g\|_\infty.$$

Given $\varepsilon > 0$, we'll show that $|\mu(f)| < \varepsilon$.

Let $V = \{x \in X : |f(x)| < \varepsilon/2M_K\}$; then V is open since f is continuous, and $E \subseteq V$. Clearly, $E^\sim$ is an open set containing the closed set $V^\sim$.

By Urysohn's lemma (Appendix I.1), there is a continuous function $h: X \to [0, 1]$ such that $h = 1$ on $V^\sim$ and supp $h \subseteq E^\sim$.

Note that $E \cap \operatorname{supp} fh = \varnothing$ and hence $\mu(fh) = 0$. Further, $f = fh$ on $K \cap V^\sim$ and $|fh| \leqslant |f|$ on X. Consequently, since $f = 0$ on $K^\sim$,

$$\|f - fh\|_\infty = \sup_{x \in V \cap K} |f(x)(1 - h(x))| \leqslant 2 \sup_{x \in V} |f(x)| \leqslant \varepsilon / M_K.$$

Therefore, noting that supp $(f - fh) \subseteq K$,

$$|\mu(f)| = |\mu(f - fh)| < M_K \|f - fh\|_\infty < \varepsilon.$$

q.e.d.

Using this result and the Hahn–Banach theorem it is not difficult [18, pp. 70–71] to prove the following approximation theorem.

Theorem III.7 *Let X be locally compact and let $\mu \in M(X)$. There is a net $\{\mu_\alpha\} \subseteq M(X)$ such that* card supp $\mu_\alpha < \infty$, supp $\mu_\alpha \subseteq$ supp μ,

$$\forall f \in C_0(X), \qquad \lim_\alpha \mu_\alpha(f) = \mu(f),$$

and $\qquad \forall \alpha, \qquad \|\mu_\alpha\|_1 = \|\mu\|_1$.

In order to get a feeling for this result, let $\mu \in M(\mathbf{R})$ be a measure, for the measurable space $(\mathbf{R}, \mathscr{B})$, and take $f \in BV(\mathbf{R})$ so that in the notation of (5.22), $\mu = \mu_f$. For simplicity, let f increase from 0 to 1 so that $\mu(\mathbf{R}) = 1$. We write

$$f_n(x) = (1/n)[nf(x)],$$

where $[y]$ is the largest integer less than or equal to y. f_n is a monotone step function. The measure μ_n, which corresponds to f_n as in (5.22), has finite support contained in supp μ and $\mu_n(f) \to \mu(f)$ for each $f \in C_0(\mathbf{R})$.

If $\mu \in M(X)$, X locally compact, then supp μ is the smallest closed set on which μ is concentrated. For perspective, take

$$\mu = \sum_{\gamma \in D} a_\gamma \delta_\gamma, \qquad \sum |a_\gamma| < \infty,$$

where D is a countable dense subset of the Cantor set $C \subseteq [0, 1]$. Then μ is concentrated on both C and D whereas supp $\mu = C$.

III.6 Haar measure

An additive group G with a locally compact topology is a l o c a l l y c o m p a c t g r o u p if the function $G \times G \to G$, $(x, y) \mapsto x - y$, is continuous. Similarly, a complex vector space X which is also a topological space is a t o p o l o g i c a l v e c t o r s p a c e if the functions $\mathbf{C} \times X \to X$, $(c, x) \mapsto cx$, and $X \times X \to X$, $(x, y) \mapsto x + y$, are continuous. Let X be an Hausdorff topological vector space. $K \subseteq X$ is absorbing if

$$\forall x \in X \qquad \exists \varepsilon > 0 \quad \text{such that} \quad 0 < |c| \leqslant \varepsilon \Rightarrow cx \in K;$$

and $K \subseteq X$ is balanced if

$$|c| \leqslant 1 \Rightarrow cK \subseteq K$$

(in both cases, $c \in \mathbf{C}$).

A fundamental result in harmonic analysis is that if G is a locally compact group then there is a Borel measure m_G on G such that

$$\forall B \in \mathscr{B} \text{ and } \forall x \in G, \qquad m_G B = m_G(B + x)$$

(where $B + x = \{y + x : y \in B\}$). Thus, the crucial feature of translation invariance for Lebesgue measure on the line extends to locally compact groups. We shall prove this existence for G compact and abelian using the Markoff–Kakutani fixed point theorem, which we'll also prove. Such an m_G for G is Haar measure. It is easy to show that there is only one such m_G which ensures that $m_G G = 1$ (for G compact).

Theorem III.8 (Markoff–Kakutani) *Let X be an Hausdorff topological vector space, take a compact and convex set $K \subseteq X$, and let $\{T_\alpha\}$ be a family of continuous linear maps $T_\alpha : X \to X$ which satisfies*

$$\forall \alpha, \qquad T_\alpha(K) \subseteq K$$

and

$$\forall \alpha, \beta, \qquad T_\alpha \circ T_\beta = T_\beta \circ T_\alpha. \tag{III.15}$$

Then there is $k \in K$ such that

$$\forall \alpha, \qquad T_\alpha(k) = k.$$

Proof. Let

$$T_\alpha^{(n)} = \frac{I + T_\alpha + \cdots + T_\alpha^{n-1}}{n},$$

where $I : X \to X$ is the identity map and T_α^j is the composition $T_\alpha \circ \ldots \circ T_\alpha$, j times. Clearly, each $T_\alpha^{(n)} : X \to X$ is continuous and linear; we write $T = \{T_\alpha^{(n)}\}$ and let $\tilde{T}$ be the set of finite products, under composition, of elements from T.

Note that for each $u \in \tilde{T}$, $u(K) \subseteq K$; this follows by the convexity of K.

Thus $u \circ v(K) \subseteq u(K)$ for $u, v \in \tilde{T}$. Further, because of (III.15) $u \circ v = v \circ u$ for $u, v \in \tilde{T}$; and hence $v(K) \supseteq v \circ u(K)$ implies $v(K) \supseteq u \circ v(K)$.

Let $\tilde{K} = \bigcap \{u(K) : u \in \tilde{T}\}$ and note that

$$\forall u, v \in \tilde{T}, \qquad u(K) \cap v(K) \neq \varnothing;$$

in fact,

$$u(K) \cap v(K) \supseteq u(K) \cap u \circ v(K) \supseteq u \circ v(K) \cap u \circ v(K) = u \circ v(K).$$

Since K is compact, $u(K)$ is compact for each $u \in \tilde{T}$. Consequently, because $u(K) \cap v(K) \neq \emptyset$ for $u, v \in \tilde{T}$, since each $u(K) \subseteq K$, and because K is compact, there is $k \in K$ such that for all $u \in \tilde{T}, k \in u(K)$; therefore $k \in \tilde{K}$.

In particular, for each α and n, $k \in T_\alpha^n(K)$, so that there is $t \in K$ (depending on α and n) such that $T_\alpha^{(n)}(t) = k$. Thus

$$T_\alpha(k) = \frac{T_\alpha(t) + \cdots + T_\alpha^n(t)}{n},$$

and hence,

$$T_\alpha(k) - k = -\frac{t}{n} + \frac{T_\alpha^n(t)}{n} \tag{III.16}$$

since $T_\alpha^{(n)}(t) = k$.

Note that the function $K \times K \to X$, $(s, t) \mapsto s - t$, takes $K \times K$ into a compact set E.

Because of (III.16) and the fact that $T_\alpha^n(t) \in K$ we have

$$T_\alpha(k) - k = \frac{T_\alpha^n(t) - t}{n} \in \frac{1}{n} E. \tag{III.17}$$

We shall show

$$\bigcap_n \left(\frac{1}{n} E \right) = \{0\}, \tag{III.18}$$

so that since (III.17) is true for each n, $T_\alpha(k) - k = 0$, and we are done.

Let $V \subseteq X$ be a balanced set for which $0 \in \text{int } V$.

By the definition of a topological vector space there is a balanced and absorbing open set $W \subseteq X$ containing 0, such that $W + W \subseteq V$.

$\{x + W : x \in X\}$ is an open cover of E so that by the compactness there are points x_i, $i = 1, \ldots, m$, such that

$$E \subseteq \bigcup \{x_i + W : i = 1, \ldots, m\}.$$

Since W is absorbing, $rx_i \in W$ for $i = 1, \ldots, m$ and for some $r \in (0, 1]$. Therefore

$$r(E \cap (x_i + W)) \subseteq W + rW, \qquad i = 1, \ldots, m; \tag{III.19}$$

and because the right-hand side of (III.19) is independent of i,

$$rE = \bigcup_{i=1}^m r(E \cap (x_i + W)) \subseteq W + rW.$$

Thus, $E \subseteq (1/r)V$, noting that $W + rW \subseteq W + W \subseteq V$; and, so, if $1/n < r$,

$$\frac{1}{n}E \subseteq \frac{1}{nr}V \subseteq \frac{r}{r}V = V.$$

Consequently, $\bigcap_n (1/n)E \subseteq V$; so that since X is Hausdorff and V is arbitrary we have (III.18).

q.e.d.

Theorem III.9 *Let G be a compact abelian group. Then there is a Haar measure m_G on G.*

Proof. Let $M_1(G) = \{\mu \in M(G): \|\mu\|_1 \leqslant 1\}$. By the Alaoglu theorem, $M_1(G)$ is weak * compact in $M(G)$.

Let $M_1^+(G) = \{\mu \in M_1(G): \mu(1) = 1\}$.

Note that μ is positive if $\mu \in M_1^+(G)$; to prove this we use the facts that G is compact and $\|\mu\|_1 = \mu(1)$ (e.g. [18, p. 101]).

If $M(G)$ is taken with the weak * topology the map $M(G) \to \mathbf{C}$, $\mu \mapsto \mu(1)$, is continuous; and hence $\{\mu \in M(G): \mu(1) = 1\}$ is weak * closed.

Thus, $M_1^+(G)$ is weak * compact; and it is easy to check that $M_1^+(G)$ is convex.

For $x \in G$ and $\mu \in M(G)$ we define the translation $\tau_x \mu$ as

$$\tau_x \mu(f) = \int_G f(y - x) \, d\mu(x),$$

where $f \in C(G)$.

Then for each $x \in G$ we define the map $T_x: M(G) \to M(G)$, $\mu \mapsto \tau_x \mu$.

Note that T_x is continuous with the weak * topology on both domain and range, linear, and

$$\forall x, y \in G, \qquad T_x \circ T_y = T_{x+y} = T_y \circ T_x$$

(since G is abelian).

It is also trivial to check that for each $x \in G$,

$$T_x(M_1^+(G)) \subseteq M_1^+(G).$$

Therefore, by Theorem III.8, there is $m_G \in M_1^+(G)$ such that $\tau_x m_G = m_G$, the required translation invariance.

Further, $\|m_G\|_1 = 1$, $m_G(1) = 1$, and m_G is positive.

q.e.d.

Bibliography

[1] Abian, A.: An example of a nonmeasurable set. Boll. U.M.I. **1** (1968) 366–368

[2] —: The outer and inner measures of a nonmeasurable set. Boll. U.M.I. **3** (1970) 555–558

[3] Arzelà, C.: Sulla integrazione per serie. Atti Acc. Lincei, Roma **1** (1885) 532–537, 596–599

[4] Asplund, E. and Bungart, L.: A first course in integration. New York 1966

[5] Austin, D.: A geometric proof of the Lebesgue differentiation theorem. PAMS **16** (1965) 220–222

[6] Banach, S.: Sur une classe de fonctions d'ensemble. Fund. Math. **6** (1924) 170–188

[7] Banach, S. and Kuratowski, C.: Sur une généralisation du problème de la mesure. Fund. Math. **14** (1929) 127–131

[8] Banach, S. and Saks, S.: Sur la convergence forte dans les espaces L^p. Studia Math. **2** (1930) 51–57

[9] Beesley, E. M., Morse, A. P. and Pfaff, D. C.: Lipschitzian points. AMM **79** (1962) 603–608

[10] Benedetto, J.: Harmonic analysis on totally disconnected sets. Lecture Notes in Mathematics, 202, Berlin–Heidelberg–New York 1971

[11] Besicovitch, A. S.: Concentrated and rarified sets of points. Acta Math. **62** (1934) 289–300

[12] Bledsoe, W. W.: An inequality about complex numbers. AMM **77** (1970) 180–182

[13] Boas, R. P.: A primer of real functions. Carus monographs, 1960

[14] Boardman, E.: Universal covering series. AMM **79** (1972) 780–781

[15] Bochner, S.: Additive set functions on groups. Ann. Math. **40** (1939) 769–799

[16] Bochner, S. and Chandrasekharan, K.: Fourier transforms. Princeton, N.J. 1949

[17] Borel, E.: Fonctions monogènes. Paris 1917

[18] Bourbaki, N.: Intégration (Chap. 1–4, 2nd ed.) Paris 1965

[19] —: Intégration (Chap. 5, 2nd ed.) Paris 1967

[20] Brooks, J. K.: On the Vitali–Hahn–Saks and Nikodym theorems. PNAS **64** (1969) 468–471

[21] Brooks, J. K. and Mikusinski, J.: On some theorems in functional analysis. Bull. Acad. Polon. Sci. **18** (1970) 151–155

[22] Bruckner, A. M.: Differentiation of integrals. AMM Slaught Paper, No. 12, 78 (1971)

[23] Bruckner, A. M. and Leonard, J. L.: Derivatives. Papers in analysis AMM **73** (1963) 24–56

[24] Carleson, L.: Representations of continuous functions. Math. Z. **66** (1957) 447–451

[25] Cesari, L.: Variation, multiplicity, and semicontinuity. AMM **65** (1958) 317–331

[26] Choquet, G.: Un ensemble paradoxal en théorie de la mesure. Bull. Sci. Math. **94** (1970) 247–250

[27] Cohen, L. W.: A new proof of Lusin's theorem. Fund. Math. 9 (1927) 122–123
[28] —: The fundamental theorem of the calculus and Borel measure
[29] Darst, R. B.: Perfect null sets in compact Hausdorff spaces. PAMS 16 (1965) 845
[30] Dieudonné, J.: Sur la convergence des suites de mesures de Radon. An. da Acad. Bras. Ci. 23 (1951) 21–38
[31] —: Sur les espaces de Köthe. J. d'Analyse 1 (1951) 81–115
[32] Dressler, R. E. and Kirk, R. B.: Non-measurable sets of reals whose measurable subsets are countable. Isr. J. Math. 11 (1972) 265–270.
[33] Dubins, L.: An elementary proof of Bochner's finitely additive Radon–Nikodym theorem. AMM 76 (1969) 520–523
[34] Dunford, N. and Pettis, B. J.: Linear operations on summable functions. TAMS 47 (1940) 323–392
[35] Edwards, R. E.: Fourier series. Vol. 1 and 2, New York 1967
[36] —: Integration and harmonic analysis on compact groups. London 1972
[37] Egoroff, D. T.: Sur les suites de fonctions mesurables. CRAS, Paris 152 (1911) 244–246
[38] Fatou, P.: Séries trigonométriques et séries de Taylor. Acta Math. 30 (1906) 335–400
[39] Fefferman, C.: A Radon–Nikodym theorem for finitely additive set functions. Pac. J. Math. 23 (1967) 35–45
[40] Frank, R.: Sur une propriété des fonctions additives d'ensembles. Fund. Math. 5 (1924) 252–261
[41] Gerver, J.: The differentiability of the Riemann function at certain rational multiples of π. PNAS 62 (1969) 668–670
[42] Gillis, J.: Some combinatorial properties of measurable sets. Quart. J. Math. 7 (1936) 191–198
[43] —: Note on a property of measurable sets. J. Lond. Math. Soc. 11 (1936) 139–141
[44] Gleason, A.: Fundamentals of abstract analysis. Reading, Mass. 1966
[45] Goldberg, R.: Fourier transforms. London 1961
[46] Grothendieck, A.: Sur les applications linéaires faiblement compactes d'espaces du type $C(K)$. CJM 2 (1953) 129–173
[47] de Guzmán, M. and Rubio, B.: Remarks on the Lebesgue differentiation theorem, the Vitali lemma and the Lebesgue–Radon–Nikodym theorem. AMM 79 (1972) 341–348
[48] Hahn, H.: Über den Integralbegriff. Anz. Akad. Wiss. Wien. 66 (1929) Nr. 2
[49] Halmos, P.: Measure theory. New York 1950
[50] Hardy, G. H.: Pure mathematics. London 1963
[51] Hardy, G. H. and Wright, E. M.: Theory of numbers (4th ed.) London 1965
[52] Hausdorff, F.: Dimension und ausseres Mass. Math. Ann. 79 (1919) 157–179
[53] —: Set theory. New York 1962
[54] Hawkins, T.: Lebesgue's theory of integration. Madison, Wisc. 1970
[55] Herz, C.: The Hardy–Littlewood maximal theorem. Warwick 1968
[56] Hewitt, E. and Stromberg, K.: Real and abstract analysis. Berlin–Heidelberg–New York 1965
[57] Hörmander, L.: Linear partial differential operators. New York 1963
[58] Hurewicz, W. and Wallman, H.: Dimension theory. Princeton, N.J. 1948
[59] Kahane, J.-P.: Lacunary Taylor and Fourier series. BAMS 70 (1964) 199–213
[60] Kahane, J.-P. and Salem, R.: Ensembles parfaits et séries trigonométriques. Paris 1963

[61] Katznelson, Y.: Introduction to harmonic analysis. New York 1968
[62] Kaufman, R. and Rickert, N.: An inequality concerning measures. BAMS **72** (1966) 672–676
[63] Kelley, J.: General topology. New York 1960
[64] Kestelman, H.: Modern theories of integration (2nd ed.) New York 1960
[65] Kirk, R. B.: Sets which split families of measurable sets. AMM **79** (1972) 884–886
[66] Koksma, J. and Kuipers, L. editors: Asymptotic distribution modulo 1 NUFFIC summer session, 1962
[67] Lebesgue, H.: Intégrale, longueur, aire. Annali di Mat. **7** (1902) 231–358
[68] —: Leçons sur l'integration et la recherche des fonctions primitives. 1st and 2nd ed., Paris 1904 and 1928
[69] —: Leçons sur les séries trigonométriques. Paris 1906
[70] —: Sur l'intégration des fonctions discontinues. Ann. École. Norm. Sup. **27** (1910) 361–450
[71] —: Measure and the integral (ed. by K. O. May), San Francisco 1966
[72] Letta, G.: Une démonstration élémentaire du théorème de Lebesgue sur la derivation des fonctions croissantes. L'Enseignement Math. **16** (1970) 177–184
[73] Lipiński, J.: On zeros of a continuous nowhere differentiable function. AMM **73** (1966) 166–168
[74] —: On derivatives of singular functions. Bull. l'Acad. Pol. Sci. **20** (1972) 625–628
[75] Lusin, N.: Sur les proprietés des fonctions mesurables. CRAS, Paris 154 (1912) 1688–1690
[76] —: Sur la notion de l'intégral. Ann. di Mat. **26** (1917) 77–129
[77] Luxemburg, W. A. J.: Arzelà's dominated convergence theorem for the Riemann integral. AMM **78** (1971) 970–979
[78] Mazurkiewicz, S.: Sur les suites de fonctions continues. Fund. Math. **18** (1932) 114–117
[79] Munroe, M. E.: Measure and integration. Reading, Mass. 1953
[80] Natanson, I. P.: Theory of functions of a real variable. New York
[81] Osgood, W. F.: Non-uniform convergence and the integration of series term by term. Amer. J. Math. **19** (1897) 155–190
[82] Oxtoby, J. C.: Measure and category. Berlin–Heidelberg–New York 1970
[83] Pesin, I. N.: Classical and modern integration theories. New York 1970
[84] Polya, G. and Szegö, G.: Aufgaben und Lehrsätze aus der Analysis. Berlin–Heidelberg–New York 1964
[85] Radó, T.: Length and area. AMS Colloquium Publication, Vol. 30, 1948
[86] Rainwater, J.: Weak convergence of bounded sequences. PAMS **14** (1963) 999
[87] Ràjchman, A.: Sur l'unicité du developpement trigonométrique. Fund. Math. 3 (1922) 287–302, **4** (1923) 366–367
[88] Riemann, B.: Sur la possibilité de répresenter une fonction par une série trigonométrique. Bull. des Sciences (1873) 20–96 (publ. by Richard Dedekind in Gött. Abh. **13** (1867) and presented by Riemann to the faculty of philosophy for his Habilitationsschrift)
[89] Riesz, F. and Sz.-Nagy, B.: Functional analysis. New York 1955
[90] Rogers, C. A.: A linear Borel set whose difference set is not a Borel set. Bull. Lond. Math. Soc. **2** (1970) 41–42

[91]   —: Hausdorff measures. London 1970

[92]   Royden, H. L.: Real analysis, 2nd edition. New York 1968

[93]   Rudin, W.: Continuous functions on compact sets without perfect subsets. PAMS **8** (1957) 39–42

[94]   —: Real and complex analysis. New York 1966

[95]   Saks, S.: On some functionals. TAMS **35** (1933) 549–556

[96]   —: Addition to the note: On some functionals. TAMS **35** (1933) 965–970

[97]   —: Theory of the integral. New York

[98]   Salem, R.: Oeuvres mathématiques. Paris 1967

[99]   Schaefer, H. H.: Weak convergence of measures. Math. Ann. **193** (1971) 57–64

[100] Schrader, K.: A pointwise convergence theorem for sequences of continuous functions. TAMS **159** (1971) 155–163

[101] —: A generalization of the Helly selection theorem. BAMS **78** (1972) 415–419

[102] Serrin, J. and Varberg, D. E.: A general chain rule for derivatives and the change of variables formula for the Lebesgue integral. AMM **76** (1969) 514–520

[103] Sierpiński, W.: Sur la question de la mesurabilité de la base de M. Hamel. Fund. Math. **1** (1920) 105–111

[104] —: Sur une problème concernant les ensembles mesurables superficiellement. Fund. Math. **1** (1920) 112–115

[105] —: Démonstration de quelques théorèmes fondamentaux sur les fonctions mesurables. Fund. Math. **3** (1922) 314–321

[106] —: Démonstration d'un théorème sur les fonctions additives d'ensembles. Fund. Math. **5** (1924) 262–264

[107] —: Sur une problème conduisant à un ensemble non mesurable. Fund. Math. **10** (1927) 177–179

[108] —: Remarque sur les suites infinies de fonctions. Fund. Math. **18** (1932) 110–113

[109] —: Hypothèse du continu. Warszawa 1934

[110] —: On the congruence of sets and their equivalence by finite decomposition. Congruence of sets and other monographs. New York

[111] Singh, A. N.: The theory and construction on non-differentiable functions. Squaring the circle and other monographs. New York

[112] Solovay, R. M.: A model of set-theory in which every set of reals is Lebesgue measurable. Ann. of Math. **92** (1970) 1–56

[113] Taibleson, M.: Fourier coefficients of functions of bounded variation. PAMS **18** (1967) 766

[114] Titchmarsh, E. C.: Theory of functions (2nd ed.) London 1960

[115] Ulam, S.: Zur Maß theorie in der allgemeinen Mengenlehre. Fund. Math. **16** (1930) 140–150

[116] —: Problems in modern mathematics. New York 1964

[117] Varberg, D. E.: On absolutely continuous functions. AMM **72** (1965) 831–841

[118] Vitali, G.: Sulle funzioni integrali. Atti. Acc. Sci., Torino **40** (1904–05) 1021–1034

[119] —: Una proprietà delle funzioni misurabili. Rend. Ist. Lombardo **38** (1905) 599–603

[120] —: Sul problema della misura dei gruppi di punti di una retta. Bologna (1905)

[121] —: Sull'integrazione per serie. Rend. Circ. Matem. Palermo **23** (1907) 137–155

[122] —: Sui gruppi di punti e sulle funzioni di variabili reali. Atti. Acc. Sci., Torino **43** (1907) 229–246

[123] —: Sulle funzioni continue. Fund. Math. **8** (1926) 175–188

[124] Volterra, V.: Sui principii del calcolo integrale. G. Battaglini **19** (1881) 333–372

[125] Wells, B. B.: Weak compactness of measures. PAMS **20** (1969) 124–130.

[126] Wesler, O.: An infinite packing theorem for sphere. PAMS **11** (1960) 324–326

[127] Zaanen, A. C.: Some examples in weak sequential convergence. AMM **69** (1962) 85–93.

[128] Zygmund, A.: Trigonometric series. London 1959

[129] —: On certain lemmas of Marcinkiewicz and Carleson. J. of Approximation Theory **2** (1969) 249–257

The items in the following indices are referenced either according to the problem (e.g., P2.10), or the section (e.g., 3.2) in which they are found.

Index of proper names

Lebesgue is mentioned so often that we do not list him below.

Index of terms

Terms which are not defined in the text are not included in the index.